Cartographic Encounters

The Kenneth Nebenzahl, Jr., Lectures in the History of Cartography

PUBLISHED FOR THE HERMON DUNLAP SMITH CENTER
FOR THE HISTORY OF CARTOGRAPHY
THE NEWBERRY LIBRARY

SERIES EDITOR, DAVID WOODWARD
    Maps: A Historical Survey of Their Study and Collecting by R. A. Skelton (1972)
    British Maps of Colonial America by William P. Cumming (1974)
    Five Centuries of Map Printing edited by David Woodward (1975)
    Mapping the American Revolutionary War by J. B. Harley, Barbara Bartz Petchenik, and Lawrence W. Towner (1978)

SERIES EDITOR, DAVID BUISSERET
    Art and Cartography edited by David Woodward (1987)
    Monarchs, Ministers, and Maps edited by David Buisseret (1992)
    Rural Images edited by David Buisseret (1996)

SERIES EDITOR, JAMES AKERMAN
    Envisioning the City edited by David Buisseret (1998)
    Cartographic Encounters edited by G. Malcolm Lewis (1998)

Cartographic Encounters

Perspectives on
Native American Mapmaking
and Map Use

Edited by

G. Malcolm Lewis

THE UNIVERSITY OF CHICAGO PRESS
CHICAGO AND LONDON

G. MALCOLM LEWIS is a research fellow (formerly reader) in the department of geography at the University of Sheffield, England. He is coeditor with David Woodward of *Cartography in the Traditional African, American, Arctic, Australian, and Pacific Societies,* volume 2, book 3 of *The History of Cartography,* forthcoming from the University of Chicago Press.

The University of Chicago Press, Chicago 60637
The University of Chicago Press, Ltd., London
© 1998 by The University of Chicago
All rights reserved. Published 1998

08 07 06 05 04 03 02 01 00 99   1 2 3 4 5
ISBN: 0-226-47694-4 (cloth)

Library of Congress Cataloging-in-Publication Data

Cartographic encounters : perspectives on Native American mapmaking
   and map use / edited by G. Malcolm Lewis.
       p.   cm. — (The Kenneth Nebenzahl, Jr., lectures in the
   history of cartography)
   Includes index.
   ISBN 0-226-47694-4 (cloth : alk. paper)
   1. Indian cartography—North America. 2. Indians of North
America—Maps. 3. Inuit—Maps. 4. Indians of North America—First
contact with Europeans. 5. America—Discovery and exploration—
Maps. I. Lewis, G. Malcolm. II. Series.
E59.C25C37 1998'912.7'08997—dc21   97-49064
                                                              CIP

⊚ The paper used in this publication meets the minimum requirements of the American National Standard for Information Sciences—Permanence of Paper for Printed Library Materials, ANSI Z39.48–1992.

To Margaret
with love and gratitude

Contents

	List of Illustrations	xi
	Series Editor's Note	
	James Akerman	xiii
	Editor's Preface	
	G. Malcolm Lewis	xv
	Introduction	
	G. Malcolm Lewis	1
PART 1 *The Four-Hundred-Year First Encounter*	CHAPTER ONE Frontier Encounters in the Field: 1511–1925	
	G. Malcolm Lewis	9
	CHAPTER TWO Encounters in Government Bureaus, Archives, Museums, and Libraries, 1782–1911	
	G. Malcolm Lewis	33
	CHAPTER THREE Hiatus Leading to a Renewed Encounter	
	G. Malcolm Lewis	55

PART 2
The Ongoing Second Encounter

CHAPTER FOUR
Recent and Current Encounters
G. Malcolm Lewis — 71

CHAPTER FIVE
Maps of Territory, History, and Community in Aztec Mexico
Elizabeth Hill Boone — 111

CHAPTER SIX
Inland Journeys, Native Maps
Barbara Belyea — 135

CHAPTER SEVEN
Native Mapping in Southern New England Indian Deeds
Margaret Wickens Pearce — 157

CHAPTER EIGHT
Eighteenth-Century Arkansas Illustrated: A Map within an Indian Painting?
Morris S. Arnold — 187

CHAPTER NINE
Indian Maps of the Colonial Southeast: Archaeological Implications and Prospects
Gregory A. Waselkov — 205

CHAPTER TEN
Debriefing Explorers Amerindian Information in the Delisles' Mapping of the Southeast
Patricia Galloway — 223

PART 3
Future Encounters in New Contexts

CHAPTER ELEVEN
Orientations from Their Side:
Dimensions of Native American
Cartographic Discourse
  Peter Nabokov — 241

CHAPTER TWELVE
Future Encounters in New Contexts
  G. Malcolm Lewis — 273

About the Contributors — 285
Index — 287

Illustrations

1.1 Wingenund map incorporated in a painting on the trunk of a blazed tree, as redrawn by or for William Bray in 1782   12

1.2 The missionary lithograph on the verso of which the Quebec Inuit Wetalltok drew a map of the Belcher Islands   16

1.3 Wetalltok, map of the Belcher Islands (1910 or earlier)   17

1.4 Lawrence van den Bosh, map of the lower Mississippi River and lands to the west   20

1.5 Powon, "Map of West Shore of Hudson Bay and Vicinity" (1894)   25

3.1 The descent and survival states of spatially arranged information originally communicated by Indians and Inuit   59

5.1 Mapa Local of Tochpan   115

5.2 Reconnaisance map, from Bernardino de Sahagún, *General History*   116

5.3a Mapa Sigüenza   119

5.3b Detail of the Aztecs leaving Aztlan, Mapa Sigüenza,   120

5.3c Detail of the founding of Tenochtitlan, Mapa Sigüenza   120

5.4 Cuauhtinchan Map 2   121

5.5 Cuauhtinchan Map 1   122

5.6 Map of Cuauhtinchan, from Historia Tolteca-Chichimeca   124

5.7 Lienzo de Zacatepec   125

5.8 Founding of Tenochtitlan, from *Codex Mendoza*   129

5.9 Page 1 of the *Codex Féjérváry-Mayer*   130

6.1 Detail of Jacques-Nicolas Bellin, *Carte de l'Amérique septentrionale*   136

6.2 Title cartouche of Jacques-Nicolas Bellin, *Carte de l'Amérique septentrionale*   137

6.3 Andrew Graham, *A Plan of Part of Hudson's Bay and Rivers, Communicating with York Fort and Severn*   138

6.4 Philippe Buache, "Carte physique des terreins les plus élevés de la partie occidentale du Canada"   143

6.5 Detail of Aaron Arrowsmith, *Map Exhibiting all the New Discoveries in the Interior Parts of North America ... additions to 1802*   145

7.1 The Mattatuck deed related to a nineteenth-century topographic map   164

7.2 The Ridgefield deed related to a nineteenth-century topographic map	168
7.3a The Sepecan deed	170
7.3b The Sepecan deed as redrawn in J. W. Barber, *Historical Collections . . . Relating to the History and Antiquities of Every Town in Massachusetts*	171
7.3c The Sepecan deed as redrawn in *Records of the Colony of New Plymouth in New England*	171
7.4 The Sepecan deed related to a nineteenth-century topographic map	172
7.5 The Weantinock deed as redrawn in Edward R. Lambert, *History of the Colony of New Haven*	175
7.6 The Weantinock deed related to a nineteenth-century topographic map	176
7.7 The Wecabaug testimony	180
7.8 The Wecabaug testimony related to a nineteenth-century topographic map	181
8.1 The Quapaw/Arkansas painted buffalo hide	188
8.2 Detail of dancers and villages, Quapaw/Arkansas painted buffalo hide	189
8.3 Detail of French village, Quapaw/Arkansas painted buffalo hide	189
8.4 Cartographic interpretation of Quapaw/Arkansas buffalo hide	196
9.1 Thomas Kitchin, "A New Map of the Cherokee Nation"	206
9.2 *Nations amies et ennemies des Tchikachas*, the Chickasaw/Alabama map of the Southeast, 1737, as redrawn by Alexandre de Batz	208
9.3 Detail of John Smith's *Virginia*	210
9.4 *Plan et scituation des villages Tchikachas*, the Captain of Pacana's map of Chickasaw towns as redrawn by Alexandre de Batz	215
10.1 Catawba Indian map collected by Francis Nicholson, ca. 1721	225
10.2 Chickasaw Indian map collected by Francis Nicholson, ca. 1723	226
10.3 Creek and Choctaw "paintings" reproduced by Bernard Romans, ca. 1770	227
10.4 Some examples of Indian house symbols from early maps of the Americas	229
10.5 Detail of lower Mississippi valley, Louis Jolliet, *Nouvelle decouverte de plusieurs nations dans la Nouvelle France*	230
10.6 Detail of lower Mississippi valley, Guillaume Delisle, sketch of the Mississippi-Alabama region	231
10.7 Reference map of Mississippi-Alabama region, drafted by Julie Smith	232
10.8 Detail of lower Mississippi valley, Guillaume Delisle, *Carte des environs du Mississipi*	234
10.9 Detail of lower Mississippi valley, Guillaume Delisle, *Carte du Mexique et de la Floride*	236
11.1 Amos Bad Heart Bull, map of Black Hills	245
11.2 Symbolism of Southeastern Indian ceremonial ground	251
11.3 Jemez Pueblo depiction of the tribal universe	252
11.4 "Mythological landscape" of Santo Domingo Pueblo, New Mexico	254
11.5 Cosmology and topography depicted on floors by Southern California Indians	256
11.6 Map of the world by a Thompson River Indian, British Columbia	257
11.7 Eliza Spalding, Protestant ladder chart, ca. 1842	261
11.8 Drum head drawn by Beaver Indian shaman	262
11.9 Tipi painting by Rock Thunder	263

Series Editor's Note

I am delighted to present this latest volume spawned by the Kenneth Nebenzahl, Jr., Lectures in the History of Cartography. In 1966, when R. A. Skelton honored the Newberry by presenting the first lectures in this series, "The Study and Collecting of Early Maps: A Historical Survey," the history of cartography was little known and appreciated outside a small circle of librarians, antiquarian scholars and collectors, geographers, and historians. Few books had been written on the subject, and fewer still strayed far from the traditional interest in the history of European discoveries and Western topographical and maritime mapping inherited from nineteenth-century scholarship. Skelton's lectures were published in 1972 by the University of Chicago Press as a small but seminal book, *Maps: A Historical Survey of Their Study and Collecting*. It was the first to take stock of the field in many years, and it spurred the successive organizers of the Nebenzahl Lectures to dedicate the series to the goals of identifying and fostering comparatively new and promising avenues of research in cartographic history. We like to think that the series has played no small role in stirring interest in the history of cartography among scholars in dozens of fields ranging from literary studies and art history to the history of science and anthropology.

The present volume continues this tradition of expanding the field of vision of the history of cartography by casting light on the underappreciated history of native North American mapping. The volume emerged from the eleventh series, which was given on June 25 and 26, 1993, to an audience that included many who had been attending the Fifteenth International Conference on the History of Cartography at the Newberry during the previous week. Though these volumes traditionally have been edited by the presiding director of the Smith Center, we are honored and grateful instead to have been able to rely on the talents of G. Malcolm Lewis, now retired from the University of Sheffield. Mr. Lewis delivered the keynote lectures at the 1993 symposium and played a key role in selecting the balance of its panel of speakers, Elizabeth Boone, Patricia Galloway, and Peter Nabokov. Since then, he has made the project very much his own quest. He has greatly expanded his contribution to the volume, an erudite and comprehensive historiographic survey of this subject, its cross-cultural pitfalls, and its potential. These enlargements have been balanced by several new essays solicited by him that broaden the disciplinary, theoretical, and geographical scope of the volume. The resulting book, though somewhat larger than previous volumes in this series, nevertheless conforms with its traditional goals of both providing a solid platform for future research and suggesting profitable directions for this future.

As series editor, I have thus played a smaller role in shaping this volume than my predecessors have in previous instances; as a result I am all the more indebted to those who have helped bring the work to fruition. I am first of all much indebted to Malcolm Lewis, who has truly blessed us with both his willingness to serve as editor at a time when he was much taken with other projects, and his willingness to share knowledge gathered over decades of diligent work. My predecessor David Buisseret got this project off the ground and guided it to its present form. We are indebted to the University of Chicago Press for its encouragement throughout the evolution of the project, and especially for its efforts to produce this attractive volume in a timely manner. But of course, the efforts of these individuals and institutions rests on the continuing enthusiasm and support of Mr. and Mrs. Kenneth Nebenzahl, founders of the Nebenzahl Fund, whose generosity has already yielded three decades of provocative scholarship and public programs.

James Akerman

Preface

Writing a preface to a multiauthored book presents its editor with an almost insoluble dilemma: how to write on behalf of all the contributors and at the same time convey to readers the essence of the book as a whole. For me, the dilemma is particularly acute because although only one of eight authors, I have written almost half of the text. I hope, therefore, that my collaborators will forgive me if in this preface the first person singular appears far more frequently than the first person plural. It is a consequence of placing the readers first and trying to give them a perspective from which to evaluate the whole rather than its several parts; in particular, how it emerged and how, as an editor, I influenced its form. I have deferred until the Introduction a statement of the book's several objectives.

In telling the White Rabbit to "begin at the beginning," the King in Alice's Wonderland was not being very helpful. In retrospect most developments can be seen to have had several beginnings. For me, those leading up to this book had at least three.

I remember clearly when and where I first became aware that native North Americans made maps. It excited me, though at the time I had no idea that the topic would later become a personal preoccupation. It was a hot summer's afternoon in central London in the early 1970s and I was working without much enthusiasm in the great round Reading Room of what had formerly been the British Museum and only recently been renamed the British Library, Reference Division. I had on my desk books relating to my interest in the emergence of ideas about the North American Great Plains. As a geographer I was particularly interested in the spatial ideas. In this context I was already aware that some of the earliest of these were manifest in manuscript maps made by two white fur traders during the last quarter of the eighteenth century. Although not a primary concern, I had idly wondered at the ability of Alexander Henry the Elder and Peter Pond to draw apparently original maps of extensive areas hitherto virtually unknown to whites; in the case of Henry's on the basis of only one season's travels.

The probable explanation came to me in a footnote in a recently published secondary source that I could equally well have been reading in the library at my own university: Alice M. Johnson, ed., *Saskatchewan Journals and Correspondence 1795–1802* (London: Hudson's Bay Record Society, 1967), 26: 253 n.1. The note referred to five Indian maps in a manuscript journal of Peter Fidler, one of the Hudson's Bay Company's early-nineteenth-century employees, only one of which had then been published and that rather obscurely. The possibility that Henry, Pond, and other traders might have incorporated maps made by Indian guides struck me immediately. It was a hot afternoon, I needed fresh air, and Fidler's manuscript journal was less than two miles away from Bloomsbury, near St. Paul's Cathedral in the Hudson's Bay Company's Archives at Beaver House. I walked there, enjoyed the fresh air, experienced for the first time the smell of pelts, was invited to wait for the archivist's attention while sitting on an old but sumptuous sofa from which, when the time came, I found it difficult to rise, and was then politely refused permission to see the journal until I could produce a letter of introduction. A few weeks later, armed with a letter from someone unknown to the company but whose credentials could be confirmed in an appropriate directory, I revisited Beaver House and at a very small desk was allowed to see the Fidler journal. My excitement at seeing the Indian maps was, if anything, heightened by the sound of a fur auction taking place elsewhere in the building.

The Beaver House experience might have been a false start rather than a beginning had I not participated in the November 1973 meeting of the Society for the History of Discoveries in Washington, D.C. I was taking part for the wrong reasons. For some time I had wanted to visit the Geography and Map Division of the Library of Congress in order to see and describe Alexander Henry's unpublished manuscript "A Map of the North West Parts of America." No research funds were available but, from a different fund, the University of Sheffield paid for me to attend the society's meetings, conditional of course on reading a paper. This was concerned with late-eighteenth-century attempts to map the boundaries between three of North America's great biogeographic regions: the grasslands of the northern Great Plains, the forests of what would eventually be the northern parts of Canada's three Prairie Provinces, and the tundra beyond these still further to the north. Some of my evidence was based on Indian maps, one of which I had first seen at Beaver House on a second visit. At the end of the session I was approached by Herman R. Friis of the National Archives. We had met briefly on a previous occasion and he had from time to time supplied me with research materials. Addressing me in his usual friendly manner, but at first routinely and with something less than his usual enthusiasm, he said, "Malcolm, I enjoyed your presentation." With great emphasis he then added, "But why did you not say far more about those Indian maps? No one in this country has shown any interest in that topic for decades. You should take it seriously." In my reply I mumbled something about only having mentioned the maps because they afforded a small part of a much larger body of diverse evidence and that I made no claim to being a historian of cartography.

The brief conversation with Herman Friis

might have been another false start had it not been for another chance event later that day. Attending the society's annual dinner, I sat with one of the very few people already known to me, David Woodward, then the first director of the recently established Hermon Dunlap Smith Center for the History of Cartography at the Newberry Library, Chicago. David's comments on my paper were similar to those of Herman Friis, but instead of just advising me to develop a serious interest in the maps of native North Americans he added, "Why not come to the Newberry Library and conduct a feasibility study? We have most of the reference sources you would require. I'm sure something could be arranged." In 1977 I did precisely that. It was not a false start but the real beginning of a preretirement research interest that in 1990 was to phase into a postretirement preoccupation.

Very quickly after 1977, I was embarrassed to find myself being approached and referred to as the authority. It was not a mark of achievement but confirmation that no one had been seriously interested in the topic. By means of research visits to archives, museums, and libraries, in the course of participating in conferences, and through a series of published papers, I seemed in the 1980s to revive interest in a topic that had stagnated since about the time of World War I. It had not been my intention to do this. At the time I was not even aware that I might be doing so. Even now, I feel that the revival might have begun if I had never got beyond my first false start. The 1980s was, after all, a decade of ferment in the attitude of whites toward native peoples, not least in social anthropology, ethnohistory, and matters relating to native artifacts hitherto in the public domain.

I had always intended to write a book about Indian and Inuit maps but the excitement of searching for new items and researching their provenances repeatedly drew me away from home and desk. By 1991 I no longer thought of myself primarily as a geographer but as one who was opening up a backwater in the history of cartography. Ironically, it was the death in that year of one of the most eminent scholars in the wider field that resulted in my involvement with not one but two books.

When Brian Harley died in December 1991 at the peak of an active career in the history of cartography he inevitably left much work that was incomplete and some projects that had only just been conceived. His involvement with David Woodward in coediting the volumes in the University of Chicago Press's ongoing series *The History of Cartography* caused David to rethink the strategy for future volumes. I had already agreed to write a chapter on maps and mapmaking by native North Americans. David's decision to include this in a separate volume about cartography in traditional African, American, Arctic, Australian, and Pacific societies was a major change of plan. In inviting me to coedit the volume with him he exposed me for the first time to traditional cartography outside North America. The invitation, which I had readily accepted, brought me into regular contact with a global network of anthropologists, geographers, archeologists, and historians of cartography, virtually all hitherto unknown to me. In the process of helping to shape their contributions to the volume I gained a new and valuable perspective on what we have come to refer to as traditional cartography.

Cartographic Encounters was also a consequence of Brian Harley's death. He had agreed to give the eleventh Kenneth Nebenzahl, Jr., Lectures in the History of Cartography at the Newberry Library. The topic was to have been "Colonial Cartographies." No attempt was made to find a replacement for what would doubtless have been a brilliantly conceived, lively, and controversial series given in Brian's inimitable style. Instead, David Buisseret, then director of the Smith Center at the Newberry Library invited me to organize a series of lectures on a quite different topic. In June 1993, Elizabeth Boone, Patricia Galloway, Peter Nabokov, and I gave five lectures in a series entitled "Cartographic Encounters: Studies in Native American Mapping." In the tradition of most of the previous Nebenzahl Lectures it was hoped from the outset that the series would be the basis for a book. But the five forty-five-minute lectures did not, and indeed could not, open up a field that was by then being recognized as having many facets, as having significance in a number of different contexts, and as having just begun to recover from decades of dormancy and neglect. Clearly there was need for a book, but if it was to be an integrated statement, I had to expand my essentially background and coordinating contributions. Equally important, we had to seek additional substantive contributions. Our failure at that stage to find one or more contributors able and willing to write about Inuit maps, mapmaking, and map use seems even more regrettable now than it did in 1993. On the other hand, the commissioned contributions of Morris Arnold, Barbara Belyea, Margaret Pearce, and Gregory Waselkov have added immeasurably to the time scale, cultural range, and contexts of the original lectures. Even more important, they have manifested different epistemologies and introduced additional ideas. Without them the book would doubtless have been better than the lectures, but would not have been as stimulating in its diversity as I now believe it to be.

The first part of this Preface describes how my interest originated and how the book emerged. It does not, however, declare an epistemology, partly because at no stage in the writing did I adopt one consciously but also because to have done so would have involved a degree of introspection likely to have confused rather than helped the perceptive reader. I will, however, enumerate a few personal factors that I know to have influenced the thrust of my contributions.

My background was in the natural sciences and not, as my current interest might lead readers to suspect, in either the humanities or social sciences. As an academic geographer I have always been more interested in space than place. Hence, my interest in maps and their referents. My acquired epistemology, though not as Darwinian as it once was, is still mildly developmental. Although a frequent visitor to North America during more than forty years, nine tenths of my life has been lived in England. Hence, I view most things North American as a reasonably well informed outsider. My formal knowledge of native North Americans' history and culture is minimal, verging on the inadequate. I came to their maps via an interest in cartography and not via cultural anthropology. Therefore, though my knowledge of these maps may well be historically, regionally, and thematically wider than that of any of my predecessors, the frameworks within which I here present it and the methods by means of which I examine it have been adopted and adapted somewhat randomly in the course of my personal encounter.

All readers should note the unlikely background from which I have approached my tasks in this book. Perceptive readers will become aware of—and I hope be stimulated by—the differences between my approaches and those of the other authors.

At the risk of offending some of the hundreds of individuals who have helped me in specific ways, I name those whose influence I feel to have been vital. I do so more or less chronologically.

Long ago, Arthur Robinson made me aware that maps were worthy of study in their own right. Then, Margaret Wilkes helped me to realize that some of the most interesting were manuscript maps made by individuals who possessed little or no formal background in cartography. It was Herman Friis who first suggested that I should develop a serious interest in maps made by native North Americans and David Woodward who then helped me to begin implementing that advice with work at the Newberry Library. While there I met my first Indianists, Francis Jennings and Helen Hornbeck Tanner. I also began to develop a network of correspondents. Of these, Norman Thrower was perhaps the first, Ed Dahl certainly the longest serving, and Charles Martijn the last but most spontaneous.

I am also indebted to the Department of Geography at the University of Sheffield. Its secretarial and technical staffs gave me practical help and invaluable advice over many years. My academic colleagues showed remarkable tolerance toward an aging colleague who seemed to be escaping more and more into what must have been seen by them as an obscure backwater; and one that drew me away to North America with increasing frequency.

Research in the United States and Canada was financed from so many sources and over so many years that I apologize if I have inadvertently omitted any from the following list: the Social Science Research Council and its successor the Economic and Social Science Research Council; University of Sheffield Research Fund; British Academy; Canadian High Commission (London) for four awards; Newberry Library, Chicago; Ontario Heritage Foundation; and the History of Cartography Program, University of Wisconsin, Madison. I was also involved as a consultant in National Endowment for the Humanities programs at the University of Wisconsin, Milwaukee.

To all these individuals and bodies I express my sincere gratitude. Without them my contributions to this book would probably never have been begun.

I have one regret that is also an apology. I regret that I have not made more opportunities to discuss my interests with native North Americans. I beg their corporate forgiveness and hope that they will find in this book little from which they dissent, much with which they can agree, some things that are new to them, and others that they will find helpful.

I am indebted to others who have influenced the final form of this book. Jim McNeil, my research assistant, made an invaluable contribution in transforming disorganized records into a working cartobibliography. Kenneth and Jossy Nebenzahl made possible the lectures from which it stemmed. David Buisseret first suggested that the lectures could be the basis of a book and gave initial advice as to how this might be achieved. The seven coauthors each accepted the invitation to collaborate, responded enthusiastically to my early suggestions, and then showed remarkable tolerance on being put through at least three cycles of revision. I hope that they will consider my own contributions, and in particular chapter 4, to have integrated theirs into a book that does justice to both its title and its objectives as announced in the Introduction. To all these I also express my sincere thanks. Without them there would not have been a publishable manuscript.

Particularly since his appointment in 1996 as director of the Smith Center at the Newberry Library, Jim Akerman has been increasingly involved in bibliographic checking, giving me sound advice, and liaising with the University of Chicago Press. From my home in suburban Sheffield, England, the first and most demanding of these would have been quite impossible. Thank you, Jim, and thanks also to the Newberry Library, which, as both an institution and collection of scholars, has helped me in numerous different ways over so many years.

Jim Akerman and I, together with the seven other contributors to this volume, wish to express our gratitude to Carol Saller, manuscripts editor at the University of Chicago Press, for her patience, perceptive observations, and constructive advice.

Finally I wish to thank my wife, Margaret. For almost four years, she has assisted, typed, acted as secretary, listened, criticized, checked my grammar, given commonsense advice, and repeatedly irritated me by reminding me of deadlines. Without her, my twin roles as editor and author would not have been completed and, notwithstanding the advice and cooperation of others mentioned and unmentioned here, this book would never have reached publication. I dedicate *Cartographic Encounters* to her in gratitude and with love.

G. Malcolm Lewis

Introduction

G. Malcolm Lewis

The extended encounter that began in North America at the beginning of the sixteenth century and has continued ever since was between its native peoples and successive incursions of aliens from elsewhere. There were fewer than five million natives living in various parts of the continent when, in 1501, Miguel Corte-Réal kidnaped fifty Micmacs or Beothuks on the coast of what was to become Maine, and sold them as slaves in Lisbon.[1] In the five hundred years since then, the number of natives has declined to fewer than three million, whereas the descendants of the many alien settlers are fast approaching three hundred million.

The encounter involved far more than demographic changes. Extending through half a millennium, at various times it has been played out on different parts of a more than seven-million-square-mile stage. In many respects, the native population was almost as diverse as the alien.[2] Initial encounters occurred in different places at different times. Aspects of the encounter were sometimes experienced before face-to-face contact: infectious diseases, changes in indigenous trading patterns, perhaps the beginnings of some native trade languages, and "domino" displacements of native groups by neighbors. Once underway, the consequences of the encounter ramified and the ramifications continued for a long time; indeed, for the native peoples forever, or until extinction. Its nature varied from one historical period to another and between one region and the next. But the encounter was far more than a series of events in time and space. It was a complex mix of many processes: face-to-face and indirect, intentional and consequential, immediate and delayed, conscious and intuitive, bilateral and unilateral, unique and universal, and so forth. Together, the sundry processes affected the spiritual, moral, intellectual, cultural, social, medical, and material worlds of all who were directly or even indirectly involved; especially those of the native peoples.

Among the face-to-face encounter processes, native-white communications were very important. Indeed, in two contexts they were paramount: negotiations of all kinds, and information seeking; the latter almost always at the request of whites. Native spoken languages were significantly different from those of the whites. Even with the emergence of white interpreters and the teaching of European languages to natives, speech continued to be an inadequate mode via which to communicate certain types of information. For reasons of syntax and inflection, the effective communication of relationships of almost any kind was particularly difficult. These included relationships between things and events in space.

In the early-encounter period, the need to communicate information about celestial and cosmographic space was unimportant. Interest in these among whites came later with intellectual attempts to understand native worldviews. In contrast, from the very beginning whites needed to know about terrestrial space. Their near *terrae incognitae*, at least, were the *terrae cognitae* of the natives with whom they had direct encounters. Some of the latter had considerable knowledge of terrain, waterways, and coasts far beyond areas directly experienced by the encountered whites. This was particularly so of tribes occupying extensive hunting and fishing territories; likewise, of native traders, especially so if they spoke one of the trade languages. At most times and in most contexts the encountered natives were willing to communicate their knowledge of rivers and lakes, portages and passes, watering places and mineral locations, distances and directions, friends and enemies, and all manner of other geographical information about the world beyond that already experienced by the whites. Almost inevitably, for this kind of purpose speech proved to be an inadequate mode of discourse. Writing, the alternative mode used by most of the white groups, did not exist among native North Americans, but a shared nonlinguistic mode for communicating spatial information emerged quickly and spontaneously: graphics of a kind that Europeans intuitively described by words derived from the Late Latin *mappa* (in English, Spanish, and Portuguese, for example) or from the Late Latin *carta* (in French and German, for example).

Employed by natives to communicate with whites, and much less frequently vice versa, maps were usually supplemented by gesture and speech. There is evidence of maps being used in every part of the continent from the earliest encounter time onward,[3] in the remoter regions of the Arctic and sub-Arctic until well into the twentieth century. Native adults of both sexes, probably all tribes, and many different positions within society possessed the ability to communicate by means of maps. Indeed, among the sane and the sentient it may well have been a universal trait. Natives made their maps on or with all kinds of organic and inorganic materials and mapped the patterns in many different ways. In doing so they were almost certainly drawing on an indigenous pictographic method for leaving messages and recording cultural traditions.

The speed with which, after the initial encounter, some natives began to make maps for whites suggests that they were not merely adapting a pictographic tradition but drawing on a more specialized one that had some of the characteristics of cartography. Native North American languages did not begin to include equivalents of *map* until the nineteenth century or later and even then the concepts embodied in the noun elements suggest European influences. Nevertheless, from a fairly early stage in the encounter, whites in several parts of the continent reported finding indigenous examples of what were recognizably and functionally maps.[4] Quite clearly, the indigenous pictographic tradition included many items that to whites neither looked like maps nor served any of the functions of maps, some that were in part maplike, and others that had many of the characteristics and functions of the whites' own maps. For how long before the encounter these had been made is still unclear. The evidence of rock art is as yet ambiguous and difficult to date. Nevertheless, this was undoubtedly the pictographic subtradition intuitively used by natives in making the earliest post-encounter maps for whites.[5] Thereafter, acculturation was inevitable, though its extent and nature varied; not least with the immediate context in which a map was made.

Some indigenously made maps have survived, as have many more examples of maps made by natives to communicate with whites, and a considerable body of contemporary non-native literature relating to specific maps, how they were made, and the contexts in which they were used. The evidence is dispersed but has been drawn on by authors contributing to an ever increasing body of secondary literature. Very little of this, however, has focused explicitly and consistently on the primary encounter process in the field, in which so many of the maps were discovered by or made for whites. Likewise, the earlier secondary literature has never been reviewed as a body of evidence relating to the quite different encounter, in which scholars of all kinds tried to interpret evidence found in museums, archives, and libraries. Nor has there been a recent attempt to consolidate what we now know about the nature of the native maps encountered during the past five hundred years. These are the subjects of the three chapters in part 1 of this volume.

By placing maps and mapmaking in the encounter process the contributors to this volume have focused on one of the many thematic contexts in which native and white thoughtworlds coexisted. In so doing, we feel that we have contributed to an understanding of native/non-native history seen as "the process of two thoughtworlds that at the time were more often than not mutually unintelligible."[6] This, however, was not our primary objective. We began with a shared interest in maps, mapmaking, and the spatial understandings that they revealed. By situating these in the encounter context we have shifted the center of gravity of native-map studies away from the gross but innocent Euro-American centrism of an earlier era and redressed the scientism that seemed likely to introduce a new centrism in more recent times.[7] It remains to be seen whether a future generation of scholars will try to shift the center of gravity still further by focusing attention exclusively on truly indigenous maps and mapmaking in entirely native contexts.

Whereas part 1 is concerned with past encounters and conclusions that can be drawn from them concerning the nature of the maps, part 2 is concerned with recent and current encounters. Chapter 4 reviews recent scholarly encounters in three contexts: historical (including archaeological), anthropological, and map-creating. It is, however, far more than just a review. Interwoven in it are contextual introductions to the core of this volume, the seven contributions (chapters 5 to 11) by invited authors. To write more here about each contribution would be to anticipate and, indeed, duplicate chapter 4.

What is appropriate at this point is to stress their distinctiveness as a set. First, they are original contributions, written by specialist authors. Second, though no less important, although no constraints were placed on the contributors, together their chapters reflect and make a significant contribution to the second (post–approximately 1970) encounter by scholars with native maps and mapping. Third, from the outset of this project in a small room in the Newberry Library, Chicago, and almost before potential authors began to be named, it was envisaged that these chapters would together constitute the volume's vital core, as examples of the current scholarly encounter and indicators of some of the ways ahead. At no stage was there an attempt to impose on the authors a working definition of "map," to define objectives, or to restrict epistemologies. Each wrote more or less independently of the others. I hope that the few but inevitable contradictions will serve to stimulate further thinking. The mix of authors was a consequence of our awareness of latent interests and work then in progress. Our one attempt to achieve a degree of representativeness failed when, very regrettably, we could not persuade anyone to write about aspects of Inuit maps and mapmaking. My task as editor was to place the core chapters in wider historical, geographical, and ideological (in the "science of ideas" sense) contexts. Once that task was underway it at times began to develop a life of its own. Readers who are in any doubt about the balance should ignore my frame and concentrate on what it frames.

The frame is not symmetrical. Indeed, it could not be. Future encounters are notoriously difficult to predict. Viewed retrospectively, proximate predictions are often found to have had at least some validity, but mediate predictions can be wrong, and ultimate predictions woefully so. Yet it would be harder to justify an incomplete frame than to present an asymmetrically framed picture. Hence, in part 3 an attempt is made to anticipate the nature of future encounters within what seem likely to be the main contexts, not as an abstract futurological exercise but in an attempt to give some direction to a still youthful field of inquiry. It is a field that to date has attracted very few full-time researchers, has interested rather more part-time and once-only investigators, and has failed abysmally to announce its existence within many communities of potentially interested scholars and scientists: men and women, including graduate students, who would have much to give, share, and take if they only knew of its existence.

Only in one respect has writing part 3 been easy: because the future is an empirical unknown, the chapter contains far fewer notes, quotations, illustrations, and other scholarly apparatus than any other part of the volume. Yet writing it has been troublesome. Not to have written it would have been a dereliction of responsibility. Doing so has been to risk leading others toward unsuccessful scholarly encounters. To suggest several ways ahead is to risk unintentionally detracting attention away from established ways and others of which I am at present unaware. To assume that initiatives will continue to come from non-natives is to risk underestimating or even insulting the native population. There are, of course, lesser risks and perhaps other major ones as yet unanticipated by me.

When this book is reviewed, I will be particularly apprehensive concerning part 1, and we will each look with a mixture of hope and anxiety for opinions concerning our respective contributions to part 2. For my part, however, I will be particularly interested in readers' opinions twenty years hence. It would be conceited to hope for a retrospective review, but I would like still to be around to undertake my own evaluation of part 3. How valid were its predictions? What, if any, were the unanticipated encounters? Did it signal any false directions and, if so, with what consequences for those who followed them? I apologize in advance to any who by then feel that they were misled. False directions, however, often lead to unexpected intersections. I suspect that each contributor to this volume can look back at a point or brief period in time when, from a noncartographic professional perspective, he or she reached such an intersection, became aware of traditional native American mapping for the first time, and developed their interest in the newly encountered.

I reached my unexpected intersection many years ago in the great domed Reading Room of the British Library, London. Perhaps surprisingly, I recognized the intersection for what it was and altered direction accordingly. We must not be frightened by unexpected encounters. Sometimes they lead to exciting exchanges of ideas and information.[8] Conversely, they may lead ourselves and others toward quite different categories of encounters from those once anticipated. Hence, the scruples that I have about publishing part 3 are overridden by the sincere hope that directly or indirectly it will prepare some "outsiders" to recognize unexpected intersections and lead them to share and further the understanding of cartographic encounters between native and non-native peoples in North America. Twenty years should be a sufficiently long period over which to test that hope.

Notes

1. William H. Goetzmann and Glyndwr Williams, *The Atlas of North American Exploration* (New York: Prentice Hall, 1992), 20–21.

2. Although alien peoples from almost every other part of the world ultimately entered North America, whites— essentially Europeans and their descendants— were by far the largest and most powerful group to encounter the native peoples. Hence, from this point in the text, "white" is frequently used to refer collectively to "alien" and "non-native" peoples.

3. Although earlier occurrences can be inferred (e.g., from Peter Martyr's map of 1511), the first certain case of a map being made by a native for a European dates from September 1540, when a Yuma (probably Halchidhoma) Indian made a "charte" of the lower Colorado River for Hernando de Alarcón, the first white man to ascend the river and within one year of Francisco de Ulloa having discovered its delta from the sea. Interestingly, in this earliest of known cartographic encounters, the Indian asked the Spaniard to make a map of the country that he came from. One was made, though whether of Spain or Mexico is not clear. Richard Hakluyt, *The Voyages, Navigations, Traffiques, and Discoveries of the English Nation* (London: George Bishop, Ralfe Newberie and Robert Barker, 1600), 3:438.

4. The first recorded finding of an indigenous map dates from July 1540. Although there is some internal ambiguity in both the Ramusio and Hakluyt published compilations of travel writings, in that month Francisco Vásquez de Coronado reached Granada (the first of the so-called Seven Cities of Cibola; now generally supposed to have been Hawikuh). It had already been abandoned by the Zuni Indians, but a painted skin (probably buffalo) was found that was interpreted by the Spaniard as a map of all seven pueblos, six of which he had not yet visited. They were distinguished from each other according

to some measure of size and linked by a route or routes. George P. Hammond and Agapito Rey, eds. and trans., *Narratives of the Coronado Expedition 1540–1542*, Coronado Cuarto Centennial Publications, 1540–1940, vol. 2 (Albuquerque: University of New Mexico Press, 1940), 171 and 173–76. Neither of the original versions indicates on what principles the Spaniards recognized the painting to be a map; though seven components must have been an important characteristic. Nor is there any indication that they subsequently tested the information in the field in order to verify the painting's status as a map.

5. The nature of native maps and mapmaking traditions, their possible cognitive and cultural roots, and their postcontact acculturation are considered in greater detail in later parts of this volume, especially in chapter 4.

6. Calvin Martin, "The Metaphysics of Writing Indian-White History," in Calvin Martin ed., *The American Indian and the Problem of History* (New York: Oxford University Press, 1987), 33.

7. The "earlier era" of native-map collecting and studies ended in the first quarter of the twentieth century and is reviewed in chapters 1 and 2. The developments of "more recent times" are traced in chapter 4. The word "scientism" is not used there but is introduced here to embrace growing but rarely explicit criticism of the work of those who supposedly consider the sciences to be "more important than the arts for the understanding of the world in which we live," adopt the position that only "a scientific methodology is intellectually acceptable," and see all "philosophical problems" as "scientific problems" that "should only be dealt with as such." Paul J. P. Noordhof, "Scientism," in Ted Honderich, ed., *The Oxford Companion to Philosophy* (Oxford: Oxford University Press, 1995), 814. Though I have not, indeed could not have, adopted any of these extreme positions, the relative silence with which one of my recent papers was received seemed to imply that among a significant group of people with established interests in native maps and mapmaking the methods and evidence of metrics, geometries, semantics, and linguistics were either unappreciated or unwelcome: G. Malcolm Lewis, "Metrics, Geometries, Signs, and Language: Sources of Cartographic Miscommunication between Native and Euro-American Cultures in North America," in Robert A. Rundstrom, ed., "Introducing Cultural and Social Cartography," *Cartographica* 30, 1 (1993): 98–106.

8. For example, a few hours before writing this sentence, I received an unanticipated letter from a complete stranger. His interest is in Ice Age parietal images in western Europe. His unexpected intersection occurred when he suspected that these images have some of the characteristics of maps. This led him to seek information about maps made in traditional cultures during historical times, in the course of which he became aware of several North American examples described in this volume. I will, of course, reply, but doing so would be easier and far more satisfactory if *Cartographic Encounters* were already in print.

PART 1

The Four-Hundred-Year First Encounter

Frontier Encounters in the Field: 1511–1925

G. Malcolm Lewis

Like every chapter in this book, this one is a report of a recent encounter between an author (the subject) and one or more aspects of mapmaking and map use by native North Americans (the object or objects) in the course of a historically extended encounter with non-natives. The two aspects on which I focus here are not, of course independent of self. I selected them and, having done so, brought to bear a unique combination of perspectives, skills, experiences, and biases. The latter combine limitations and attitudes of which I am at least in part conscious with lacunae or silences of which I am for the most part not. I have tried to summarize all these in the Preface, and the perceptive reader should be aware of them in approaching this and my other chapters. They have inevitably influenced the chapters' respective focuses, structures, emphases, interpretations, and objectives.

This chapter is concerned with maps, mapmaking, and map use among indigenous peoples as described and sometimes transcribed by whites in a more than four-hundred-year encounter on the North American continent. As with the Industrial Revolution, very few who participated in the Encounter were aware of doing so. It is the name given retrospectively to include the events associated with, processes involved in, and consequences of the meeting of white with indigenous peoples; an extended event paralleling, often interacting with, and sometimes difficult to separate from other extended events such as frontier expansion and agricultural settlement. Though interactive, evidence of it is predominantly that of contemporary white observers and retrospective interpretations by white archaeologists, ethnohistorians, and anthropologists; so much so that it is difficult to consider any aspect of it from a non-white perspective. Regrettably, but inevitably, this is so for the aspects of the encounter considered in this chapter.

The possibility that some prehistoric rock art incorporates maps is now debated and evaluated almost exclusively by white specialists, with perhaps a very few white-educated and -trained indigenous specialists. Contemporary accounts of maps, mapmaking, and map use were almost exclusively written by white observers. Likewise, extant map artifacts were for the most part collected, described, and interpreted by whites. Similarly, contemporary transcripts of maps were made, annotated, and selectively modified almost exclusively by whites. To a considerable extent, categorizations of artifacts as maps, of procedures as mapmaking, and of functions as map use were likewise imposed by whites. Yet, except for the recent past, these alien perspectives on indigenous cartography are almost all we have in the way of evidence. This chapter reviews that evidence from 1511 to 1925, the latter a somewhat arbitrary date,[1] selected because it marks the approximate beginning of a hiatus in the recording and collecting of traditional-style evidence.

Not only was the evidence reviewed in this chapter mediated by whites, it has been mediated again by me in my role as subject. Although with every intention of enhancing what was thought to be known already, I have inevitably distorted. In focusing on artifacts, processes, and functions that are of considerable interest within cartography and to its history I will be seen by some to have detracted from—even trivialized—the traditionally recognized major encounter processes of which they were but a part: for example, geographical displacement of indigenous peoples, the spread of diseases, and acculturation. My reply to such criticism is that every focus distorts. Whether a distortion is justifiable depends on the extent to which new insights lead to different and richer interpretations of the object which was temporarily distorted.

This chapter has two objectives. Drawing mainly on documentary and printed evidence, I have tried first to establish how, why, and in what contexts the indigenous peoples of North America made and used maps within their own societies before, at, and immediately after they were encountered by whites. To enumerate all the distorting factors inherent in this task would be unproductive, but one of them does need to be stated. I began with the received assumption that at least some indigenous North Americans made maps at, and therefore before, first contact. Whether these peoples were aware of maps as a functional category of cultural products or of mapmaking and map use as discrete activities did not at the time concern me. Much as I would now like answers to these questions I do not feel that there is as yet conclusive evidence for or against. Recreating the cognitive styles of peoples who left no written records of their own is a difficult task and leads to tentative conclusions that are usually unverifiable.

The second objective of this chapter was somewhat easier to achieve than the first. Using artifactual as well as documentary and printed evidence, I have examined maps, mapmaking, and map use as recorded in a range of encounter contexts. Whereas the focus is almost exclusively on maps made by the indigenous peoples, virtually all the evidence originated with whites. Extant examples of truly indigenous pre-nineteenth-century map artifacts are exceedingly rare. Pre-nineteenth-century accounts are often ambiguous, but one form of ambiguity should be noted. Most accounts use the noun "map" or the verb "to map" or their equivalents in other European languages. For the sixteenth, seventeenth, and even eighteenth centuries this raises the question, to what kinds of European and Euro-American maps were the indigenous artifacts being compared? Similarly, to what kinds of mapmaking procedures? Such questions cannot be answered, except occasionally in specific cases and on the basis of considerable research. In being aware of them, however, we must not forget that those who recorded their observations were the exceptions among the whites involved in the encounter. They were literate and, therefore, better educated than most of their white contemporaries. The contexts in which they recorded their observations almost all involved the need or desire for information. In the earlier phases of the encounter the needs included personal and group survival, political and military strategies, and a variety of entrepreneurial aspirations. Later on, the needs were somewhat less immediate and usually less vital. This was particularly so among nineteenth-century scientists. Curiosity led them to seek information about almost every aspect of indigenous peoples, including their maps, mapmaking, and map use. Scientists also learned to use these maps in facilitating their own systematic surveys.

What we do not know is the extent, if any, to which white artisans involved in the encounter—foot soldiers, French-Canadian *voyageurs*, and emigrants on the western wagon trails, for example—observed, solicited, understood, or were even aware of maps made by indigenous peoples. If they had recorded any of these things, the conclusions we might have drawn from their observations might have been different from the ones I have reached here. It could be, of course, that whites who did not themselves make and use maps elicited information from indigenous people only in noncartographic modes.

Pre-encounter and Indigenous Post-encounter Mapmaking

Native peoples throughout North America made maps indigenously, certainly from early contact times and probably long before that. They were made in a number of contexts, of which the following were important: as messages or instructions to others; as interactive planning; in order to reconstruct past events and record them for posterity; to make sense of the world beyond that of direct experience and relate it to the known world; and to divine. The list is not exhaustive.

At their simplest, maps were ephemeral gestures. Because these were fugitive, they are known of only in encounter contexts. For example, in 1761 the chief of the Pookmoosh Band of Micmac, in what was to become New Brunswick, whose people had hitherto been allied to the French, encapsulated their changing geopolitical situation in one dynamic hand gesture. It was a case of what, almost two centuries later, and in a quite different context, Winston Churchill called "The Closing of the Ring." The French Acadians had been deported or taken flight and the English, some Scots, and, most of all, people from the New England colonies were about to move in. The Micmac perceived this to be a threat. "Their chief made almost a circle with his forefinger and thumb, and pointing at the end of his forefinger, said there was Quebec, the middle joint of his finger was Montreal, the joint next [to] the hand was New York, the joint of the thumb next [to] the hand was Boston, the middle joint of the thumb was Halifax, the interval betwixt his finger and thumb was Pookmoosh, so the Indians would soon be surrounded, which he signified by closing his finger and thumb."[2]

Far more important than gesture maps were the message maps left at strategic locations by departing persons for the information of others expected to arrive or pass by. Particularly common on birchbark sheets and somewhat less so on the exposed white wood of conspicuously blazed trees, the practice was most characteristic among the hunting peoples of the Northeast woodlands. The best known and probably oldest extant indigenous example of the former is that found in 1841 on the Ottawa River–Lake Huron watershed. Almost certainly made by Ojibwa, its geographical content is minimal; just sufficient to indicate who had left it, the route via which they had arrived, the location of the previous night's camp, and the intended route ahead.[3] Such maps are known to have been used from the late seventeenth century onward.[4] Maps on blazed trees were almost certainly less common. They seem usually to have been associated with military activities. Prominently placed, they were symbols of power, resistance, success in battle, or aggressive intent. That described and illustrated by Bray in 1782 is a good example (fig. 1.1).[5] It incorporates stylized plans of three forts (certainly Detroit [9] and Fort Pitt [10], and perhaps Fort Loudoun [8]) and a map of the confluence of the Allegheny and Monongahela Rivers at Fort Pitt to form the Ohio River, together with the adjacent civil settlement of Pittsburgh (11). The whole is certainly not a map but, like many pictographic messages, some of the components are recognizably carto-

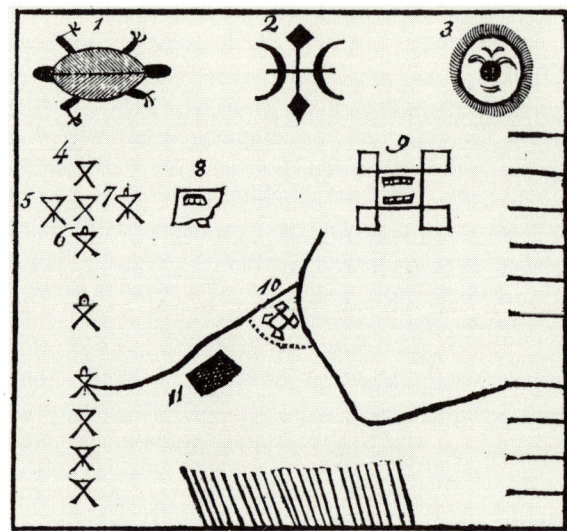

1.1 Wingenund (a Delaware warrior), map incorporated in a painting on the trunk of a blazed tree seen adjacent to the Muskingum River (Ohio) in or a little before 1781, as redrawn in William Bray, "Observations on the Indian Method of Picture Writing," *Archaelogia* 6 (1782): 159. Courtesy of the Newberry Library.

graphic. In the absence of written words and, more specifically toponyms, such composite pictographs conveyed to the initiated information about places, either by means of identifiable plans or distinctive networks.

An event involving mapmaking for instructional purposes was reported by Col. Richard Dodge on the basis of an account of events which occurred almost fifty years before. Sometime in the 1820s, Comanche braves were about to undertake a journey of approximately one thousand miles from central Texas to near Monterrey, Mexico, and back. As none of the braves had been there before, they were instructed in advance by older men who had. The route to be taken was divided into days of travel. For each day a map was made on the ground. This was memorized by all the young men before the map for the next day's journey was made.[6] Military activity was also the context for an extremely good example of mapmaking done as a basis for interactive planning. Sometime between 1860 and 1865, warriors from three Nootkan villages on the west coast of Vancouver Island were planning to attack a fourth village some ninety miles to the north. Only one man knew the village well, because he had courted his wife there. He was asked to make a map of it. The warriors all went down to a beach where the chosen man modeled a map three-dimensionally in the sand. It was very detailed, showing critical topographic features, tracks, and individual houses. "All this time the warriors . . . stood round the delineator in a large circle . . . questions were asked and eager conversation held." Only then was the general plan of attack, which had already been proposed, finally accepted.[7]

Made for use at a federal government council in Washington, D.C., in 1837, an Iowa chief's map was produced, apparently spontaneously, in trying to resolve a dispute between two groups of Indians concerning land. It shows "the route of my (Ioway) forefathers—the land that we have always claimed" and does so against a topological but detailed representation of the upper Mississippi and Missouri River basins; in all, more than one quarter of a million square miles. Though apparently lacking a beginning or an end, the dotted line zigzags across much of the western Middle West. Dots in circles appear to represent locations where the ancestors had settled in the course of a long and complicated migration.[8]

Maps preserving traditions for later generations are the least known but most indigenous. It is impossible to estimate how many are still preserved by native people. In *Maps and Dreams*, Hugh Brody described a public hearing held in 1979 by the Northern Pipeline Agency for the Beaver Indians of northeast British Columbia. The first day's proceedings were almost over. They had involved the submission of evidence in map form: hunting, trapping, and berry-collecting territories carefully and systematically plotted for the occasion on modern topographic maps. Then, as the meeting was about to break up, two Indians produced a large moosehide bundle. When opened, it was as large as the table top and was revealed to be a dream map. No one knew how old it was. It showed heaven, the trail to it, a false trail, and animals. All its content had been discovered in dreams. A corner of it was missing. The detached part had been buried with someone who would not otherwise have found it easy to get to heaven. Dream maps are unpacked only on very special occasions. Their intricate routes and meanings are difficult to understand. When the owner of a dream map dies, it is buried with him. Not surprisingly, they are little known.[9] Likewise, little was known of the Southern Ojibwa midé migration scrolls until Selwyn Dewdney made a partial inventory and wrote about them in the 1970s. He was, incidentally, able to locate only eight certain and seven possible examples.[10] The Pawnee sky chart on buckskin at the Field Museum, Chicago, is another example of a map made to preserve tradition.[11] But we must be careful. William Gartner, a graduate student at the University of Wisconsin, cautioned that it "is not merely a map of the celestial sky. Its direct uses are as a beacon for heavenly forces, as an earthly guide, as a symbol of cosmological unity, and as a flag of identity during the Thunder and/or Great Washing ceremonies."[12] This is a level of interpretive understanding rarely attained by non-natives.

All the indigenous maps referred to thus far were based on information derived via experience, by tradition, or in dreams. Some patterns considered to be maps were, however, produced by chance. Especially among the Naskapi of Quebec and Labrador, divination involved inducing cracks on the scapula bone of certain mammals. Patterns induced by heat or percussion were often interpreted as maps and sometimes related to real-world features; especially rivers, lakes, and trails.[13] The same tribe also made decorative patterns by biting folded pieces of birchbark. Patterns were sometimes random or emerged by error in the course of trying to create something else. Deliberately or accidentally made, patterns were usually explained or given names. Frequently, these indicate an ability to read patterns as maps of trails.[14]

Regrettably, there are no pre-encounter artifacts reliably identifiable as maps.[15] Even if there were, it is doubtful whether it would ever be possible to determine the contexts in which they were made and used. Some rock paintings and glyphs do contain patterns which, from a modern Euro-American perspective, appear maplike. The purposes for which they were made, when, and by members of which cultures are, however, too much in doubt to afford bases for firm conclusions. Nevertheless, the indigenous uses of maps observed in the course of the encounter were almost certainly rooted in earlier practices; perhaps much earlier. It is very unlikely that maps made by Indians and Inuit in early-encounter contexts were spontaneous innovations made to satisfy alien whites.

Native-White Cartographic Encounters in the Field

No systematic study has ever been made of what was probably a universal and undoubtedly very old process: the acquisition of geographical information by alien cultures from the indigenous peoples whose territories they were invading. In North America, as elsewhere, Europeans did this from the very beginning—apparently, instinctively so.[16] Many, probably most, of the occurrences were unreported. Furthermore, where reports have survived, it is sometimes impossible to distinguish cartographic from gestural and oral transmissions. Nevertheless, there are several hundred unambiguous accounts of Indians and Inuit making maps, and a similar number of extant examples. Even so, as a proportion of the total literature of early European discovery and subsequent exploration, accounts of Indians and Inuit making maps are small. Likewise, extant maps make up a very small proportion of Indian and Inuit artifacts. The accounts give the impression that, even from a white perspective, mapmaking was a natural practice. For this, if for no other reason, most of the accounts are brief, conveying less information about the maps and mapmaking procedures than later scholars would have wished.

Occasionally, Indians were reported to have used maps spontaneously in communicating geographical information. On September 11, 1540, four young St. Lawrence Iroquoisan men "shewed" Jacques Cartier three of the Lachine

Rapids on the St. Lawrence River just above the site of what was to be Montreal, together perhaps with the lower Ottawa River. They did so with "little stickes, which they layde upon the ground in a certaine distance [representing the River(s)] and afterward layde other small branches between both, representing the Saults."[17]

Though not necessarily with "little stickes," spontaneously made maps were to be used on many occasions during the next 450 years, in the course of informing and guiding Europeans and, later on, Euro-Americans. The most widely known reports are of Inuit mapmaking in the nineteenth century, but examples can be found from every culture region in North America. Although historically late, one of the best documented is from the Moisie River, eastern Quebec, where in 1861, Henry Youle Hind led an expedition to the headwater region and the sources of the eastward-flowing Churchill River beyond. Louis and Pierre, his Montagnais guides, became increasingly unsure of the way ahead. They contacted another Montagnais, Domenique, and an unnamed Naskapi youth. Hind was advised to give the two Indians a good meal and allow them to rest. This was in preparation for soliciting geographical information, when the following events took place:

> "Where are you going to Louis?" someone enquired, as the Indian was rolling off into the woods with a torch of birch-bark, about an hour after supper.
> "Get birch-bark for map."
> "What map?"
> "Domenique going to make map of portages to show us the way. Tomorrow," continued Louis, with a knowing leer, "I speak to Domenique about young Nasquapee; Domenique well pleased—like supper, like tobacco, like everything. Think he will let young Nasquapee go."
> We sat by the fire till a late hour talking to Domenique and the young Nasquapee. The lad appeared to be very intelligent, and apparently knew the upper country well. He and Domenique together constructed a map of the Moisie and the old Montagnais route, as far as the dividing ridge—showing the point where the Ashwanipi River took its rise, and began its long course of several hundred miles to Hamilton Inlet, on the Atlantic Coast of Labrador.
> He put in all the portages, and explained the map to Louis and Pierre. The latter took charge of the map, and before we rose went over every little detail to see if he understood it perfectly.[18]

These events were painted by Hind's brother, William.[19] They are particularly interesting because the soliciting of information was by Indians from Indians; albeit in the context of guiding Europeans. Long after the first encounter, mapping was still an indigenous procedure among Indians and Inuit and not merely a means of communicating with white aliens.

Very occasionally, existing Indian maps were found and then used by Europeans. Three months before the stick map was made for Cartier, but more than two thousand miles to the southwest of the upper St. Lawrence River, Francisco de Coronado found at the Zuni pueblo of Hawikuh a skin (probably a buffalo hide) on which would already seem to have been painted the tribe's other six pueblos (the so-called Seven Cities of Cibola) and a route or routes between them. It is not clear how maplike the painting was or how Coronado recognized it to be a map, but the relative sizes of the seven pueblos were shown: one "somewhat larger" than Hawikuh; "another

1.2 The missionary lithograph on the verso of which the Quebec Inuit Wetalltok had drawn a map of the Belcher Islands, Hudson Bay, in or before 1910. From the American Geographical Society Collection, University of Wisconsin–Milwaukee Library.

of the same size as" Hawikuh; and "four ... somewhat smaller."[20]

In 1910 Robert Flaherty, on a geological expedition prospecting for iron ore but now remembered as a film director and particularly for the documentary *Nanook of the North*, met an Inuit, Wetalltok, who told him of islands in Hudson Bay that had been virtually forgotten by whites.

From a litter, odds and ends of tools, old carvings of ivory, harness toggles, harpoon heads, and the like Wetalltok drew out an old coloured lithograph, tattered and torn. On the back of it, in pencil and crudely drawn, was a map, obviously handiwork of his own.[21]

Like the Zuni skin, the map already existed. Unlike the skin, it is still extant (figs. 1.2, 1.3).[22] It is a remarkably good representation of the Belcher Islands; an exceedingly complex series of islands that have since been given as important toponyms the names of the two participants: Wetalltok Bay and Flaherty Island. Flaherty's manuscript notes contain significant additional information about this encounter conducted through Johnny, an interpreter. Wetalltok indicated to Flaherty that

> his country is far out from the coast—on islands out at sea.
>
> "Where are they" I [Flaherty] asked, pulling out my map and spreading it before Wetalltok.
>
> For a long time Wetalltok studied the map—at a loss, I thought to understand. Finally he pointed to a group of little islands in dotted outline—the dots showing, of course, that they were unexplored [by whites]—marked on the map as the 'Belcher Islands.'
>
> "Wetalltok says his land is here, sir," says Johnny, "where these islands are; but the way the white man draws them is crazy. It's a big land that's there, Wetalltok says."

At this point, Wetalltok produced his map and Flaherty asked to look at it.

> It was an old missionary picture of the Good Samaritan.
>
> "No, it's this side he wants you to look at," said Johnny, turning it over. A map had been drawn on its blank side in pencil. "It's Wetalltok's map."
>
> It was well done. The land was leaded in with pencil; the sea was simply the blank paper.

FRONTIER ENCOUNTERS IN THE FIELD: 1511–1925   17

1.3 Wetalltok drew this map of the Belcher Islands in pencil on the verso of an old missionary lithograph, apparently retrospectively, some time before giving it to Robert Flaherty in 1910. From the American Geographical Society Collection, University of Wisconsin–Milwaukee Library.

"He says this is what his island really looks like," said Johnny.

"Is there any scale to it?" I asked.

"I think the only way we can get a scale to it," said Johnny, "is to ask Wetalltok the travel time from one point to another. That's the only way I know of getting a scale from a native map."

When Johnny asked Wetalltok how long it would take to travel from the southern extreme of the land mass he had drawn to the northern end, Wetalltok said:

"If I go fast by Kayak, two days."

"Two days!" I exclaimed.

"Yes," said Johnny. . . .

"Good Lord!" I said. "If that's true, this land of his must be upwards of a hundred miles long!"

But of course I knew what Wetalltok said was preposterous. It was impossible that there could be a group of islands upwards of a hundred miles long outlying the east coast of Hudson Bay, and not known to the white man. I thought of the once-a-year ships which had been sailing into the Bay for more than two hundred years. Certainly many a time they must have passed not far from this group of islands. I looked again at my own map, at the place where Wetalltok claimed there was land. My map showed soundings! Johnny told Wetalltok it was impossible that the land could be so big; but Wetalltok, laughing a little, said yes, it was possible.

"The white man doesn't know everything," he added with a grin.

Before I left, I asked Wetalltok if he would lend me his map so that I could make a tracing of it.

"Here, take it," he said. "I have others."[23]

There are three noteworthy aspects of the Flaherty-Wetalltok encounter. The Inuit already had in his possession several maps made by himself or other Inuit. One can only speculate about their roles but reinforcement of nostalgia seems to have been one of them. Secondly, with some effort, the Inuit was able to locate his island homeland on a very inadequate Euro-American map. Finally, as late as 1910, an intelligent white man (albeit with little experience at that stage of either Indians or Inuit) assumed that the concept of linear scale could be applied to maps made by native people.

Increasingly, after the early contacts, Europeans requested information in cartographic or, at least, graphic form. In 1540, almost one year before the stick map had been made for Cartier, but twenty-five hundred miles away on the lower Rio Colorado, the Spaniard Hernando de Alarcón asked an elderly, probably Halchidhoma, man to "set me downe in a charte as much as he knew concerning that River, and what maner of people those were which dwelt upon the banckes thereof on both sides: which he did ... [according to a later translation] with pleasure ... on a piece of paper" Interestingly, the Indian then asked Alarcón to "describe" his "country" (Spain or Mexico?) "in the same way he had done his." This could be interpreted to confirm that, on his first contact with Europeans, the Indian had a well-developed concept of map. Alarcón's response, however, could equally be interpreted to indicate that the man was incapable of distinguishing between cartographics and other forms of graphics: "In order to please him, I ordered a picture drawn of a few things."[24] The evidence concerning this important point is tantalizingly ambiguous, hinging as it does on the word "things" and, more specifically, whether or not these were topographic features arranged spatially.

The originals of maps solicited before the eighteenth century have rarely survived. Those that have are, for the most part, either contemporary transcripts or incorporations on European-compiled maps. The earliest undisputed transcript is a map drawn on paper in 1602 in the City of Mexico; not by a native Mesoamerican but by a Plains Indian, who probably originated from somewhere within what are now Kansas, Oklahoma, and Texas. Miguel had been captured in 1601 in the course of the Oñate expedition northeastward or eastward from New Mexico. The map was made by him during a long and carefully recorded interrogation, one of several attempting to establish the circumstances of and reconstruct events occurring on the expedition.[25] Regrettably, the original map was not preserved, but the transcript is remarkable and deserves far more attention than it has received from scholars.[26] Routes, rivers, salines, and Indian settlements are shown; the latter, apparently, crudely differentiated according to relative size or cultural importance. The area represented is either the central or southern Great Plains.[27] An insert shows a reported route across a great river (probably the Rio Grande) to a distant shallow lake, where gold was traded (probably Lake Texcoco), on the shore of which Miguel made the map.

Several aspects of Miguel's map are noteworthy. It is very unlikely that the Indian could have been influenced either directly or indirectly by Europeans before his capture only seven months prior to making the map. Furthermore, the transcript shows no evidence of instruction or influence by the Spaniards. Yet, even excluding the Mexican inset, the map seems to embrace a vast territory. Apart from San Gabriel, in the upper Rio Grande valley, no ground referents have been recognized with certainty. Even so, the area mapped probably extends over more than one hundred thousand square miles. Its content was for the most part derived through experience, but the Mexican inset was based on what Miguel had heard. The Indian appears to have had no difficulty in understanding and responding to the request "to mark with pen and ink on a sheet of paper . . . the pueblos of his land."[28] Evidently he had at least a latent concept of map. Perhaps because he was asked to indicate the pueblos, he started with these, inserting the patterns of rivers and roads later. Whatever the reason, he began by placing (rather than plotting in relation to a base) the nodes, after which he added the two networks.

Most noteworthy in the context of the exchange were the reasons why Miguel was asked to make the map. Once it was made, he was asked to use it to indicate "the place where he was born, how he was taken prisoner and carried away by the enemy [Indians] to other lands where he grew up, how he came to fight the Spaniards and was taken prisoner."[29] Under further interrogation, Miguel gave the distances between specific places on the map in number of days of travel. These ranged between fifteen and forty-four days.

Once it had been made, Miguel's interrogators used the map to establish intelligence about possible sources of several metals, of which they produced specimens. The only one he knew of was gold. The source of placer gold was at the Mexican inset. Miguel moved golden articles over his evidently large map to indicate the trade patterns.

Quite clearly, the Spanish officials' interest in Miguel's map was not as an aid to navigating what was still a *terra incognita,* but as a means of obtaining reasonably reliable information about the political relationships between groups of frontier Indians and about the sources of and native trade in precious metals. In obtaining this information the map was the most important facilitator, supplemented with sign language and material specimens.

By the late seventeenth century, maps of often very large *terrae incognitae* began to be solicited for strategic reasons relating to intercolonial competition. Originally, England, France, and Spain had had widely separated toeholds on the North American continent but, as the intervening spaces began to be exploited by traders, colonial administrators and entrepreneurs sought geographical information about very large territories. In 1687, for example, the Sieur de la Salle reached the great village of the Cenis in what is now east Texas.

With a "mind to make an Attempt upon the [Spanish-controlled] Mines of St. Barbe in New Mexico," he questioned the Indians about the Spaniards to the west. In response, the Cenis made a map for him with charcoal on bark. Though not extant there are contemporary accounts of it.[30] Later, the survivors of the ill-fated expedition had made their way up the Mississippi River to the nearest French settlement: Fort St. Louis on the Illinois River. Seven years later, Lawrence van den Bosh of Maryland compiled a manuscript map of the lower Mississippi River

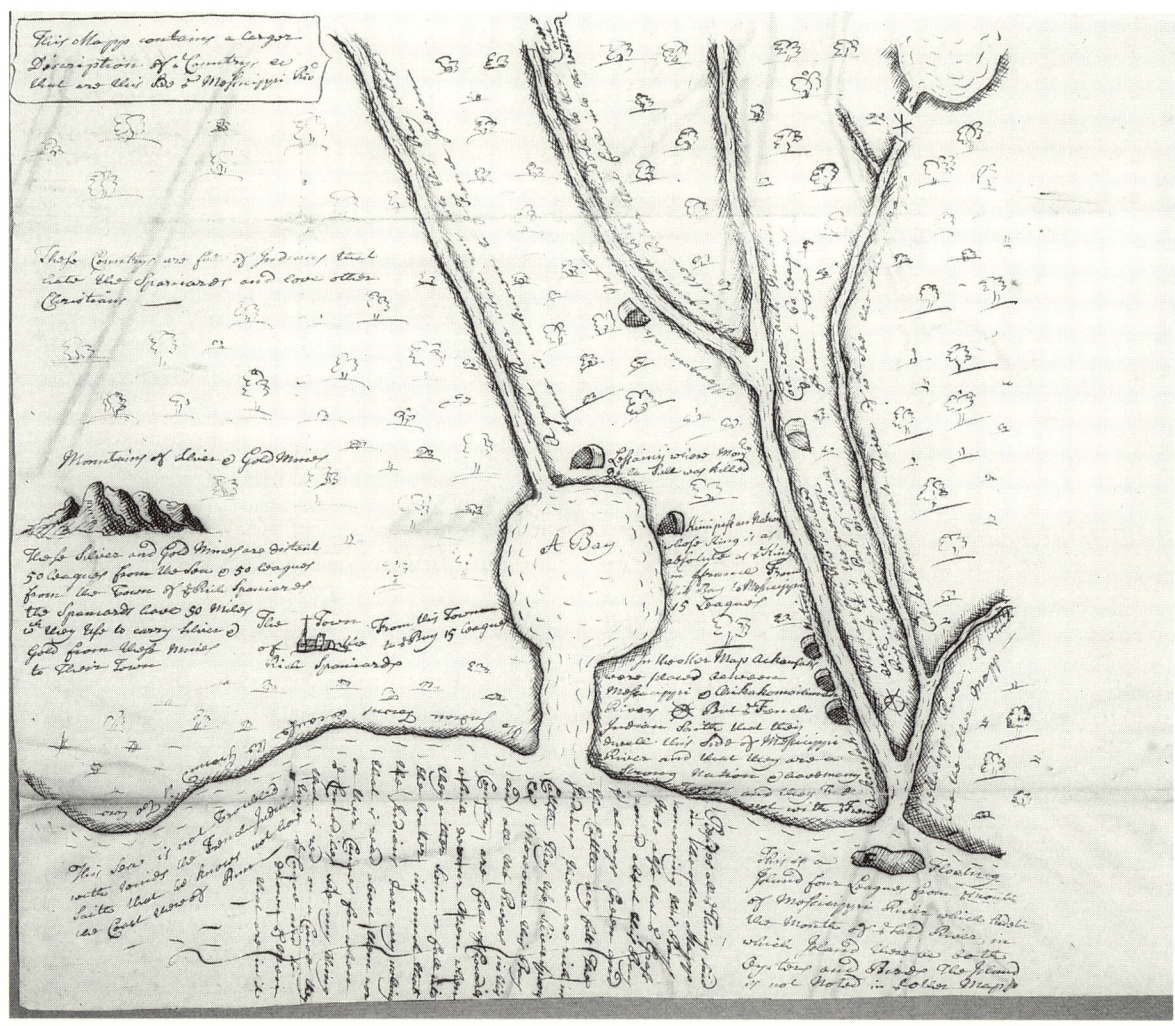

1.4 Lawrence van den Bosh, map of the lower Mississippi River and neighboring coast, ca. 1693, apparently based in part on an Indian map, probably Shawnee, in turn based on a Southeast Indian cartographic tradition. Courtesy of the Edward E. Ayer Collection, Newberry Library.

and lands to the west of it (fig. 1.4).[31] In a letter of transmittal to Francis Nicholson, then governor of the colony, he acknowledged that the content of the western part had been "lately reced of the French Indian," probably an Illinois or Shawnee from the Fort St. Louis region.[32] The Cenis' map as described by Hennepin, Douay, and Cavelier has so much in common with the French Indian's component of the van den Bosh map to suggest a common origin, or at least a common Indian tradition concerning the spatial relations of the events and crucial locations of a few years before. In particular the van den Bosh map shows "Lessainy where Monsr De la Sale was killed," and "Mountains of Silver & Gold Mines . . . distant . . . 50 leagues from the Town of ye Rich Spaniards."

Miguel's map and that of the Cenis were both of large territories. Indeed, their respective areas may well have overlapped. Our knowledge of the maps, however, differs considerably because of the contrasting circumstances in which they were made. Miguel's was a captivity document, made by request, under close scrutiny, and apparently with the aid of an interpreter. Attempts were made by the inquisitors to clarify content, and the whole process was officially and professionally recorded for posterity. The Spaniards were in virtually complete control of the procedures, though language was a barrier. In contrast, although responding to La Salle's request, the Cenis made their map on a traditional medium, presumably using their traditional techniques, and certainly in their own cultural environment over which they exerted complete control. The three retrospective accounts, though confirmatory, are each brief and lack detail. La Salle was certainly not in charge of the information exchange, and his party "did not understand the [Cenis] language." Interestingly, however, the cartographic exchange was two-way, to the extent that a few days later, at another Cenis village, someone in La Salle's party depicted for the Indians "on the bark of a tree by means of a piece of charcoal, the place we were coming from and that where we were going."[33]

Not only were the circumstances in which the maps were made quite different, so were their purposes. Miguel's was made in response to the Spaniards' attempt to discover more about the native peoples and environmental conditions to the east and northeast of New Mexico and to establish what, if anything, they knew about gold. In 1602 the Spanish authorities in Mexico had very little concern about incipient French and English activities far to the northeast in the St. Lawrence valley and to the east on the Atlantic seaboard. In contrast, by 1687 the inter-European *terra incognita* had shrunk. La Salle, in what is now south-central Texas, knew that Spanish territory was only a few hundred miles to the west and he was intent to discover more about it from the Indians. Anastasius Douay's account of the context in which the Cenis made their map makes this patently clear:

> They [the Cenis] have intercourse with the Spaniards through the Choumans, their allies, who are always at war with New Spain [where Spain had had a tenuous presence for approximately one hundred years]. The Sieur de la Salle made them draw on bark a map of their country, of that of their neighbors, and of the river Colbert, or Mississippi [which La Salle had descended five years before and Jacques Marquette partly so nine years before that], with which they [the Cenis] are acquainted. They reconned themselves six days' journey from the Spaniards, of whom they gave us

so natural a description, that we no longer had any doubts on that point, although the Spaniards had not yet undertaken to come to their villages, their [the Cenis'] warriors merely joining with the Choumans to go to the war on New Mexico.[34]

Governor Francis Nicholson, to whom van den Bosh had forwarded his map in 1694, was to become foremost among colonial officials as a solicitor of maps from Indians for strategic reasons. In a letter of 1699 to Governor Joseph Blake of South Carolina, he expressed concern about French activities in the lower Mississippi valley and on the eastern Gulf Coast. He urged Blake to develop a plan of action for submission to the Lords of the Council for Trade and Plantations in London. This was to be based, in part, on maps of the *terra semicognita* between coastal South Carolina and the lower Mississippi: some to be obtained "under oath" from Indian traders whose veracity could be relied on and others from "Indians, which you can rely upon: . . . let them draw out the Country as Henepin Says one of the Shavana [Shawnee] Indians did for him, for I think at least 400 leagues, which he found to be true."[35] Nicholson sent with this letter a copy of Hennepin's book (almost certainly *A New Discovery of a Vast Country in America*, published in London in the previous year) but Blake died soon afterward and there is no evidence that such maps were solicited by him or on his behalf.

More than twenty years after writing to Blake (by which time he himself had been appointed governor of South Carolina and as a consequence moved much closer to the French sphere of influence in the lower Mississippi valley), Nicholson was involved in obtaining and transmitting to London two important and, in contemporary copied forms (transcripts), still extant Indian maps. Both the originals were in color and done on skin. Each covered an extensive part of the very region for which Nicholson had urged Blake to solicit maps. The earlier (ca. 1721) and slightly less well known of the two was made by Catawba Indians and is extant in two contemporary transcripts (see fig. 10.1).[36] Like Miguel's map of 120 years earlier, it represents paths by approximately straight lines and Indian settlements or tribes by circles, the relative sizes of which appear to be proportional to population size or some other measure of demographic or cultural importance.

It is not clear whether Nicholson solicited the Catawba map in the manner he had advised Blake to do more than twenty years earlier. Indeed, the titles of both versions indicate that the original was "presented" which, taken literally, excludes soliciting. He probably received it soon after assuming the governorship of South Carolina when he met the chiefs of the Piedmont tribes loosely affiliated to the English and known by the omnibus name "Catawbes." This map would not have been as important strategically as a second one "presented" to him by the Chickasaw approximately two years later, of which a transcript is extant (see fig. 10.2).[37] Similar in cartographic characteristics, the Chickasaw map covers a much larger area, extending to the west into and beyond the French settlements and French-allied Indians of the lower and middle Mississippi valley. Indicating by circles thirty-two Indian tribes and settlements, it differentiates by an "F" those allied to the French. For this reason alone, it must have been of much greater significance to Nicholson than the earlier Catawba map. It was an up-to-date geopolitical statement of intercolonial relationships in the wider Southeast; much of which was still a *terra semicognita* to colonial officials.

During the latter part of the eighteenth century, exploration by whites became more scientific. Increasingly, in the course of these encounters, maps were commissioned and the chance of their survival for posterity improved, albeit, as before, usually as transcripts. Furthermore, the circumstances and characteristics of the maps were described with greater precision so that more is known about the nonsurvivors than their earlier equivalents. More than ever before, map content was used by whites in developing strategies and in adding systematically to geographical knowledge. In 1770, for example, Samuel Hearne of the Hudson's Bay Company began a third attempt to cross more than a thousand miles of sub-Arctic and Arctic Canada between Hudson Bay and the lower Coppermine River. This time the attempt was successful and his account of it became a classic of exploration, accompanied by a remarkable map, of which the manuscript original is extant.[38] The expedition was planned on the basis of a map that had been made a few years before by two Chipewyans: Meatonabee and Idotlyazee. The original, on skin, has not survived, but a contemporary transcript has.[39] It was the first of more than thirty maps and topographic sketches known to have been made for the Hudson's Bay Company by native peoples during the next eighty years.[40] Its strategic message for Hearne was clear: the direct route must be avoided in favor of a much longer one that remained to the south, within the shelter of the woods for all but the summer months. Hearne's own experience was essentially along one outward and one return route, closely spaced and in places identical. Yet his own map represented features as much as a hundred miles beyond these. Although transformed to conform to Hearne's graticule, there is little doubt that some of his map's content was derived from that of the two Indians.

Like Hearne, on their transcontinental expedition of 1804–6, Meriwether Lewis and William Clark obtained information in map form from Indians. One by the Cut Nose, probably a Nez Percé Indian, was of the "principall watercourses West of the Rocky mountains."[41] When incorporated in the postexpedition cartographic compilation, it gave rise to one of the grossest errors in the mapping of the region: a Multnomah (now Willamette) River that rose not as it in fact does in the Cascade Range of what is now west-central Oregon but in the Rocky Mountains, more than eight hundred miles to the southeast, near to the source of the Rio Grande.[42]

Maps were solicited from Indians on almost all the great post–Lewis and Clark surveys of the American West. Likewise, the Inuit made maps for several of the Sir John Franklin and post-Franklin expeditions to the Arctic.[43] The true scientific surveys, though less well known, also engaged natives in mapping and the resulting maps have often survived. For example, in the early 1840s, the Hudson's Bay Company undertook a full-scale field experiment. It was an attempt to establish a beaver preserve on uninhabited Charlton Island at the south end of James Bay. In 1839 two East Cree Indians, one of whom, Cauc-chi-chenis, had first suggested the idea, were sent to the island to hunt the natural predators of the beaver and to make a map of the coast, rivers, and lakes. Thereafter, several female beavers and one male, Old George, were released to breed in the wild. From time to time during the next few years the beaver population was mapped by Indians specially sent out to the island.[44] Altogether, eight maps were made. The population survey for 1839–40 was recorded by alphabetically and numerically coded lakes,[45] in much the same way that we now use minor civil divisions for enumerating human populations.

Geologists, in particular, solicited information from native peoples in map form. William E. Logan, the founder and first director of the Geological Survey of Canada, was one of the first to do this. On July 30, 1858, he probably established a precedent for the survey when he copied into his own field notebook an Algonkin Indian's map of the Rivière Rouge in southwest Quebec.[46] It was the first of almost 150 known to have been made for the survey between then and 1914 in regions as far apart as southern Quebec and the Yukon Territory.[47] The most accessible collection of these is in the Tyrrell Collection in the Thomas Fisher Rare Book Library in the University of Toronto.

A distinguished geologist in the early part of his life, Joseph Burr Tyrrell spent all but one of the summers between 1892 and 1896 conducting surveys in Manitoba and the Districts of Saskatchewan and Keewatin. Tyrrell frequently used Chipewyan—and occasionally Inuit—guides, and his papers contain more than fifty of their maps. Mainly done in pencil on large sheets of coarse paper, most merely indicate the rivers and lakes to be canoed during the next few hours or, at most, days. An Inuit example from the Dubawnt Lake region is typical in its overall sparseness but also in its emphasis of critical details such as trees and large trees (conspicuous on the otherwise open tundra), and in its gross stereotyping of a peninsula.[48] A few of the maps, however, are rich in detail and cover enormous areas. One made in October 1894 at Fort Churchill by the Chipewyan Powon covers a vast territory of rivers, lakes, boreal forest, and tundra to the west of Hudson Bay (fig. 1.5).[49] Significantly, it represents the same "Edge of Woods" that had been so strategic in Meatonabee's route planning for Samuel Hearne 120 years before. Equally significantly, both were made at Fort Churchill, which had remained an important center of information exchange.

The National Anthropological Archives, Washington D.C., likewise holds corpora of native maps. In 1882, James Owen Dorsey, working on behalf of the newly formed Bureau of American Ethnology, obtained from the Omaha, Ponca, and Kansa of the east-central Great Plains toponyms of hydrological and other features of their respective regions. The respondents made their own base maps, to which Dorsey added phonemic transcriptions. There are at least ten such maps in the Dorsey Papers.[50] These are outnumbered by approximately thirty surprisingly comparable maps in the same archive's collection of papers by the German-American anthropologist Franz Boas.[51] Boas spent 1883–84 on Baffin Island, where he used a technique very similar to Dorsey's to compile from the Baffinland Eskimo both a gazetteer and a vocabulary. Twenty years later, by which time he was curator of anthropology at the American Museum of Natural History, New York, Boas was responsible for acquiring further maps from the same region—this time, however, as a consequence of commissioning whaling captains to make collections of artifacts. A map collected by the Scottish captain James Mutch is remarkably like those collected by

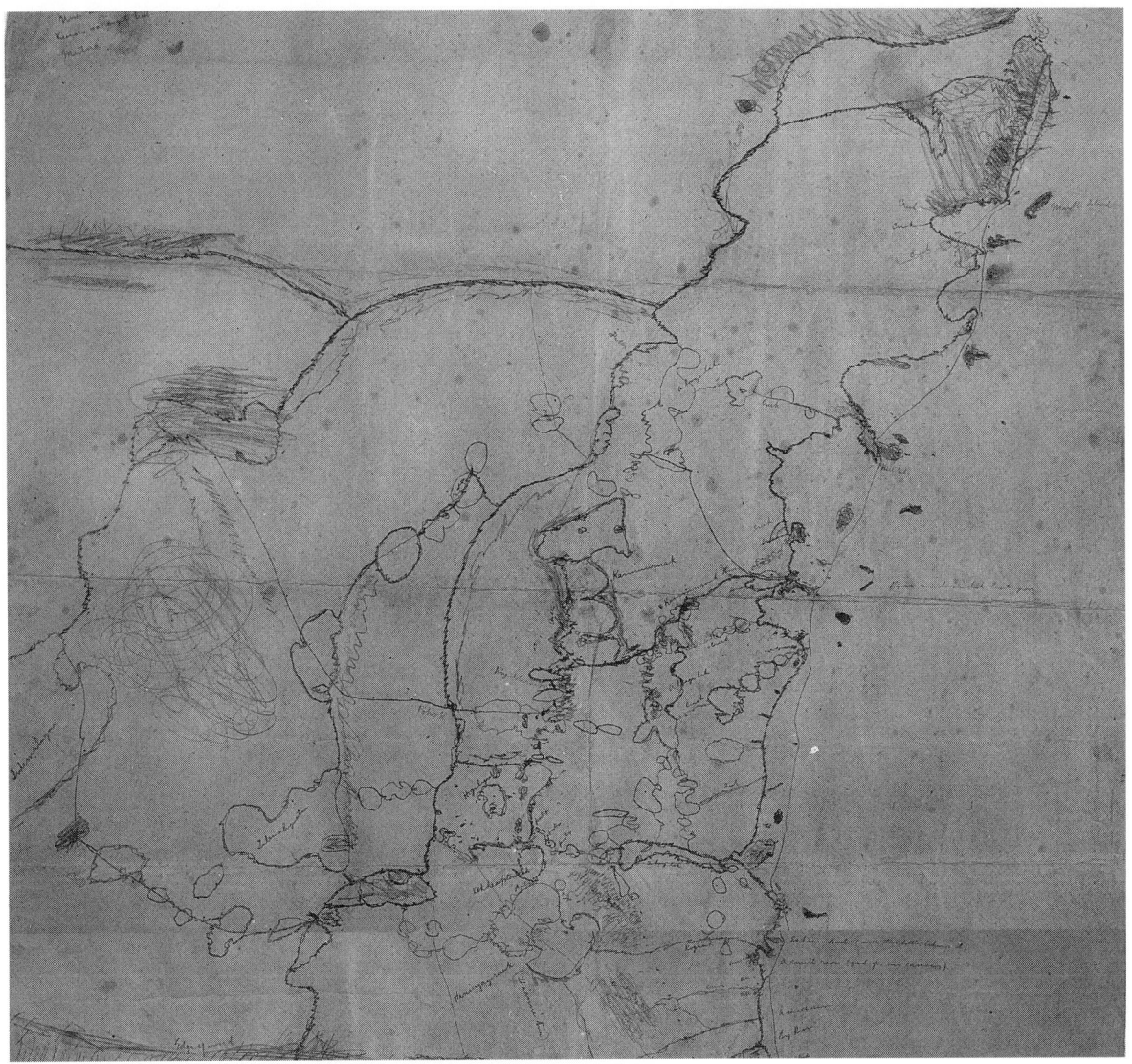

1.5 Powon (a Chipewyan guide), map of the western shore of Hudson Bay and vicinity made for James B. Tyrrell in the fall of 1894. Annotated by Tyrrell, the line work is probably Powon's. Courtesy of the Joseph B. Tyrrell Papers, Thomas Fisher Rare Book Library, University of Toronto Library.

Boas.[52] Another, collected by the American whaler Capt. George Comer, is from the Canadian mainland in the District of Keewatin. It is an Iglulik map of musk-ox hunting in several seasons from 1893–94 onward.[53] Routes are marked, with dots at intervals to show "where Igloes were built." Presumably, therefore, the intervals between adjacent dots represent days' journeys.

Apart from its zoological interest, the map contains at least one mythological component: the island in Wager Bay that, according to the Iglulik, "was once a whale." This is a late example of what was quite common in earlier encounter maps: the coexistence of real and mythically endowed features.

By the late nineteenth century, much ethnographical material was being commissioned for museums, European as well as North American. These included maps on paper and indigenous map artifacts. Regrettably, usually because they are not listed as single items or because they are not catalogued as maps, they are not easily discovered in museum collections. One suspects that many examples remain virtually unknown. Intuition can sometimes lead to remarkable finds, for example, fifty-seven maps by Inuit of the Canadian eastern Arctic in the National Museum of Denmark, collected between 1921 and 1925 on the Fifth Thule Expedition.[54] Sometimes such intuition has delayed results. An unsuccessful search for a map made by a Chilkaht chief in 1869 for the Californian scientist George Davidson involved contacting the Bancroft Library at the University of California-Berkeley. The response was efficient and polite, but negative.[55] Three years later, the library acquired it, or rather, them—two maps were found—together with a manuscript account of how they were made.[56] The chief and his wife had been dissatisfied with their first effort done on standard letter paper and asked for a larger sheet. Given the plain verso of a printed coastal chart, they then spent three days producing a magnificent map of a more than six-hundred-mile return journey to Fort Selkirk and back, involving very difficult terrain and undertaken seventeen years before. It was a remarkable feat of memory as well as an exhibition of skillful mapmaking.

Concluding Observation

The significance of the somewhat more than four-hundred-year frontier encounter beginning in 1511 is that the accounts of maps, mapping procedures, and mapmaking contexts, together with extant map artifacts and surviving transcript maps, constitute a considerable, though dispersed, body of data. Only a very small proportion of this was drawn on in the first phase of scholarly encounters extending from the late eighteenth century to the early twentieth century (chapter 2). All the contributors to part 2 of this volume have drawn heavily on it for evidence and ideas. Because twentieth-century Indians and Inuit have lost so many of their traditional skills and their indigenous modes of world making, it will remain the main body of evidence for future scholarly interpretations (part 3). For the latter reason alone, every opportunity must be taken to conserve, add to, and systematize the pre-1925 data.[57]

Notes

1. The first date is not arbitrary. A case can be made for the Floridian part of the untitled woodcut map included in some copies of the first (1511) edition of Petrus Martyr de Anglerius, *Oceani Decas,* having been based on information received from Arawak or Lucayo Indians: Louis De Vorsey, Jr., "Native American Maps and World Views in the Age of Columbus," *Map Collector* 58 (Spring 1992): 26. If the case is accepted, then 1511 almost certainly marks the earliest cartographic encounter between native American and European cul-

tures. The terminal date of 1925 is more arbitrary, being the last year in which maps were collected from Polar Eskimos on the Fifth Thule Expedition.

2. Gamaliel Smethurst, *A Narrative of an Extraordinary Escape Out of the Hands of the Indians in the Gulph of St. Lawrence* (London: J. Bew and A. Grant, 1774), 14.

3. Map on birchbark, 9.0 x 38.0 cm, mounted on paper and framed behind glass. An inscription on the paper reads: "Map drawn by Indians on Birch-Bark and attached to a tree to shew their route to others following them, found by Capt. Bainbrigge Rl. Engineers at the 'ridge' between the Ottawa [River] and Lake Huron. May 1841." RUSI (Misc.), fol. 2, Map Room, British Library, London.

4. On the basis of experiences in the St. Lawrence and upper Mississippi valleys between 1683 and 1692, Louis Armand de Lom d'Arce, Baron de Lahontan, described "Chorographical Maps . . . drawn upon the Rhind of your Birch Tree." Although observed fairly soon after the first encounters, they were certainly, and apparently widely, made and used in indigenous contexts, because "when the old Men hold a Council about War or Hunting they're always sure to consult them." Louis Armand de Lom d'Arce, Baron de Lahontan, *New Voyages to North America* (London: H. Bonwicke and others, 1703), 2:13–14.

5. William Bray, "Observations on the Indian Method of Picture Writing," *Archaeologia* 6 (1782): 159.

6. Richard I. Dodge, *The Hunting Grounds of the Great West* (London: Chatto and Windus, 1877), 414.

7. Gilbert M. Sproat, *Scenes and Studies of Savage Life* (London: Smith, Elder and Co., 1868), 191–92.

8. Non-Chi-Ning-Ga (Iowa), untitled manuscript map of the upper Mississippi and Missouri drainage systems between Lake Michigan and the Rocky Mountains, showing "the route of my (Ioway) forefathers—the land that we have always claimed," presented at a council between Indians of the Mississippi and Missouri Rivers in Washington, D.C., October 7, 1837, 104 x 69 cm, Record Group 75, Map 821, Tube 520, Cartographic Branch, National Archives of the United States, College Park, Maryland; reproduced in G. Malcolm Lewis, "Indian Maps: Their Place in the History of Plains Cartography," in Frederick C. Luebke et al., eds., *Mapping the North American Plains: Essays in the History of Cartography* (Norman: University of Oklahoma Press, 1987), figs. 4.2 and 4.3.

9. Hugh Brody, *Maps and Dreams: Indians of the British Columbia Frontier* (London: Jill Norman and Hobhouse, 1982), 266–67.

10. Selwyn Dewdney, *The Sacred Scrolls of the Southern Ojibway* (Toronto: University of Toronto Press, 1975), 183–84.

11. Celestial chart on tanned buckskin, originally belonging to the Skiri band of Pawnee, 66.0 x 46.0 cm; collected at Pawnee, Oklahoma, 1906, when it was already old; Artifact 71898-10, Department of Anthropology, Field Museum of Natural History, Chicago.

12. William G. Gartner, "Pawnee Cartography," Term paper, Geography 570 (University of Wisconsin, Madison, December 1992), 40.

13. Frank G. Speck, *Naskapi: The Savage Hunters of the Labrador Peninsula* (Norman: University of Oklahoma Press, 1935), 145 (fig. 12A), 148 (fig. 14D), and 149.

14. Frank G. Speck, "Montagnais Art in Birch-Bark: A Circumpolar Trait," *Indian Notes and Monographs*, 11, 2 (1973): pl. xiii, patterns a, b, d, e, f, i, and j. Pattern h was intended to be of trails but owing to an error in biting or folding, came out looking like a star.

15. There is one intriguing but not entirely convincing exception: the engraving on a conch shell from Spiro, Oklahoma (a Mississippian site from the period A.D. 700–1540). The pattern has been described as "Concentric cross-in-circle motifs in a connected grid": Philip Phillips and James A. Brown, *Pre-Columbian Shell Engravings from the Craig Mound at Spiro, Oklahoma* (Cambridge, Mass.: Peabody Museum Press, 1978), pt. 1:, plate 122.3. The pattern has been likened to a Chickasaw map of 1737: "Nations Amies et Ennemies des Tchikachas," signed manuscript in black and red, 51.0 x 34.5 cm, transcribed by Alexandre de Batz from an original by Mingo Ouma, C/13/a/22, fol. 67, Archives d'Outre Mer, Aix-en-

Provence (fig. 9.2 of this volume). Having been likened to a map, the engraving was then hypothesized to be a map of Mississippian sites in what is the now the South Central United States: Robert H. Lafferty III, "Prehistoric Exchange in the Lower Mississippi Valley," in Timothy G. Baugh and Jonathon E. Ericson eds., *Prehistoric Exchange Systems in North America* (New York: Plenum Press, 1994), 202–5, including figs. 9–11.

16. For a review of the noncartographic aspects of this process as it operated within North America, see G. Malcolm Lewis, "Native North Americans' Cosmological Ideas and Geographical Awareness: Their Representation and Influence on Early European Exploration and Geographical Knowledge," in John L. Allen, ed., *North American Exploration*, vol. 1, *A New World Disclosed* (Lincoln: University of Nebraska Press, 1997), 71–126.

17. Richard Hakluyt, *The Voyages, Navigations, Traffiques, and Discoveries of the English Nation* (London: George Bishop, Ralfe Newberie and Robert Barker, 1600), 3:235.

18. Henry Y. Hind, *Explorations in the Interior of the Labrador Peninsula: The Country of the Montagnais and Nasquapee Indians* (London: Longman, Green, Longman, Roberts, and Green, 1863), 83, 88.

19. "Indian Preparing Birch Bark Map for Professor Hind," oil on board, 27.9 x 40.6 cm, John Ross Robertson Collection, Metropolitan Toronto Central Library. "Montagnais Chief Explaining His Map on Birch Bark," watercolor and graphite with scraping, on a commercially prepared colored ground, 13.9 x 22.3 cm, Bert and Barbara Stitt Family Collection, Art Gallery of Hamilton, Hamilton, Ontario. William Hind was the expedition artist.

20. The interpretation is based on two translations of Francisco Vásquez de Coronado's report to Antonio de Mendoza, Viceroy of New Spain, dated August 3, 1540, concerning events of just before July 7 of that year: George P. Winship, "The Coronado Expedition, 1540–1542," *Fourteenth Annual Report of the Bureau of Ethnology to the Secretary of the Smithsonian Institution 1892–1893* (Washington, D.C.: Government Printing Office, 1896), pt. 1:558, 562; and George P. Hammond and Agapito Rey, eds. and trans., *Narratives of the Coronado Expedition 1540–1542*, Coronado Cuarto Centennial Publications, 1540–1940, vol. 2 (Albuquerque: University of New Mexico Press, 1940), 171, 173–76.

21. Robert J. Flaherty, *My Eskimo Friends, "Nanook of the North"* (Garden City, N.Y.: Doubleday, Page and Co., 1924), 18.

22. Untitled map of the Belcher Islands in pencil on the verso of a colored missionary print, 35.4 x 31.1 cm, RARE 772, 3-c. B44 A-19, American Geographical Society Collection, University of Wisconsin–Milwaukee Library.

23. Robert J. Flaherty, "The Islands That Were Not There," manuscript on file in the Robert J. Flaherty Collection, Butler Library, Columbia University, New York. Either Flaherty's memory of the missionary print was wrong or he was referring to another of Wetalltok's maps. The print on the verso of which the extant map was drawn (fig. 1.2) is certainly a biblical scene. Equally certainly, it is in an urban setting and does not represent any aspect of the parable of the good Samaritan.

24. Hammond and Rey, eds. and trans., *Narratives of the Coronado Expedition,* 153; and Hakluyt, *The Voyages,* 438.

25. Part of the "Questioning of the Indian Miguel Brought from New Mexico by the Maese de Campo, Vincente de Zaldivar," itself part of the "Official Inquiry Made by the Factor, Don Francisco de Valverde, by Order of the Count of Monterrey, Concerning the New Discovery Undertaken by the Governor Don Juan de Oñate toward the North beyond the Provinces of New Mexico," George P. Hammond and Agapito Rey, eds. and trans., *Don Juan de Oñate, Coloniser of New Mexico 1595–1628*, pt. 2, Coronado Cuarto Centennial Publications, 1540–1940, vol. 4 (Albuquerque: University of New Mexico Press, 1953), 871–77.

26. "Pintura q- Por man^{do} de don Fran^{co} Valverde de mercado factor de su mag^d hizo myguel yndio de las pro vincias del nuevo mex^{co} . . ." (endorsement) ["Sketch which, by order of don Francisco Valverde de Mercado,

factor of His Majesty, was made by Miguel, an Indian, native of the provinces of New Mexico, of the relative position of the towns of the said provinces and the statement on them as he said it and gave to understand by signs and the names of the towns pronouncing them as they are and how populated. Don Francisco Valverde de Mercado/Hernando Esteban—This transcript is (certified) correct and truthfully (by) Hernando Esteban"], transcript in ink on paper. 31.0 x 43.0 cm, May 11, 1602 (from original of April 22, 1602), estante 1, cajon 1, legajo 22, ramo 4, folio 172, Patronato, Archivo General de Indias, Seville, Spain; reproduced in Lewis, "Indian Maps," fig. 4.8.

27. Opinion hitherto has been that Oñate's 1601 expedition was northeastward from the Pecos valley, across the upper Canadian River to the Arkansas River, perhaps in the region of what is now Arkansas City, Kansas: e.g., Waldo R. Wedel, "Archaeological Remains in Central Kansas and Their Possible Bearing on the Location of Quivira," *Smithsonian Miscellaneous Collections* 101 (1942): 18–20. All interpretations, however, have been made without reference to Miguel's map. With reference to it, a strong but not conclusive case can be made for it being of the southern rather than central Great Plains.

28. Hammond and Rey, *Don Juan de Oñate*, pt. 2:872–73.

29. Ibid., 873.

30. Louis Hennepin, *A New Discovery of a Vast Country in America* (London: M. Bentley and others, 1698), pt. 2:26; Anastasius Douay, "Narrative of La Salle's Attempt to Ascend the Mississippi in 1687," in John D. G. Shea, *Discovery and Exploration of the Mississippi Valley* (New York: Redfield, 1852), 204–5; and Jean Delanglez, *The Journal of Jean Cavelier: The Account of a Survivor of La Salle's Texas Expedition* (Chicago: Institute of Jesuit History, 1938), 101, 103, 105.

31. Untitled supplemented manuscript copy by Lawrence van den Bosh of a ca. 1693, probably Shawnee version of a Southeast Indian map of the country extending west of the lower Mississippi River toward the Rio Grande, 1694; Edward E. Ayer Collection manuscript map no. 59, Newberry Library, Chicago.

32. The verso of the map contains a "Copy of a letter to Collo Nicholson Govr. of Maryland" from "Lawrence Vanden Bosh" and dated "From North Sassifrix ye 19th Day of October 1694." Although the content is ambiguous, it is clearly a letter of transmittal and refers to "a larger description of the Countrys River etc that are on the left [west] Side of the Messacippi River, which description I lately reced of the French Indian."

33. Delanglez, *The Journal of Jean Cavelier*, 105.

34. Douay, in Shea, *Discovery and Exploration of the Mississippi Valley*, 204–5.

35. Extract from a letter by Francis Nicholson to Joseph Blake, James Town, Virginia, September 2, 1699, CO5/1311/10 liv (10), Public Record Office, Kew, London.

36. "This Map describing the Scituation of the Several Nations of Indians to the NW. of South Carolina was coppyed from a Draught drawn & painted on a Deer-Skin by an Indian Cacique and presented to Francis Nicholson Esqu^r . . . ," paper, 81 x 112 cm, ca. 1721, Additional MS 4723, British Museum, London. A second, almost identical manuscript copy also exists: C.O.700, North American Colonies General no. 6 (1), Map Room, Public Record Office, Kew, London (fig. 10.1 in this volume; often referred to as the Catawba Deerskin Map). For comprehensive descriptions and explanations of both these contemporary transcripts see Gregory A. Waselkov, "Indian Maps of the Colonial Southeast," in Peter H. Wood et al., *Powhatan's Mantle: Indians in the Colonial Southeast* (Lincoln: University of Nebraska Press, 1989), 295, 320–24, and figs. 3, 11.

37. "A Map Describing the Situation of the several Nations of Indians between South Carolina and the Missisipi; was Copyed from a Draught Drawn upon a Deer Skin by an Indian Cacique and Presented to Francis Nicholson Esqr. Governour of Carolina [1725]," paper, 114 x 145 cm, C.O. 700, North American Colonies General no. 6 (2), Map Room, Public Record Office, Kew, London (fig. 10.2 in this volume; often referred to as the

Chickasaw Deerskin Map). For a comprehensive description and explanation of this contemporary transcript see Waselkov, "Indian Maps of the Colonial Southeast," in Wood et al., *Powhatan's Mantle*, 295, 324–29, and figs. 4, 12. The date of 1725 is in a different hand from the rest of the inscription and appears to have been inserted later. The original was almost certainly presented to Nicholson in 1723.

38. "A Map exhibiting Mr. Hearne's Tracks in his two Journies for the Copper Mine River, in the Years 1770, 1771, and 1772 . . . ," in Samuel Hearne, *A Journey from Prince of Wales's Fort in Hudson's Bay, to the Northern Ocean* (London: Printed for A. Strahan and T. Cadell, 1795). The manuscript map from which the printed map was derived is "A Map of part of the Inland Country to the Nh Wt of Prince of Wales's Fort Hs; By, Humbly Inscribed to the Govnr Depy Govnr and Committee of the Honble, Hudns, By Compy By their Honrs, moste obediant humble servant. Sam,l Hearne; 1772," paper, 76.7 x 82.5 cm, manuscript map G2/10, Hudson's Bay Company Archives, Provincial Archives of Manitoba, Winnipeg, Manitoba, Canada.

39. "An Explanation of a Draught Brought by Two Northern Indians Leaders Call'd Meatonabee & Idotlyazee, of Y^e Country to Y^e Northward of Churchill River Viz^t Hudsons Bay," paper, 139.5 x 69.5 cm, 1767–68, manuscript map G2/27, Hudson's Bay Company Archives, Provincial Archives of Manitoba, Winnipeg, Manitoba, Canada. For a detailed explanation of this map see June Helm, "Matonabbee's Map," *Arctic Anthropology* 26, 2 (1989): 28–47.

40. Richard I. Ruggles, *A Country So Interesting: The Hudson's Bay Company and Two Centuries of Mapping, 1670–1870* (Montreal: McGill-Queen's University Press, 1991), 266, app. 9 and 10.

41. In the spring of 1806 several Indians told about or made maps of the Multnomah River. Of the maps, the last would seem to have been the most influential: "Sketch given to us May 8th 1806 by the *Cut Nose* and the brother of the *twisted hair*," paper, 34.3 x 24.4 cm, 1806. William Robertson Coe Collection, Yale University, New Haven; reproduced in *The Journals of the Lewis and Clark Expedition*, vol. 1, *Atlas of the Lewis and Clark Expedition,* edited by Gary E. Moulton (Lincoln: University of Nebraska Press, 1983), map 98.

42. "A Map of part of the Continent of North America . . . Compiled from the information of the best informed travellers . . . By William Clark," paper, 73.7 x 137.2 cm, 1810, William Robertson Coe Collection, Yale University, New Haven; reproduced as map 125 in Moulton, ed., *The Journals of the Lewis and Clark Expedition*, vol. 1.

43. For a review of printed examples from the eastern Arctic see John Spink and D. W. Moodie, *Eskimo Maps from the Canadian Eastern Arctic*, Cartographica Monograph no. 5 (Toronto: University of Toronto Press, 1972), including 24 figures.

44. Ruggles, *A Country So Interesting*, 86.

45. Probably Robert Miles, "Sketch Simpson's Beaver-preserve Charlton Island as originally laid down by the Indians 1838/39 Red Ink additions thereto by them 1839/40," paper, 52.8 x 65.3 cm, probably 1842, G1/68, Hudson's Bay Company Archives, Manitoba Provincial Archives, Winnipeg, Manitoba, Canada.

46. "Copy [by W. E. Logan] of a map by Shushuk [an Algonquin]," simple, annotated inked map of rivers and lakes, 16.0 x 9.5 cm in Logan's field notebook of surveys in the valley of the Rivière Rouge, southwest Quebec, during the period June 4–August 25, 1858, facing p. 117, RG45, vol. 159, notebook 1983, National Archives of Canada, Ottawa.

47. At least thirty-five are in the notebooks of field officers. Most are annotated transcripts. Officers include Robert Bell, Charles Camsell, George M. Dawson, Albert P. Low, and A. R. C. Selwyn. The notebooks are contained in the R.G.45 Group, National Archives of Canada, Ottawa.

48. Ahyont (a Caribou Inuit), "Maps by Ahyont Aug. 23 Doobaunt Lake," Pencil on two sheets of brown paper, August 23, 1894, 61.0 x 46.0 and 46.0 x 30.0 cm; represents Dubawnt Lake, District of Keewatin, in some detail and its relationship to the Kazan River; "Trees" marked in

several places, Tyrrell Papers, Thomas Fisher Rare Book Library, University of Toronto, Toronto, Canada. A typed transcript of Tyrrell's notebook entry for that day does not specifically mention the map but, referring to Ahyont and his son as the two "Huskies [who] spent a good deal of their time in our tent, [Tyrrell] endeavoured to improve the time a little by getting a vocabulary and an account of the adjoining country. . . . They say that there is a large lake lying east of Doobaunt lake, but smaller than it discharging into the River [Kazan] some distance below." J. B. Tyrrell, Field Note Books (typed transcripts) for 1894, University of Toronto Manuscript Collection 26, item 47, bk. 3, pp. 23–24, for August 23, 1894.

49. Powon (Chipewyan), "Map of West Shore of Hudson Bay & vicinity by Powon. Churchill. Oct. [1894]" (endorsement on verso). 91.0 x 61.0 cm on brown paper. Tyrrell Papers, 1894 014 B, Thomas Fisher Rare Book Library, University of Toronto. The annotations were almost certainly added by J. B. Tyrrell, including "Edge of wood." The pencilwork is probably Powon's original, including the line delimiting the edge of the woods and the shading of the wooded areas.

50. Ten large pencil, crayon, and ink manuscript maps of east-central Plains drainage systems drawn by Omaha-Ponca and Dheigiha-Kansa Indians as base data for recording their names for rivers; all presumed to date from 1882. Two similar Indian maps of the Yaquina and Siuslaw Rivers, Oregon, 1884, 4800 Dorsey Papers, National Anthropological Archives, Washington, D.C., files 129, 130, 241, 393, and 394.

51. Anonymous Baffinland Inuit and Franz Boas, thirty-four pencil maps and four ink maps of parts of Baffin Island. Line work is apparently Inuit, except for that in ink, which appears to be by Boas. Most of the maps are not identified as to the precise area they represent except by Inuit toponyms. Notes on the maps include short texts in Inuit-Inupiaq and notes in German and English. Most of the notes are illegible because Boas used a shorthand system of his own in writing many of them, and his handwriting is very difficult to read even when not abbreviated. The maps are mounted on linen and quite well preserved. Each less than 61.0 x 79.0 cm, 1883–84, 129,270 Eskimo, gift of Franz Boas, c/o BAE, through O. I. Mason, February 25, 1895, USNM acc. 29,060, National Anthropological Archives, Washington, D.C. Five similar maps from the same expedition are in the Museum für Völkerkunde, Berlin.

52. Unknown Baffinland Inuit, with a numbered list of Inuit toponyms added by James Mutch, untitled manuscript map of a section of the fjord coast of east Baffin Island; pencil with land shaded in an apparent attempt to represent relief. Mutch's numbered locations and list of forty-eight toponyms in ink; three adhered sheets of lined paper, 29.5 (max.) x 46.0 cm 1903–4; purchased for the museum at the request of Dr. Franz Boas from Captain James Mutch of Peterhead, Scotland; uncatalogued (except as negative 317,380 in the museum's Photography Collection), Department of Anthropology, American Museum of Natural History, New York. Franz Boas, "The Eskimo of Baffin Land and Hudson Bay," *Bulletin of the American Museum of Natural History* 15 (1907): 4–570, is subtitled "From Notes Collected by Captain George Comer, Captain James S. Mutch, and Rev. E. J. Peck." Whereas Comer and Mutch were whalers, Edmund James Peck was a Church Missionary Society missionary who served at Cumberland Sound from 1894 to 1905. Five manuscript Baffinland Inuit maps on paper drawn in 1902 are preserved in file 32, Peck Papers, Anglican Church of Canada Archives, Toronto. File 21 contains a letter of June 1, 1903, sent to Peck at an address in England supplied by Mutch, in which Boas stated: "I value the [Inuit] material you obtained very highly indeed, and I beg to ask you if you would be inclined, during your coming sojurn in Cumberland Sound, to add some more to the details collected by you." Quite clearly, Inuit maps were a small part of a much larger ethnographic collection initiated by the museum but assembled by a loose association of commercial and church agents in the course of, but incidental to, their main employment.

53. Probably in the hand of George Comer, "Melikis Map of musk ox hunting different seasons," Aivilingmiut manuscript map of the west coast of Southampton

Island, Roes Welcome Sound, and the Keewatin mainland between Depot Island and Repulse Bay; linework in pencil with names and legends in ink, 41.5 x 56.5 cm, ca. 1898, 60/2842/E, Department of Anthropology, American Museum of Natural History, New York.

54. The maps collected on this expedition are little known and scattered. In addition to the fifty-seven in the Department of Ethnography, National Museum of Denmark, there is one other known collection. In 1979, at least seventeen maps collected by the expedition leader, Knud J. V. Rasmussen, were given by his daughter to the small museum in Jacobshavn, Greenland, in the house where Rasmussen was born. Photographic copies of these are held in the Arktisk Institut, Charlottenlund, Denmark. Several of the maps are reproduced in publications arising from the expedition.

55. "I regret to inform you that I have been unsuccessful in finding the original pencil sketch map made by Chief Kohklux" (R. Philip Hoen, Map Librarian, The Bancroft Library, University of California, Berkeley, letter to the author, April 11, 1977).

56. "Eureka! . . . the Kohklux map for which you have long been searching has turned up in the hands of one of George Davidson's heirs, and The Bancroft Library has just acquired it" (Hoen to the author, March 6, 1980). Kohklux, a chief of the Chilkaht, and his two wives, "This map was drawn by Kohklux in 1869 at his village. It is the first time he ever used a pencil." Shows the route from the Chilkat River, northern British Columbia, to Fort Selkirk, Yukon Territory, as taken by Kohklux and his father in 1852 on their way to burn Fort Selkirk. Some, at least, of the mountain ranges appear to be represented in profile. Manuscript in pencil on the verso of a printed nautical chart with native toponyms transliterated and entered in ink by George Davidson, 109 x 67 cm, August 1869, G4370, 1852, K6D, Map Library, Bancroft Library, University of California, Berkeley. A smaller, incomplete manuscript map is endorsed, "Kohklux started from his place at Kluwan and drew all around the paper for want of room. He asked for a big sheet of paper and I gave him the back of an old map [the nautical chart], on which he and his wife drew their route etc. in 1852": 26 x 20 cm, K4370, 1852, K61A, Map Library, Bancroft Library, University of California, Berkeley. The circumstances of the mapping, together with a transcript of the larger map, are described in George Davidson, "Explanation of an Indian Map," *Mazama* (April 1901): 1–8.

57. Arguably, maps and accounts thereof originating after 1925 are too acculturated by formal education to afford evidence of traditional forms and methods. There will, of course, be exceptions to this, but, by analogy with spoken language, the vast majority are best considered as pidgins.

2

Encounters in Government Bureaus, Archives, Museums, and Libraries: 1782–1911

G. MALCOLM LEWIS

Introduction

Field encounters with native North Americans' maps almost always involved either contact with those who made them or, at least, with the cultures in which they were made. As chapter 1 makes clear, almost as much was learned about the production process as about the maps produced. But many early encounters with native maps did not occur in the field. Almost always retrospective and rarely involving direct contact with the native mapmakers, most of these were made in one of two broad contexts.

From the early sixteenth century onward, European mapmakers used native maps as sources with which to extend into their *terrae incognitae* maps of areas they had already explored. Sometimes the process was acknowledged or manifest in the resulting map: *incorporations* akin to distinctive older components inserted by architects into new buildings. More frequently, it was unacknowledged and not manifest: *assimilations* detectable, only in retrospect, on the basis of historical research.

Much later in time and for a variety of reasons, native maps began to attract the interest of scholars: at first merely as individual curiosities; then because as a category they seemed to fit into European structures and theories of knowledge; and finally as evidence of native ways of worldmaking, how they differed from European ways, and the manner in which they had changed in the course of contact with non-natives.

Incorporations and Assimilations

There is a spectrum of European maps containing native supplementations, though one extreme of the spectrum probably never existed. It is difficult to think of a fully acknowledged incorporation of one truly indigenous native map at the periphery of a European sketch or survey map. Apart from that extreme, there are examples of almost every form of mix, although there must be many examples of subtle but unacknowledged assimilations still to be detected. Until far more case studies have been undertaken, the full significance and consequences of Indian and Inuit inputs on the European mapping of North America will be unknown. They were certainly much greater than once suspected.

Occasionally, the importance of native inputs forces itself upon the intuitive scholar. There are now guidelines for those who wish to search and a growing number of case studies.[1] Chapters 6 and 10 of this volume add significantly to the latter.

Most incorporating and assimilating was by mapmakers working in the central bureaus or colonial offices of European governments or, from the late eighteenth century onward, by those with access to official records. Although this use of native sources began early, knowledge of it is very largely a consequence of recent research. For that reason, further consideration will be deferred until chapter 4 of this volume. Even so, it should be stressed that the origins of the practice and procedures preceded scholarly awareness of them by approximately four hundred years.

The Emergence and Early Development of Awareness among Scholars

The first generation of scholars and scientists to mention in their writings maps made by native North Americans were men who, in their earlier careers, had participated in exploration, engaged in scientific fieldwork, or had some form of frontier experience. Some made brief references to specific maps in the course of writing major syntheses and a few even published papers about particular maps. During the last quarter of the eighteenth century and first half of the nineteenth, these authors contributed to a slowly emerging awareness among the growing post-Enlightenment intelligentsia that native North Americans had made and were probably still making maps. Mainly European and predominantly German, or writing in a Germanic tradition, their influence in this respect was almost certainly not then as great as the publications of contemporary explorers, including Meriwether Lewis and William Clark, the mainly British searchers for a Northwest Passage, and others of the period. Their significance is that they heralded a more remarkable development after the creation of the German Empire in 1871: the recognition and study of the maps of native peoples as a distinctive category of cartography. These studies were rarely specific to North America, but almost always included North American examples.

The absence of a colonial tradition and of a ready-made colonial empire may in part have accounted for the global perspective of the new Germany after 1871, but intellectual and institutional factors were also important. The interest in the traditional cartography of native peoples stemmed from initiatives and developments in the academic fields of geography and cartography. In 1874, the Prussian government decided to establish a chair of geography in each of the Prussian universities. By 1880 ten had been filled and three more were about to be. Furthermore, the Prussian initiative was being followed elsewhere in the newly unified Germany.[2] This was all the more remarkable because it preceded, in some cases by several decades, the emergence of a new geography in France, Britain, the United States, and elsewhere in the Western world. Furthermore, most university departments of geography quickly adopted the traditions of the seminar and doctoral research, the latter often conducted outside Europe. New geographical journals were established, definitive texts were written, and there was vigorous and sustained debate about the nature and methods of geography as an academic discipline.

At first, physical geography seemed to be dominating Germany's new geography, but the publication in 1882 and 1891 of the two volumes of Friedrich Ratzel's *Anthropogeographie* marked the coming of age of a new global, systematic, and evolutionary-based human geography.[3] The new human geography had close links with academic cartography. For several decades before Ratzel, cartography in the states that were to form the German Empire had been university—or education—based and relatively less preoccupied with the principles and practice of topographic mapping than in France, Britain, and other European countries with national topographic surveys. Geographer-cartographers from Adolf Stieler through Heinrich and Hermann Berghaus to August Petermann had created an awareness of a wider range of map types than in perhaps any other Western country.[4] Many of their maps were of the whole world, presenting global distributions of systematically defined phenomena.

The coexistence of the somewhat older thematic cartography with the new geography provided in Germany the institutional and intellectual milieu for many new developments. Relatively minor among these but important in the context of this chapter was the beginning of the systematic and comparative study of the traditional cartography of native peoples at a global scale and usually from a somewhat crude evolutionary perspective. Between 1877 and 1910, it was the milieu from which stemmed the published work of Richard Andree, Georg Frauenstein, Wolfgang Dröber, and Bruno Adler. Indirectly, if not always directly, it was also the milieu that influenced late-nineteenth- and early-twentieth-century German field scientists to observe and record maps and mapmaking in traditional societies in sundry parts of the world.

As will be shown, serious scholarly interest in Indian and Inuit maps and mapmaking began earlier within North America than in Germany. Not surprisingly, it was stronger. It was also more diverse, reflecting the opportunities and interests of a variety of mainly scientists, but without ever in the pre–World War I period showing signs of become a field of study. What is surprising is the relatively few studies or even scattered accounts published during this period by Canadians in Canada. Given the considerable number of examples and pertinent records that are now known to have existed there by World War I, this paucity must have been a consequence of lack of interest in the subject. Whereas the United States had some exposure through its own publications to pertinent work by German scholars from Georg Kohl to Bruno Adler, Canadian scholarship was still dominated by that in Great Britain, within which a consolidated interest in traditional cartography had not developed.

Scholarly Encounters by Germans

Native maps were collected, described, and used by Europeans from the early sixteenth century onward, but evidence of scholarly interest did not emerge until the late eighteenth century. When it did, German scholars were particularly influential. An early account was by Johann Reinhold Forster, who a few years before had served as naturalist on Capt. James Cook's second voyage to the South Pacific. In 1784 he published in German a book that was to appear two years later in an English translation: *History of Voyages and Discoveries made in the North*.[5] It contained the following:

> Two northern Indians, whom Scroggs had with him, and who had passed the winter in Churchill, told him of a rich mine of native copper, which was to be found on the [Arctic] coast ... They had also at Churchill drawn on parchment with charcoal, the situation of the coast from thence to the spot; and as far as the ship was then come, the sketch perfectly corresponded with the real situation of the country.[6]

This brief extract from a substantial scholarly book is of little intrinsic interest, but several of its negative characteristics were indicative of many later examples. Although it indicated the context in which it had been made, the map itself was not reproduced. Nor was there any attempt to describe its characteristics. Although it was said to "perfectly correspond with the real situation," no indication was given as to the principles and experience on which the evaluation was made. Forster did not give the source of the information, indicate whether he had seen the map, or even state whether it was extant.

In retrospect, it is virtually certain that Forster was paraphrasing from the journal of John Scroggs written in 1722 on a voyage up the west coast of Hudson Bay. According to an extract from the journal published in 1744:

> [T]wo [Northern] Indians had been entertained at [Churchill] Factory all the foregoing Winter....
> [T]hey gave us Intimation of a rich Copper Mine, that lay near the Surface of the Earth, and said they could direct the Sloop or Ship to lie by it... These Indians sketch'd out the Land with Charcoal upon a Skin of Parchment, before they left the Factory, and as far as the Sloop went, they find it agree very well.[7]

Forster's encounter was, therefore, almost certainly made in a library. Ironically, he may have been physically close to making a far more exciting encounter in a very important archive. The Hudson's Bay Company had at its London headquarters an archive of documents relating to its own affairs. Available to members of the company's committee, outsiders were only occasionally admitted. Forster may well have been one of the occasional exceptions. In the course of preparing books and articles on the fauna and flora of North America, he had drawn very heavily on the unpublished notes and specimens sent back to London by Andrew Graham, a company employee.[8] It is difficult to imagine how he could have done this without access to the archives. While there he would presumably have been very close to a remarkable manuscript map. Still extant, it is the second oldest in the Hudson's Bay Company's Archives and the company's first map of the Northwest.[9] Its history is still not completely understood, but in essence it was made by James Knight, who as governor of the company's territories,

> took every opportunity to question ... the Indians during their trade visits, as to the physical geography and human occupants of their home areas, and of the broader regions of the interior.... In the spring of 1716, Knight had asked some Indians to make several sketch maps for him. On another occasion, a Cree chief drew a map of the region

where friends of his, on a war party in 1715, had killed some fourteen tents of Copper Indians to the "WNW" of their own home area ... Then when a group of young Northern Indians arrived at York . . ., Knight "had abundance of Discourse wth. them about there Country and did gett them to lay down there Rivers along Shore to the Norward they chalkd 17 Rivers some of them very Large."[10]

Though Forster may well have been in the very room in which the manuscript sources for this explanation were preserved, it is a level of scholarly interpretation never attempted until the last quarter of the twentieth century.

Whether or not Johann Forster actually worked in the Hudson's Bay Company Archives, he probably did not know that it contained two native maps. By 1810 there were at least twenty-four and by 1843 not less than five more.[11] For the most part, however, encounters with native maps in the company archives were exceedingly rare until the second half of the twentieth century.[12] One important exception involved another German scholar, Johann Georg Kohl. A distinguished geographer, after 1854 he spent several years working and traveling in the United States. On a visit to new German settlements in what is now southeast Minnesota late in 1854, he was made aware that the settlers "used the Indians, who had a much more detailed and intimate knowledge of the country than themselves, to gather information from them regarding directions and to discover fertile patches of soil." Kohl already knew that "Indians had extensive geographical knowledge and that they were skilled in fashioning cartographical pictures of their landscapes. In order to test their knowledge [he] started a conversation with a Sioux ... about the various lakes and tributaries of the headwater regions of the Cannon River." He then asked the Indian to "put this on paper ... [whereupon the Sioux took Kohl's pencil] and designed ... a detailed map of the whole region."[13]

The significance of Kohl's unexceptional frontier experience is that, as a scholar, he was aware of Indian maps before even reaching North America and, therefore, predisposed to learn more about them. Prior to his arrival he had combed major European archives and libraries in search of information relating to the exploration of the New World. In the course of these visits he had made a large collection of transcripts of important maps.[14] These included one of an Indian map that could only have been copied in the Hudson's Bay Company Archives: "An Indian Map of the Upper-Missouri 1801."[15] Now included in the Johann Georg Kohl Collection in the Library of Congress, Geography and Map Division, it is a transcript of a manuscript map still in the company archives.[16] First brought to the attention of scholars in 1886, it was reproduced, rather than the eminently reproducible original, as late as 1957 in one of the modern classics in the history of North American cartography.[17] In chapter 6 Barbara Belyea alludes to this map as one of those embodying a stylized drainage pattern that conferred spiritual significance on the region and was easy to remember. For Kohl, it was doubtless at first of interest simply because it was made by an Indian. Soon, however, he recognized that it was one of an important category of maps. His lecture to the Smithsonian Institution in the winter of 1856–57 was an informed and detailed plea for the establishment of a federal map library. In the discussion of categories of maps to be collected, he included all those rude sketches of interior parts of America, which on different occasions, have been drawn by the Indians on skins or the bark of trees, and which sometimes were the first guides, by the help of which Europeans were enabled to find their way. Such Indian maps have often been considered as conveying very valuable information, and, consequently, have been sent home to England or France by governors of provinces, have been copied by European geographers into their works, and have then been deposited as valuable documents in the archives of state, or have been found worthy, as historical curiosities, of being preserved in the British Museum and in similar collections. Nay, there are still some parts of America, as the interior of Brazil and Labrador, and the vast territories of Hudson's Bay, which are delineated on our maps on no better authority than that of an Indian sketch or report. It is evident, then, that we cannot neglect the study of these aboriginal productions, but must give them also a place in our collection.[18]

Quite clearly, Kohl was fully aware of Indian maps as a large, distinctive, and important category, examples of which were to be found in many collections. They should be collected not as curios but because they were important to an understanding of certain stages in the European mapping of America. Kohl was almost certainly the first scholar to recognize this. Although his plea for the establishment of a national (United States) collection of maps of all kinds was not answered for another forty years, other German scholars began to give serious attention to what would now be referred to as traditional cartography. Then, however, titles almost always included "Naturvölkern" or "Primitive Peoples," terminology that persisted at least until 1943, the year in which Leo Bagrow completed *Geschichte der Kartographie*, the very brief first chapter of which was "Die Naturvölker."[19] In many respects, Bagrow's monograph marked the end of an important stage in the development of the history of cartography. Having worked in Berlin from 1918 onward, he was very aware of the literature in German. His first chapter reflects but does not acknowledge this.

After Kohl, German syntheses embraced "primitive" cultures in all parts of the world, though in many cases the North American component was the most dominant. This reflected a reliance for sources on published works of discovery and exploration. By the latter part of the nineteenth century, North America had generated more publications of this type than any of the other "peripheral" (to Europe) continents. Many mentioned specific maps made by Indians or Inuit, a few contained more detailed accounts, but virtually no examples were reproduced. This was reflected in the considerable post-Kohl German literature on traditional cartography. One of the earliest of these contributions was particularly significant because it presented the maps of native peoples in an evolutionary, or at least developmental, context.

In 1878, Richard Andree published *Ethnographische Parallelen und Vergleiche*. Seeking parallels and comparisons within the diverse social world, the book reflected the evolutionary ideas that were by then beginning to influence thinking in fields outside biology. One chapter, "Anfänge der Kartographie," attempted to present the history of cartography in a developmental context.[20] Superficial when read from a late-twentieth-century perspective, it anticipated the emergence of more deterministic ideas, in which map types were correlated with the "levels" of society in which they were made. Citing examples from many parts of the world, it implied that maps made in traditional cultures, as observed in the non-European world after the Age of Discovery, were relics of "the beginnings" (die Anfänge) of cartography. North American examples were numerous but usually drawn from secondary sources and only briefly alluded to.[21] A short section, "Eskimokarten," was illustrated with a German line drawing derived from "Chart of Coast from Cape York to Smith Channel Drawn by Kalli-herua (Alias Erasmus York). . . ."[22] It is a remarkably un-Eskimo-like map in appearance, mainly because of omissions. It is almost impossible to distinguish the Eskimo's delineation of the shoreline of northwest Greenland as derived from Kalli-herua from that superimposed onto it from the Admiralty Chart. Most of the original's many Eskimo toponyms are omitted. None of the fascinating legends are included, such as "Hares in plenty," "some natives were killed by falling from the Cliffs," "[n]o ice ever forms in this Channel [but] icebergs pass through this Channel in Winter," and "Glacier traversed in sledges by natives." We will never know the extent to which the first printed map differed from Kalli-herua's original, but the second printed version as reproduced in Andree is a diluted travesty of the first.

Although the empirical content and scholarly standards of works in the German tradition were to improve after Andree's seminal contribution, the conceptual context in which the material was presented remained implicitly developmental and the visual illustrations inadequate. Briefly mentioning, but not citing, Indian and Inuit examples, Georg M. Frauenstein ended a short paper, "Primitive map-making," with categorical statements that fell far short of conclusions arising from the evidence: "The Polynesians, the Esquimaux, and the [North American] Indians, have all thus given us marks of the different degrees of advancement they have independently made in the use of this, the most important of geographical aids [cartography] . . . maps appear among the evidences of the degree of civilization that has been reached."[23] The theme was continued by Wolfgang Dröber in his doctoral dissertation at Erlangen.[24] Though the empirical content was somewhat greater, the earlier characteristics persisted; particularly the developmental theme.

The most substantial of the German contributions was Bruno F. Adler's 350-page monograph of 1910. Published in Russian in St. Petersburg, it has never been readily accessible outside Russia. In the English-speaking world, knowledge of the contents was, and for the most part still is, through a translated abridgement. Whereas the original devoted ten columns (including twenty-three footnotes and seven figures) to maps made by North American Indians and Inuit,[25] the translation abridged it to one paragraph of less than two hundred words, devoid of citations, mentioning very few authors by name, and unsupported by figures.[26] In the Russian original, however, the seven figures were each inadequate line drawings reproduced from published sources. Indeed, for the North American section, at least, there is no indication that Adler saw either an original map artifact or a direct transcript of one. Likewise, all the cited sources were fairly well known published works and there was no indication that he saw any original manuscript sources. Monumental in scope, his monograph neither added to the concepts nor altered the thrust of the earlier works by German scholars. Yet, world events and a decline in interest in this and related aspects of North America's native peoples ensured it a reputation, reinforced somewhat by the mystique of its inaccessibility, that was not to be surpassed for more than sixty years.

Scholarly Encounters by North Americans

Whereas German scholars drew on published accounts of maps and mapmaking to provide evidence in support of developmental stages in the global history of cartography, their North American contemporaries tended to focus attention on surviving examples of the maps themselves, their production, and the contexts in which they were used. The contrast was not, of course, absolute. The Moravian missionary John Gottlieb Ernestus Heckewelder wrote one of the earliest and best accounts of a cosmographical map.[27] Conversely, Charles P. Daly, an eminent Irish-American scholar and judge, mentioned Inuit and Indian maps in a developmental context only two years after Andree had formally introduced the idea in Germany.

In 1819, Heckewelder published a detailed account of a drawing that had been made on deerskin in 1762. It had been used as a visual aid by an itinerant Delaware Indian in the course of preaching a politico-spiritual message concerning the displacement of his people by settlers of European stock who were moving westward from the middle Atlantic colonies into the upper Ohio valley. Published in the eminently scholarly *Transactions of the Historical and Literary Committee of the American Philosophical Society*, the account is in retrospect remarkable for at least three reasons: (1) it is one of the earliest records of a map made and used in an indigenous context, albeit one that had been created by the activities of aliens; (2) though not published until fifty-seven years after the event described, it appears to have been based on personal observations; and (3) it combines detailed graphic description with informed interpretation.[28] Though better in every respect, it was in the tradition of Indian-map description initiated in 1782 by the English antiquary William Bray: a published account, obtained from "a gentleman" of his "acquaintance," of a painting on the blazed trunk of a tree. Like Heckewelder's deerskin map, it was made by a Delaware Indian in the course of the ferment of events that affected the upper Ohio valley during the third quarter of the eighteenth century. Apparently indigenous in style and production, it too represented an aspect of the disruptive encounter between Indians and people of European stock. Unlike Heckewelder's account, however, it was not based on the author's own observations. The "gentleman" who had brought it to Bray's attention had found the painted tree on the banks of the Muskingum River, in what is now southeast Ohio, where it had been interpreted for him by White Eyes, a Delaware chief.[29] Reproduced as a keyed line drawing (fig. 1.1), it was a pictographic painting of military events, of which only parts were maplike: the lower Allegheny and Monongahela Rivers, their confluence to form the Ohio River, and the locations of Fort Pitt and the adjacent civil settlement in relation to these. Bray's account barely mentions them and does not use the word "map." Although a precise and important early contribution to the scholarly understanding of Indian pictography, the paper's significance in the context of cartography did not emerge until much later. Bray's failure to recognize the cartographic component doubtless reflected his lack of knowledge about the region. It was an extreme case of early scholarly encounters failing to consider ground referents.

Charles P. Daly, chief justice in the Court of Common Pleas in New York City, was for many years president of the American Geographical Society. He gave annual presidential addresses, several of which reviewed the status of and current developments in branches of geography he deemed to be important. It is doubtful whether he ever went far beyond the urbanized Atlantic seaboard or met a native American, but in his 1879 review of the early history of cartography he referred to Inuit maps on page 2 of a thirty-seven-page paper in which he "sought to bring together and arrange in something like consecutive order the principal facts in the history of cartography down to the time of Mercator [Gerardus Mercator, 1512–94]."[30] Yet the Inuit maps referred to were from the early nineteenth century, when the British Arctic explorers Parry and Ross

> were astonished to find that the Esquimaux understood their charts; that they were able to recognize upon them not only the outlines of coasts and the positions of places, but to continue the drawing of the coast lines in the delineation of portions of the Arctic unknown to the explorers. Parry published several of these Esquimaux charts, or maps, which when further explorations were made, were found to be remarkable for their accuracy. Nor is this art confined to the Esquimaux, for the North American Indians have, and have always had, maps which, however rude they may appear to us, are intelligible and serviceable to them.

This was the second part of a paragraph that had begun with the statement that "The origin of maps is involved in as much obscurity as the invention of letters" and continued with the assumption that:

> the cartographic art is probably as old, or older than the invention of the alphabet.... [Indeed] we may fairly assume that long before letters were invented ... man had sufficiently advanced in the knowledge of the arts of design, to be able to represent the position of countries, cities and towns, the course of rivers, the situation of seas, the locality of mountains, or other distinguishing features of the earth's surface, by some form of delineation, or map ... for this art has been found in use amongst races that had never advanced so far as to invent a written language.[31]

Daly did not cite the sources of his basic implication that maps made by North American Indians and Inuit in the modern era were akin to those made in early civilizations. During the 1870s, intellectual life in New York City was much influenced by German scholarship, science, and philosophy. Daly may not have read Andree's "Die Anfänge der Kartographie" of two years before, but the ideas were almost identical and must have stemmed from the same intellectual milieu. Alas, they were equally vague, made without rigorous examination of specific artifacts among the traditional cartography of the modern era, or careful comparisons between them and the maps cited from antiquity. Daly was an armchair geographer. Fortunately, these ideas were not to take root in North America, where, with a few exceptions, scholarly encounters were to assume different directions.

One of those exceptions came remarkably late. For his presidential address to the Association of American Geographers in 1926, J. Paul Goode took as his theme the map as a record of progress in geography. In a manner quite similar to Daly, he too first conflated and then equated the maps of modern traditional cultures with those of ancient times. Unlike Daly's, however, his example was a specific and truly indigenous map, of which he had been told by a friend, George A. Dorsey. It was a Pawnee "star chart on fine buckskin. More than a thousand stars were entered, and so accurately was the work done that when he was permitted to bring the chart to his colleagues in astronomy, they could make a fair estimate from the positions of the planets that the chart had been made about the time of the discovery of America."[32] There is no evidence in the text that Goode had seen this artifact, which for twenty-one years had been in the Field Museum of Natural History, close to the campus of the University of Chicago, of which he had been a faculty member for even longer.[33] Nor does Goode seem to have been aware that at the time he was giving his lecture, an astronomer was completing a study of the buckskin in what was to become probably the first published scientific analysis of a truly indigenous map artifact.[34]

The main and distinctive contributions of North American scholars to the understanding of Indian and Inuit maps during this period were sixfold: they began to make known the characteristics, contexts, and locations of specific early examples; they recognized that some of these had influenced early European maps of the Americas; they fostered the collecting of map artifacts by museums; they described examples of maps discovered in the course of field research; they recorded the methods by which maps were still being made; and they demonstrated that mapmaking had its roots not in the encounter context, but in the indigenous and continent-wide method of communicating and recording by means of pictographs.

Johann G. Kohl's activities in Europe immediately before arriving in North America in 1854 gave rise to the spread of knowledge of historically important maps. Kohl had made "well-executed hand-copies, with but occasional attempts at reproduction by fac-simile . . . of maps relating to the progress of discovery in [North, Central, and South] America."[35] The published cartobibliography of 1886 listed and described 474 items arranged by region, and contained an author list, but lacked an index. Philip Lee Phillips's index to the 1904 reprint revealed that the collection contained three "Indian map[s]."[36] Two were important but virtually unknown manuscript maps in London collections: Ac ko mok ki's 1801 map of the sources of the Missouri River as transcribed by Peter Fidler and preserved in the archives of the Hudson's Bay Company; and a map "describing the Scituation of the Several Nations of Indians to the NW of South Carolina . . . coppyed from a Draught drawn & painted on Deer-Skin by an Indian Cacique and presented to Francis Nicholson Esqu^r. Governour of South Carolina by whom it is most humbly Dedicated To His Royal Highness George Prince of Wales," preserved in the British Museum.[37]

Kohl also copied one printed map, thereby helping to draw attention to the fact that Indian maps had been redrawn for publication from the early eighteenth century onward. Now fairly well known and often reproduced, Louis Armand de Lom D'Arce's "A Map drawn upon Stag skins by

Yᵉ Gnacsitares..." has still to be adequately interpreted more than three hundred years after it was made for Baron La Hontan in the winter of 1688–89, almost certainly somewhere in the middle reaches of the Minnesota River.[38] In 1854, when Kohl asked the Sioux to make a map of the Cannon River, he was probably unaware that he was little more than a hundred miles from the eastern edge of the stag skins map, and probably just within the southeastern limit of the La Hontan map of the lower "River Longue" (Minnesota River), of which it was a part.

Several decades after Kohl copied maps of the Americas in Europe, Woodbury Lowery, a lawyer born in Washington, D.C., began making an even bigger collection of map transcripts for a considerably smaller area: former Spanish possessions that had eventually been included within the United States. Although Lowery had no means of knowing it when the copy was made in the Archivo General de Indias, Seville, one transcript would now seem to be of the oldest extant North American map. Apparently copied by Lowery himself, it is full of errors, not least the name of the Indian who made the original. Now known to have been Miguel, it was transcribed as "Nigual yndio natural" and listed by that name in the published catalogue.[39] More serious than the copying errors, however, was the failure of scholars to take much note of it. A spectacular example of this occurred as late as 1953 in the definitive study of Don Juan Oñate, on whose expedition of 1601 across the southern or central Great Plains Miguel had been captured. The "Official inquiry...concerning the new discovery undertaken by Governor Don Juan de Oñate toward the north beyond the provinces of New Mexico" contains an almost two-thousand-word detailed official record indicating how and why Miguel made the map. Yet the editors did not indicate that the map itself was extant in Seville and available in the Lowery transcript form in the Geography and Map Division of the Library of Congress.[40] The reason for this was almost certainly the physical segregation of the map from the "Official inquiry" in the Archivo General de Indias. Lowery had found the map; Hammond, Rey, or one of their assistants had found the record of the "Official inquiry"; but until 1987 no one seems to have recognized the existence of both, each enhancing the significance of the other.[41]

In 1901, Woodbury Lowery published *The Spanish Settlements within the Present Limits of the United States: 1513–1561, with Maps*. A discussion of the Cantino planisphere of 1502, perhaps the earliest map to show European discoveries in the Americas, contained several hypotheses concerning its "origin" and "what land it was intended to portray."[42] One of these is of interest for two reasons. Some of the information was said, by unstated persons, to have been "obtained from [natives of] Hispaniola or Cuba and shrewdly applied in the making of the map, for these Indians were not without a knowledge of contiguous coasts." At this point, a footnote directed the reader to "Appendix E, this volume, 'Indian Charts.'" It cited three early accounts of Indian mapmaking in North America: Alarcón's on the lower Colorado River in 1540; Archer's off the coast of Maine in 1602; and Champlain's of the St. Lawrence River in 1603. It was by no means the first time that these had been referred to in print, and they served only to prove that some native peoples (albeit North American Indians and not

Carib Taino) made maps at or very soon after their first contacts with Europeans.[43] The significance of the appendix is that, as in Phillips's index to the second edition of *The Kohl Collection*, published three years later, it was a clear indication that Indian maps were increasingly being viewed as a category by American scholars. In this respect, they were following earlier German scholars, but it was a potentially dangerous trend. None of the three maps cited by Lowery was extant and two (the lower Colorado and St. Lawrence Rivers) did not represent coasts. Even more dangerous was the assumption that because the Halchidhoma, Micmac, and St. Lawrence Iroquois had made maps, the Taino of Hispaniola (now Haiti and the Dominican Republic) and the Taino, Subtaino, and Ciboney of Cuba had been able to do likewise.

Lowery's hypothesis that information supposedly supplied by natives in map form had been "shrewdly applied" by whoever compiled the Cantino planisphere is but one of several apparently independently reached conclusions that early European maps must have incorporated geographical and cartographical information obtained from indigenous peoples. For example, in one of the earliest of a fifty-year sequence of monographs and papers on the early cartography and toponymy of the Atlantic coast of Canada, the botanist William F. Ganong asked rhetorically "whether maps made by Indians have played any part in our cartography?" His answer was general but positive: "No doubt sketches made by them for the early explorers gave much information about places the explorers could not visit, and it is not unlikely that such sketches are the originals of some features upon early maps; in fact there is no other imaginable source from which accurate information could have been derived where the explorers did not themselves visit the places."[44]

Ganong was also one of the earliest North American scholars to recognize the importance of circumstance and the mapmaker's personal background in determining the characteristics of a map. Concerning one of several Passamaquoddy maps made on birchbark about 1798, he observed, "It is very crude, as are the others in the same set, but probably no more so than would be the case if made by white men under the same circumstances."[45] Ganong possessed a map of the Tobique River, New Brunswick, that had been drawn for him by a Malecite Indian which, except for a distortion due to the size and shape of the paper, he considered to be "surprisingly accurate," perceptively adding, however, that "as the maker had been to school, it can hardly be viewed as aboriginal."[46] It was a degree of caution that all too few later scholars have exercised in discussing Indian maps and deriving evidence from them.

Increasingly from the 1860s onward, Euro-Americans who encountered native American maps and mapmaking in the field subsequently published detailed accounts and, in most cases, ensured that the maps were preserved. Three encounters mentioned in chapter 1 would have been almost equally appropriate as examples here: Sproat's with three-dimensional Nootkan mapmaking on a Vancouver Island beach sometime between 1860 and 1865;[47] Davidson's with a Chilkaht and his two wives making a map in 1869 of the route of a return journey undertaken many years before across the rugged mountains and basins of southeast Alaska and the southern part of what was to become the Yukon;[48] and Flaherty's chance discovery in 1910 of an Inuit's map of the Belcher Islands[49] (figs. 1.2 and 1.3). A fourth is less well known because it was presented in a lecture and never published. On February 10, 1870, the Canadian geologist Robert Bell delivered the first lecture in the Somerville Course, given under the auspices of the Natural History Society of Montreal. Since the course was presumably named after the British geographer Mary Somerville, whose *Physical Geography* had been first published in 1848 and in whose memory the Oxford College was later named, the title of the lecture was very appropriate: "Lecture on Exploration in the Nipigon Country." Bell, an officer of the Geological Survey of Canada, had surveyed the region north of Lake Superior during the previous summer. In the lecture, he described an incident that had occurred immediately prior to that:

> Before leaving Fort William I took the precaution to ascertain from the H.B. Comp.[s] [Hudson's Bay Company's] officers and the Indians all I could in reference to the Great Lake [Nipigon] we proposed to explore. They assured me it would occupy a whole [summer field] season, or perhaps two, simply to paddle once round its shores, hugging all the bays and points. One day a party of Nipigon Indians [Ojibwa] arrived at the fort. I invited them to my camp. I gave one of them a sheet of paper, which he spread upon the cook's baking board and went to work with a lead pencil to make a sketch of the lake; all the rest standing round him in a circle helping him by their suggestions and improvements. While this interesting work and discussion were going on some one (in an evil moment) knowing their weakness—tossed them a pack of cards. In an instant, baking board, paper and pencil were dropped and games commenced, at which they played with great eagerness all afternoon . . . I afterwards got the map finished and here is the result of their combined effort, only touched up with a little color. I should mention that the shape is distorted Indian fashion to fit the paper and to make the most of it."[50]

The map is extant and has pin holes in it that presumably date from its display at the lecture.[51] It is not known whether the lecture had an impact beyond the audience, but the account supports several aspects of the mapmaking process as described by Sproat and Davidson. As with the Nootkas, and presumably the Chilkaht and his two wives, the mapping was interactive, with one Indian taking the lead and others offering "suggestions and improvements." Likewise, as with the Chilkaht chief and wives, shape and size of paper were important. The Chilkahts aborted their first attempt at mapping because the sheet of paper that they had been given was too small. Whether they realized it or not, the Nipigons' map, according to Bell, who by the time he gave the lecture had surveyed the lake, was distorted in shape by the proportions of the paper. These accounts reveal that Nootkas, Chilkahts, and Nipigon Ojibwa all approached mapmaking with enthusiasm. These kinds of details had rarely been recorded in the earlier frontier encounters.

Davidson, Bell, and Flaherty evidently anticipated the significance that their encounters would have for future generations of scholars. In each case, the map that they described has survived.[52] Indeed the fifty-year interval between the end of the American Civil War and the outbreak of World War I was a period of most active map collecting and preserving. Many maps, but by no means all, were on paper, and to that extent not indigenous. Whereas maps made on paper had served earlier generations of explorers well and continued to be solicited by geologists, the first generation of cultural anthropologists began to observe, describe, and collect truly indigenous artifacts that they categorized as maps or which afforded evidence that Native Americans sometimes interpreted linear patterns as maps. Extant examples are scattered and varied.

In 1887, Sapiel Selmo, an elderly Passamaquoddy chief from eastern Maine whose father had supposedly commanded six hundred Passamaquoddys in the Revolutionary War, made several maps on birchbark for Garrick Mallery. As a younger man, Selmo had camped and hunted with his father at Machias Lake, Maine.[53] The birchbark maps, though not dating from that period, were almost certainly made as examples of message maps formerly made by hunters to convey information about the hunt to companions expected to pass by. Three were reproduced by Mallery as line drawings, but, notwithstanding their fragility, the originals have survived in the Department of Anthropology, National Museum of Natural History, Washington, D.C.[54] The museum also houses the National Anthropological Archives; formerly the Manuscript Division of the Bureau of American Ethnology. It contains a large number of native maps on paper, mainly made for field officers after the bureau was created in 1879.

Not all maps collected during this period were eventually placed in national collections. Indeed, some of the more interesting ones went elsewhere. In the early twentieth century, the University of Pennsylvania anthropologist Frank G. Speck deposited in the Museum of the American Indian, New York, twelve pieces of birchbark on which Montagnais women had made patterns by first folding and then biting with their teeth. Some of the patterns were intentional, but some were produced in error, and these were often described as trail maps. Among the examples deposited in the museum is an error pattern explained by an unknown Montagnais as "started to make trees, but trails came out."[55]

Publication between 1896 and 1901 of Reuben Gold Thwaites's seventy-three-volume and exceedingly well-indexed *The Jesuit Relations and Allied Documents* made generally available for the first time early accounts of native map awareness and of hitherto almost unsuspected types of indigenous maps; especially in the Northeastern culture region and adjacent parts of the sub-Arctic. In addition to accounts of bitten bark and birchbark biting, the volumes also contained early accounts of two other indigenous activities that involved perceptual and cognitive skills very similar to those used in spatial aspects of map reading: scrying and scapulimancy.[56] Both were divinatory. Scrying involved peering for long periods at a smooth surface (traditionally, almost always water) in order to "see" things that were geographically far away, sometimes apparently in plan. Its purpose was to locate enemies, strangers, and animals in the forests. Scapulimancy involved interpreting cracks induced on animal bones by heat or percussion, often as maps indicating the locations of animal herds. Though not necessarily stimulated by *The Jesuit Relations*, the early twentieth century saw the first serious anthropological studies of these in part cartographic forms of divination.[57] Examples of these and other map artifacts continued to be accessioned by museums.[58]

Whereas Adler's 1910 monograph was the culmination of a long, sustained, but for the most part undeviating German interest in the historical context of native people's maps worldwide, by that date interest among North Americans had diversified but remained confined to their own continent. The best single indicator of diversification had been published seventeen years before Adler: Mallery's "Picture Writing of the American Indians." Entitled "Notices," chapter 11 was divided into four sections: "Notices of visit, departure, and direction," "Direction by drawing topographic features," "Notice of condition," and "Warning and guidance."[59] All the examples had an environmental context and almost all had a strong spatial component. Many of those illustrated and described were maplike, but the word "map" rarely appeared in the text. Of the thirty-three figure captions only two contained the word and one other contained a "chart."[60] Yet evaluated retrospectively, nine of the figures were undoubtedly maps, two were very simple maps, and six more were topographic profiles. The chapter's pervading theme was the pictographic representation of simple spatial relationships and more complex spatial patterns. Its significance would have been clearer, therefore, if the title had contained "map," "chart," or even "topography." Nevertheless, it was in several respects innovative and drew on a wider range of examples than any previous work of scholarship.

Whatever they were called and however they were categorized, the artifacts illustrated and described by Mallery were presented as indigenous in function and form. There was very little reference to maps made on paper to assist or satisfy alien whites. The volume of which the chapter was a part was concerned with the North America–wide native method of communicating and recording that was slowly becoming recognized by scholars as a highly adaptable system common to all Indians and Inuit. First called "pictography" by Henry R. Schoolcraft in 1851, it sometimes involved the representation of features, patterns, and distributions in vertical perspective.[61] There is no evidence to suggest that native North Americans thought of vertical perspective as constituting a distinctive subset of pictography akin to the map, chart, or plan of the European tradition.[62] In "Notices," therefore, Mallery came close to creating a subcategory that most native North Americans had never had cause to recognize. His chapter established a potential basis for comparative studies of formal European-style maps and native vertical perspectives as manifest in pictography, but the potential was to remain latent until the 1980s.

Mallery's other major contribution was bringing together examples of many types of vertical-perspective pictographs as used in a diversity of indigenous contexts. Drawing on the collections and observations of the Bureau of Ethnology's staff, including not only his own among the Passamaquoddy and Malecite of northern Maine and New Brunswick, but also the published records of other scientists, including cultural anthropologists, he described or illustrated examples incised on birchbark and birchbark scrolls, drawn with charcoal on a spruce wood chip, modeled three-dimensionally on the ground, traced in earth, and carved in wood. The examples were from several quite different parts of the continent and Greenland. For many Mallery gave a native's own explanation. Important though this chapter was, its contemporary impact was less than it might otherwise have been. The genuine map artifacts were illustrated by annotated, rather crude, certainly unattractive, and probably incomplete line drawings; as a basis for further research the bibliographic standards were grossly inadequate; and, as indicated already, the organization of the chapter focused on functions rather than on mapmaking as a distinctive way of presenting certain kinds of information. The extent to which these shortcomings weakened the chapter's potential impact during the eighteen years leading up to the end of the first period of scholarly encounter cannot be determined. Almost certainly, however, Mallery had more influence than Adler on scholars involved in the post-1970 revival of interest in Indian and Inuit maps and mapping. All who initiated that revival should feel indebted to him.

Notes

The dates in this chapter's title are not arbitrary. The first, 1782, was the year in which an account and reproduction of an Indian map first appeared in a scholarly journal: William Bray, "Observations on the Indian Method of Picture Writing," *Archaeologia* 6 (1782): 159. Indian maps had been reproduced before that date but always, so far as has been established, without an informed discussion, e.g., "A New Map of the Cherokee Nation with the Names of the Towns + Rivers. They are Situated on N°. Lat. from 34 to 36" (title in cartouche); "Engraved from an Indian Draught by T. Kitchin" (note below map), *London Magazine* 29 (February 1760), opp. 96. The terminal date of 1911 is the year in which an abridgement was published in English of a global review of traditional cartography that had been published in Russian in the previous year. Thereafter, little of scholarly importance was published for approximately fifty years: H. de Hutorowicz, "Maps of Primitive Peoples," *Bulletin of the American Geographical Society* 43 (1911), 669–79; translated and abridged from Bruno F. Adler, "Karty Piervobytnyh Narodov," *Iviestia Impierator Obshshestva Lubitieleï Estiestvoznania, Antropologii i Etnografii, sostoyaszchavo pri Impieratorskom Moskovskom Universitietie*, vol. 99, *Trudy Geograficheskavo Otdielienia*, no. 2 (St. Petersburg, 1910).

1. G. Malcolm Lewis, "Indicators of Unacknowledged Assimilations from Amerindian *Maps* on Euro-American Maps of North America: Some General Principles Arising from a Study of La Vérendrye's Composite Map, 1728–29," *Imago Mundi* 38 (1986): 9–34. In it the diagnostic characteristics of information transmitted by Amerindians in *map* form are on pp. 22–31. These and other diagnostic characteristics were later employed by the author in two detailed case studies of La Vérendrye's composite map: "Misinterpretation of Amerindian Information as a Source of Error on Euro-American maps," *Annals of the Association of American Geographers* 77, 4 (1987): 542–63; and "La Grande Rivière et Fleuve de l'Ouest: The Realities and Reasons behind a Major Mistake in the Eighteenth Century Geography of North America," *Cartographica* 28, 1 (1991):54–87.

2. Preston E. James and Geoffrey J. Martin, *All Possible Words: A History of Geographical Ideas*, 2d ed. (New York: John Wiley, 1981), 165–66.

3. Ibid., 169.

4. Ibid., 144–45.

5. Johann R. Forster, *Geschichte der Entdeckungen und Schiffahrten im Norden: Mit neuen Originalkarten versehen* (Frankfurt an der Oder, 1784); English edition: *History of the Voyages and Discoveries Made in the North, Translated from the German . . . and Elucidated by Several . . . Maps* (London: G. G. J. and J. Robinson, 1786).

6. *History of the Voyages and Discoveries,* 388. Interestingly, this was almost certainly the map briefly alluded to exactly fifty years later by Alexander von Humboldt in another great work of synthesis: *Kritische Untersuchungen uber die Historische Entwicklung der Geographischen*, vol. 1, *Kentnisse von der Neuen Welt* (Berlin: Nicolai, 1836), 297–98. Humboldt, however, cited post-Forster sources.

7. An extract from the 1722 journal of John Scroggs as published in Arthur Dobbs, *Remarks upon Capt. Middleton's Defence: Wherein His Conduct during the Late Voyage for Discovering a Passage from Hudson's Bay to the*

South-Sea is Impartially Examin'd (London: J. Robinson, 1744), 114.

8. Forster had influential scientific friends, including Alexander Dalrymple, who in 1799 had become hydrographer to the East India Company. Dalrymple had been in part responsible for inviting Forster to England from Russia and is known to have used the Hudson's Bay Company Archives himself. He was also a close friend of Samuel Wegg, deputy governor of the company from 1778 and later governor.

9. Untitled sketch of rivers between Prince of Wales' Fort and the "Norther most Coper Mind," giving Indian names; by or derived from James Knight but with later additions, post-1719, 52.0 x 66.5 cm, G1/19, Hudson's Bay Company Archives, Winnipeg. There are published transcripts of this map in John Warkentin and Richard I. Ruggles, eds., *Manitoba Historical Atlas* (Winnipeg: Historical and Scientific Society of Manitoba, 1970), 86–87; and in William H. Goetzmann and Glyndwr Williams, *The Atlas of North American Exploration* (New York: Prentice Hall, 1992), 102.

10. Richard I. Ruggles, *A Country So Interesting: The Hudson's Bay Company and Two Centuries of Mapping, 1670–1870* (Montreal: McGill-Queen's University Press, 1991), 30.

11. From a list of extant native manuscript sheet maps supplied in March 1993 by Judith Hudson Beattie, Keeper, Hudson's Bay Company Archives. There were formerly more. Richard Ruggles gives details on an additional ten in *A Country So Interesting,* 241–45, cat. C. In chapter 6 Barbara Belyea discusses Aaron Arrowsmith's use of some of these in revising "A Map Exhibiting all the New Discoveries in the Interior Parts of North America" (1795). See also the indexed references to Aaron Arrowsmith in Ruggles, *A Country So Interesting,* esp. 60–72.

12. Even then there were restrictions. Richard Ruggles used the archives intensively in the course of research for his 1958 Ph.D. thesis at the University of London: "The Historical Geography and Cartography of the Canadian West, 1670–1795: The Discovery, Exploration, Geographic Description, and Cartographic Delineation of Western Canada to 1795." Access was doubtless facilitated by his eminent thesis supervisor, Sir Dudley Stamp. Even so, photography was not permitted and all the numerous maps had to be transcribed. Several of Ruggles's transcripts of Indian maps were subsequently published in *Manitoba Historical Atlas*, figs. 32–34.

13. Johann G. Kohl, *Reisen im Nordwestern der Vereinigten Staaten* (New York: D. Appleton and Co., 1857), 268. The Sioux map is not extant, but from Kohl's fairly detailed account it was of the Cannon River, and the many small lakes in its upper valley, including Cedar Lake; also the Blue Earth River, the land around it, and the track to Faribault (44°20'N, 93°15'W).

14. The general background to these events is described in John A. Wolter, "Johann Georg Kohl and America," *Map Collector* 17 (December 1981): 10–14. The collection is described in Justin Winsor, *The Kohl Collection of Maps Relating to America: With Index by Philip Lee Phillips* (Washington D.C.: Government Printing Office, 1904).

15. Transcribed by Johann G. Kohl, "An Indian Map of the Upper-Missouri 1801," ink on paper, 18.4 x 26.0 cm, Kohl Collection, Library of Congress, Washington, D.C.

16. "An Indian map of the Different Tribes that inhabit on the East & West side of the Rocky Mountains with all the rivers & other remarkable places, also the Number of Tents. Etc. Drawn by the Feathers or ac ko mok ki—a Blackfoot chief—7th Feby. 1801—reduced 1/4 from the Original Size—by Peter Fidler," ink on paper, 37.2 x 47.0 cm, G1/25, Hudson's Bay Company Archives, Winnipeg; reproduced in Ruggles, *A Country So Interesting*, pl. 19.

17. Carl I. Wheat, *Mapping of the Transmississippi West 1540–1861*, vol. 1, *The Spanish Entrada to the Louisiana Purchase, 1540–1804* (San Francisco, Institute of Historical Cartography, 1957), opp. 181, map 249.

18. Johann G. Kohl, "Substance of a Lecture Delivered at the Smithsonian Institution on a Collection of the Charts and Maps of America," *Annual Report of the Board of Regents . . . Smithsonian Institution* (Washington: Smithsonian Institution, 1857), 127.

19. Leo Bagrow, *Geschichte der Kartographie* (Berlin: Safari-Verlag, 1951). An earlier edition, which had been prepared for publication in 1944, had been destroyed by fire.

20. Richard Andree, *Ethnographische Parallelen und Vergleiche* (Stuttgart: Julius Maier, 1878), 197–221. The first part of the chapter had been published in the previous year: "Die Anfänge der Kartographie," *Globus* 31 (1877): 24–27.

21. Included among the North American examples were a Kiowa sand map described by Lieutenant Amiel Weeks Whipple, and Yuma and Paiute examples of the same sand map as described by Heinrich Balduin Möllhausen, the student of Alexander von Humboldt who had accompanied Whipple on the 1853 U.S. Army expedition in search of a railroad route to the Pacific. Other North American sources included Father Pierre-Jean De Smet (birchbark map), Josiah Gregg (map in pencil on paper), and Frederick William Beechey (Eskimo map in sand and stones).

22. "Chart of Coast from Cape York to Smith Channel. Drawn by Kalli-herua (alias Erasmus York). Partly from his own observation, and partly from what he has heard his own people say. The Eskimo names are given by himself, and are written as pronounced. The names doubly underlined are to be found on our Charts. Drawn on board H.M.S. Assistance during the Winter of 1850–51. Erasmus Ommanney, Capt[n], R. N. Note. The Coast line, ec in Red is in accordance with the Admiralty Chart," 53.0 x 12.5 cm, in Clements R. Markham, ed., "Papers on the Greenland Eskimos," *A Selection of Papers on Arctic Geography and Ethnology Reprinted and Presented to the Arctic Expedition of 1875, by the President, Council and Fellows of the Royal Geographical Society* (London, 1875), facing 184.

23. Georg M. Frauenstein, trans. from *Das Ausland*, "Primitive map-making," *Popular Science Monthly* 23 (1883): 687.

24. Wolfgang Dröber, *Kartographie bei den Naturvölkern* (Erlangen: Junge, 1903); reprinted by Meridian in Amsterdam in 1964. At the time of publication it was probably best known in a published abstract form: "Kartographie bei den Naturvölkern," *Deutsche Geographische Blätter* 27, 1 (1904): 29–46.

25. Adler, "Karty Piervobytnyh Narodov," 161–71, including figs. 80–86.

26. Hutorowicz, "Maps of Primitive Peoples," 672–73.

27. Heckewelder was Germanic in culture but not German. A missionary in the Moravian Church, his education had been in Moravian schools, and some of his publications were in German. Born in England, he lived most of his life in North America.

28. John G. E. Heckewelder, "Historical Account of the Indian Nations," *Transactions of the Historical and Literary Committee of the American Philosophical Society* 1 (1819): 286–90. For a reconstruction of the map based entirely on this account see G. Malcolm Lewis, "The Indigenous Maps and Mapping of North American Indians," *Map Collector* 9 (December 1979): 26, fig. 2. The reconstruction, with legends in German, also appears in Rainer Vollmar, *Indianische Karten Nordamerikas* (Berlin: Dietrich Reimer Verlag, 1981), 70.

29. Bray, "Observations on the Indian Method of Picture Writing," 160. Interestingly, a Captain White Eyes provided "Details geographiques" upon which Michel Guillaume Saint Jean de Crèvecoeur in part based his elegant printed map: "Esquisse des Rivières Muskinghum et Grand Castor," *Lettres d'un Cultivateur Américain* (Paris: Cuchet, 1787), 3: folded to face p. 413. The English edition, *Letters from an American Farmer*, had been published without maps in London in 1782, immediately after Crèvecoeur's brief visit there en route to France with all his papers. This was the same year in which "Observations" was published and it is quite possible that Crèvecoeur was the gentleman of Bray's acquaintance. If so, Bray's scholarly encounter was probably only one stage removed from the original encounter.

30. Charles P. Daly, "On the Early History of Cartography, or What We Know of Maps and Map-Making before the Time of Mercator," *Journal of the American Geographical Society* 11 (1879): 2, 37.

31. Ibid., 2. William E. Parry included accounts of two Iglulik maps made in 1822 in his *Journal of a Second Voyage for the Discovery of a North-West Passage* (London: John Murray, 1824), 2:196–97; printed transcripts face pp. 197–98. John Ross included accounts of three Netsilik maps made in January 1830 in his *Narrative of a Second Voyage in Search of a North-West Passage . . .* (London: A. W. Webster, 1835), 254–55, 259–60, and 261–62. None is extant, but they appear to have been incorporated by Ross in "Chart Drawn by the Natives," a map of the Gulf of Boothia, facing p. 262.

32. J. Paul Goode, "The Map as a Record of Progress in Geography," *Annals of the Association of American Geographers* 17, 1 (1927): 2.

33. Painted Pawnee celestial chart on tanned antelope or deer skin; originally belonging to a Skidi band of Pawnee, 66 x 46 cm, Artifact 71898–10, Field Museum of Natural History, Chicago; it was collected at Pawnee, Oklahoma, in 1906 as part of a sacred bundle. It may be a descendant of a precontact original. The artifact is reproduced photographically in G. Malcolm Lewis, "Indigenous Maps and Mapping," 31, fig. 8.

34. Ralph N. Buckstaff, "Stars and Constellations of a Pawnee Sky Map," *American Anthropologist* 29, 2 (1927): 279–85. In 1906, Dorsey had corresponded with the astronomer Edward E. Barnard in an attempt to have the stars and constellations identified, but nothing was published: Von Del Chamberlain, *When Stars Came Down: Cosmology of the Skidi Pawnee Indians of North America* (Los Altos, Calif.: Ballena Press for the Center for Archaeoastronomy, University of Maryland, College Park, 1982), 192 and app. 2.

35. Winsor, *The Kohl Collection*, 17.

36. Ibid., 178.

37. Additional MS 4723 (formerly Sloane MS 4723), British Museum, London. This is virtually identical to C.O.700, North American Colonies General no. 6(1), Map Room Public Record Office, Kew, London (the latter reproduced as fig. 10.1 in this volume). In one or other of its two forms, it has been referred to and described in numerous scholarly encounters since Kohl copied it; most notably by Gregory A. Waselkov in "Indian Maps of the Colonial Southeast," in Peter H. Wood et al., eds., *Powhatan's Mantle: Indians in the Colonial Southeast* (Lincoln: University of Nebraska Press, 1989), 295, 324–29, figs. 3, 11.

38. "A Map drawn upon Stag skins by Y^e Gnacsitares who gave me to know Y^e Latitudes of all Y^e places mark'd in it," in Louis Armand de Lom D'Arce, Baron de La Hontan, *New Voyages to North America* (London: H. Bonwicke et al., 1703), map partly folded. There has never been agreement as to who the Gnacsitares were. The printed version of their map is incorporated onto the western edge of La Hontan's own map of the lower Minnesota River. Almost certainly the Indian original represented the lakes at the head of that river, the steep-sided Coteau des Prairies to the west of them, and the headwaters of the Big Sioux River.

39. Philip Lee Phillips, ed., *The Lowery Collection: A Descriptive List of Maps of the Spanish Possessions within the Present Limits of the United States, 1502–1820* (Washington, D.C.: Government Printing Office, 1912), 104–5 and listed as WL 93. The original is: "Pintura q- Por man^{do} de don Fran^{co} Valverde de mercado factor de su mag^d hizo myguel yndio de las pro vincias del nuevo mex^{co} . . ."(endorsement) ["Sketch which, by order of Don Francisco Valverde de Mercado, factor of His Majesty, was made by Miguel, an Indian, native of the provinces of New Mexico, of the relative position of the towns of the said provinces and the statement on them as he said it and gave to understand by signs and the names of the towns pronouncing them as they are and how populated. Don Francisco Valverde de Mercado/Hernando Esteban—This transcript is (certified) correct and truthfully (by) Hernando Esteban"], transcript in ink on paper, 31.0 x 43.0 cm, May 11, 1602 (from original of April 22, 1602), Estante 1, Cajon 1, Legajo 22, Ramo 4, Folio 172, Patronato, Archivo General de Indias, Seville, Spain; reproduced in G. Malcolm Lewis, "Indian Maps: Their Place in the History of Plains Cartography," edited by Frederick C. Luebke et al., *Mapping the North American Plains: Essays in the History of Cartography* (Norman: Uni-

versity of Oklahoma Press, 1987), fig. 4.8.

40. "Questioning of the Indian Miguel Brought from New Mexico by the Maese de Campo, Vincente de Zaldivar," itself part of the "Official Inquiry Made by the Factor, Don Francisco de Valverde, by Order of the Count of Monterrey, Concerning the New Discovery Undertaken by the Governor Don Juan de Oñate toward the North beyond the Provinces of New Mexico," in George P. Hammond and Agapito Rey, eds. and trans., *Don Juan de Oñate, Coloniser of New México 1595–1628*, pt. 2, Coronado Cuarto Centennial Publications, vol. 4 (Albuquerque: University of New Mexico Press, 1953), 871–77.

41. Lewis, "Indian Maps," 74 and n. 27.

42. Woodbury Lowery, *The Spanish Settlements within the Present Limits of the United States* (New York: Putnam, 1901), 129 and app. E, 436–37.

43. Although not cited in the originals by Lowery, the three accounts are now well known: Hernando de Alarcón's 1540 account of the response made by an elderly man, probably of the Halchidhoma tribe, when asked to "set me downe in a charte as much as he knew concerning [the Colorado] River," in Richard Hakluyt, *The Voyages, Navigations, Traffiques, and Discoveries of the English Nation* (London: George Bishop, Ralfe Newberie, and Robert Barker, 1600), 3:438; Gabriel Archer's 1602 account of a probably Micmac map of the coastline of Maine and perhaps Nova Scotia and southern Newfoundland in Samuel Purchas, *Purchas His Pilgrimes* (London: William Stansby, 1625), 4, pt. 6, 1647; and Samuel de Champlain's 1603 accounts of two Algonquin maps of the upper St. Lawrence river now best known in Henry P. Biggar, ed., *The Works of Samuel de Champlain* (Toronto: Champlain Society, 1922), 1:153–55, 159–61.

44. William F. Ganong, "A Monograph on the Cartography of the Province of New Brunswick," *Proceedings and Transactions of the Royal Society of Canada*, 2d ser., 3 (1897): 328 n. 1.

45. Ibid. Ganong had already reproduced a transcript of this birchbark map in "The St. Croix of the Northeastern Boundary," *Magazine of American History with Notes and Queries* 26 (1891): 264.

46. Ibid.

47. Gilbert M. Sproat, *Scenes and Studies of Savage Life* (London: Smith, Elder and Co., 1868), 191–92.

48. "This map was drawn by Kohklux in 1869 at his village. It is the first time he ever used a pencil." Shows the route from the Chilkat River, northern British Columbia, to Fort Selkirk, Yukon Territory, as taken by Kohklux and his father in 1852 on their way to burn Fort Selkirk. Some, at least, of the mountain ranges appear to be represented in profile. Manuscript in pencil on the verso of a printed nautical chart with native toponyms transliterated and entered in ink by George Davidson. 109 x 67 cm, August 1869, G4370, 1852, K6D, Map Library, Bancroft Library, University of California, Berkeley. A smaller incomplete manuscript map is endorsed "Kohklux started from his place at Kluwan and drew all around the paper for want of room. He asked for a big sheet of paper and I gave him the back of an old map [the nautical chart], on which he and his wife drew their route etc. in 1852," 26 x 20 cm, K4370, 1852, K61A, Map Library, Bancroft Library, University of California, Berkeley. The circumstances of the mapping, together with a transcript of the larger map, are described in George Davidson, "Explanation of an Indian Map," *Mazama* (April 1901): 1–8.

49. Untitled map of the Belcher Islands in pencil on the verso of a colored missionary print, 35.4 x 31.1 cm, RARE 772, 3-c B44 A–19, American Geographical Society Collection, University of Wisconsin Library, Milwaukee.

50. "Lecture on 'Exploration in the Nipigon Country.' By R[obert] Bell, delivered under the auspices of the Nat. Hist. Socy. of Montreal, Feby. 10th 1870. Being the First of the Sommerville Course," unpaged manuscript, Robert Bell Collection, MG 29 B 15, vol. 36, National Archives of Canada, Ottawa.

51. "Sketch Plan of Lake Neepigon pronounced Am-Neepigon [by] Windigo (Chief) E. side of Neepigon + another Neepigon Indn. & corrected by seven other Indians and others," pencil and ink on paper, 53.0 x 65.5

cm, H2/410—(Lake) Nipigon—(1869), National Map Collection, National Archives of Canada, Ottawa. The map contains two interesting marginal comments: "Indian Chief Windigo to have present of some canoes bought of him and also a copy of map [presumably Bell's] of L. Neepigon, shewing his Lodge etc. etc."; and at the upper right "Shape of lake here limited by size of sheet of paper. Should be much larger to be in propn. to rest of lake."

52. In each case, a degree of chance was involved. The two maps by Kohklux and his two wives remained in the possession of Davidson's heirs until 1980, when they were purchased by the Bancroft Library, University of California at Berkeley, three years after I had written to map librarian Philip Hoehn in the hope that they might be there. Wetalltok's map is not, as might have been expected, in the Robert J. Flaherty Collection at the Butler Library, Columbia University. For many years it was known only through the facsimile published in Flaherty, "The Belcher Islands of Hudson Bay," *Geographical Review* 5 (1918): 440. The original evidently remained in the possession of the journal's publisher, the American Geographical Society, and in the early 1990s was rediscovered in the society's collection in the University of Wisconsin Library, Milwaukee. Windigo's map of Lake Nipigon was included in the Robert Bell Papers when they were obtained by the National Archives of Canada. As is so often the case, the map was separated from the associated documents (in particular, the lecture script) and it was a fortunate coincidence that I related the two. In each case, therefore, the scholarly encounter has been renewed by chance after an interval of many decades.

53. Jeanne O. Snodgrass, "American Indian Painters: A Biographical Directory," *Contributions from the Museum of the American Indian, Heye Foundation* (New York: Museum of the American Indian, 1968), 21, pt. 1:165.

54. Three maps incised on birchbark. Almost certainly collected by Garrick Mallery from Sapiel Selmo in 1887. The originals have dimensions of 19.7 x 22.9 cm, 20.3 x 24.1 cm, and 20.3 x 19.0 cm, and are catalogued 393,431-3 in the Department of Anthropology, National Museum of Natural History, Washington, D.C. Each is reproduced as a line drawing in Mallery, "Picture Writing of the American Indians," in *Tenth Annual Report of the Bureau of American Ethnology 1888–89* (Washington: Government Printing Office, 1893), figs. 456–58. Each is captioned "Passamaquoddy wikhegan" and represents the activities of two Indian hunters in relation to a dendritic drainage system with lakes. The hunters, animals, and lodges are represented pictographically in profile.

55. Original Montagnais bitten birchbarks, Museum of the American Indian, New York: items 19/5763–19/5771 nd 19/5773–19/5775; reproduced as simple line drawings many years later in Frank G. Speck, "Montagnais Art in Birch-bark, a Circumpolar Trait," in *Indian Notes and Monographs* 11, 2 (1937), pl. 13. For a recent, well-referenced review of this art form see Cath Oberholtzer and Nicholas N. Smith, "I'm the Last One Who Does It: Birch Bark Biting, an Almost Lost Art," in David H. Pentland, ed., *Papers of the Twenty-Sixth Algonquian Conference* (Winnipeg: University of Manitoba, 1995), 306–21.

56. Reuben Gold Thwaites, *The Jesuit Relations and Allied Documents* (Cleveland: Burrows Brothers Co., 1896–1901). Scrying is mentioned in the *Relations* 15 (1639):178; 17 (1639): 210; 33 (1648): 192–94; and 39 (1653): 20. Scapulimancy is referred to in the *Relations* 6 (1634): 214; 16 (1639): 194; and perhaps—the accounts are not clear—29 (1646): 163; and 31 (1647): 211.

57. For example: John M. Cooper, "Northern Algonkian Scrying and Scapulimancy," in William Koppers, ed., *Festschrift publication d'hommage offerte au P. W. Schmidt* (Vienna: Mechitharisten-Congregations-buchdruckerei, 1928), 205–17, esp. 210; and Frank G. Speck, *Naskapi: The Savage Hunters of the Labrador Peninsula* (Norman: University of Oklahoma Press, 1935), 138–64.

58. Surprisingly, a recent search failed to locate an example of a scapula used in divination and interpreted as a map.

59. Mallery, "Picture Writing," 329–57.

60. Ibid., figs. 452, 453, 455.

61. Henry R. Schoolcraft, "Indian Pictography,"

Historical and Statistical Information Respecting the History, Conditions, and Prospects of the Indian Tribes of the United States (Philadelphia: Lippincott, Grambo, 1851), 1:333–430.

62. An attempt to test whether native North Americans possessed a concept of map proved inconclusive. Until the recent past, European-Indian/Inuit word lists and dictionaries rarely included "map." The Iroquoisan, Siouxan, and Algonquian languages of the eastern and interior regions were, however, found to have much higher than average frequencies. This could reflect relatively early contact with Europeans and Euro-Americans; a hypothesis that is reinforced by many of the native word roots. The element for "earth" or "ground" is frequently combined with an element suggesting European culture. Of these "paper," "book," and "toy" are common.

3

Hiatus Leading to a Renewed Encounter

G. Malcolm Lewis

The Sixty-Year Hiatus

The years 1911 and 1970 marked fairly precisely the beginning and the end of a hiatus in scholarly interest in Indian and Inuit maps and mapmaking.[1] There was no single explanation for this, especially as these were decades in which real and imagined events involving native North Americans became increasingly important in the popular culture of fiction and cinema, thereby influencing consumer products, including toys and certain aspects of clothing. Increasingly popular recreational activities such as touring in ever remoter areas, camping, and canoeing, brought more and more whites into contact with native environments and the more superficial aspects, at least, of native life. Though increasingly debased, Indian and Inuit craft items were acquired by more and more whites. Indian and Inuit art was purchased by wealthy private white collectors. Some collections passed into the public domain and began to stimulate serious study. Like most maps, many of the items were flat: textiles, hide paintings, and works on paper. Though not usually representational, some had affinities with maps, but this appears not to have been recognized.

If the hiatus in interest in the traditional cartography of native North Americans coincided with an increase in popular white awareness of things Indian and Inuit, an explanation must be sought within the academic sphere. This has never been attempted. When it is, the broader factors operating on academe will need to be taken into account; for example, a major economic depression, two world wars and two major regional ones, the abandonment of many traditional interests in the face of new demands, opportunities, and isms, and the fragmentation of most traditional disciplines. Of particular relevance in explaining the hiatus will be the ways in which these and other macro factors influenced the development of pertinent traditional fields such as geography, cartography, history, and anthropology; especially the extent to which parts at least of their former concerns were taken over by and refocused within new amalgamates such as the behavioral, cognitive, and communicational sciences and regional and cultural studies. Though difficult to prove, it could be that after ca. 1970, cultural anthropology and other fields with interests in native North Americans did not renew the earlier concern with maps per se but developed a new awareness of space as a factor in so many aspects of native life. This could explain why some scholars did begin to develop an interest in native maps as evidence rather than as artifacts.

In retrospect, the depth of the hiatus in interest in native North Americans' maps and mapmaking is reflected in the paucity of primary publications, their time-worn themes, and their failure to generate a new momentum of interest. Almost without exception, the authors were aging if not elderly.[2] There was one exception, though at the time it probably attracted little attention. Canada's Dominion archivist, William Kaye Lamb, in his *Report of the Public Archives for the Year 1951,* placed first among "Other New Acquisitions": "Photostatic copies of 45 'aboriginal maps,' selected by Mr. Richard Ruggles, formerly Lecturer in Geography at McMaster University, were acquired during the year. These are quite diverse in origin, date and coverage."[3] All the "photostatic copies" were of printed maps and almost all the published sources were widely available.[4] Nevertheless, the Ruggles collection was almost certainly the first systematically assembled set of native North Americans' maps. Although no originals were reproduced in it, Ruggles's first major published work marked the end of the hiatus.[5]

Field studies by cultural anthropologists did occasionally find maplike artifacts, reveal map-reading abilities, and discover unexpected forms of spatial representation.[6] These, however, were few and seem not to have come to the attention of historians of cartography until well after 1970.[7]

In 1970, therefore, systematized knowledge of native North Americans' maps and mapmaking was essentially as it had been sixty years earlier. Hence, before examining recent and current encounters and presenting new case studies in part 2 of this volume, and before trying to anticipate future developments in part 3, 1970 is the appropriate year for which to take stock. What was it that had been encountered? More specifically, which aspects of the encountered had been perceived clearly; which only dimly; and which not at all? Finally, a question subtly different from the last, were there things still to be encountered?

Native Cartography as Perceived circa 1970

Whereas the encounter *sensu lato* had always been a two-way process, as late as 1970 the written record of it was almost entirely from the perspective of whites. This was particularly true of the cartographic encounter.

Records of native encounters with European- and Euro-American-style maps were few, mainly involved Inuit, and were almost always as perceived by whites. One rare exception was more surprising than revealing. Years after the event, the Inuit Nuligak had recollected that in the "summer of 1916 the H.B.C. [Hudson's Bay Company] built a trading post at Bernard Harbour [about 85 miles north of Coppermine]. I was hired as pilot on their schooner, the *McPherson*. I had never seen the eastern country but I found my way using those drawings that white men make on paper."[8]

White perceptions of Inuit encounters with European- and Euro-American-style maps were few and almost always lacked the necessary background information with which to make retrospective evaluations. Diamond Jenness, the anthropologist on the Canadian Arctic Expedition of 1913 to 1918, had reported negatively that among the Copper Indians:

> not a single native was encountered who had the slightest conception of map, with the sole exception of Uloksak. Even he had only a vague comprehension. He understood that certain lines represented the coast and others the rivers, and he seemed able to picture a bay as a curve, but he totally failed to comprehend the purpose of the map, and so could not reproduce on paper his own topographic knowledge.[9]

The apparent ability or inability of Inuit to comprehend maps and charts of their own territories presented to them by whites had been determined by circumstance and mood. Occasionally whites perceived this. In his ship's cabin in June 1853, Rochfort Maguire, having asked three Inuit about places and distances on charts of the coast between Point Hope and Cape Prince of Wales in northwest Alaska, observed, "It was remarkable how well they comprehended the charts when their ideas were turned to it, but if I had asked them a question at another time and brought them to the chart when their mind was else where, it would have been difficult to make them understand any thing about it."[10] In reporting an incident that had occurred in 1912, Robert Flaherty revealed none of Maguire's caution. On reaching the east coast of Hudson Bay after his return journey by sledge across the northern Labrador peninsula, he had been anxious to establish whether the coastal ice had cleared. Meeting a group of Inuit he acted and observed as follows:

> On the floor of the largest topek I spread out my maps of the coast of Hudson Bay and they traced with their fingers and told us where we might find driftwood and where there might be camps of Eskimos. What pleased us most of all was their assurance that there was no more ice—everywhere the coast was clear.[11]

From the few white observations on natives' reactions to European- and Euro-American-style maps before 1970, it was apparent that, in certain circumstances, some natives recognized in plan perspective topographic features in areas with which they were familiar on the ground. What is not clear is whether natives with no or very little formal education categorized the whites' maps they had encountered as equivalents of some of their own pictographic drawings or as a distinctive subcategory representing the spatial arrangements of things. Nuligak's categorization of them as "drawings that white men make on paper" implied the latter.[12] Conversely, his claim to have used them in navigating what to him was a strange coast implied an ability to relate maps to hitherto unknown ground referents.

Long before 1970, whites' perceptions of native pictography had begun to impose categories. Pertinent among these were rock art, painted hides, sand paintings, wampum, and incised birchbarks. It was not a systematic classification. Particularly in rock art, meaning and purpose were usually unclear and always unverifiable.

Almost intuitively, specific cases within each of the material categories of pictography had been recognized by whites as having some of the characteristics and serving some of the functions of their own maps. To be categorized as a map the elements of a pictograph had been perceived as arranged vis-à-vis each other according to something close to vertical perspective. Categorization of a pictograph as a map was confirmed in those cases where it had been observed to function in ways similar to maps in Western cultures; especially so if its elements had been related convincingly to ground referents. By 1970 the diversity of material categories had become widely known: on trees, bark, hides, ivory, and wood; inscribed on the ground in ash, sand, and snow or modeled three-dimensionally in whatever materials were at hand; stylized in wampum; and perhaps incorporated in petroglyphs. Rock-art panels created the greatest uncertainty. Some undoubtedly looked like maps, but they were old and undatable, and the local native peoples were for the most part unable to offer satisfactory explanations or indicate the purposes for which they had been made.[13] Conversely, the most convincing examples were those artifacts that were still being made and used in indigenous contexts. Of these, the best were the birchbark message maps, widely used by Indians of the northeastern forests when they needed to leave spatially structured messages for individuals expected to arrive later.[14]

Dimly Perceived Aspects of Native Cartography circa 1970

By 1970 Europeans and Euro-Americans had only dimly perceived three aspects of Indian maps that were later to be recognized as important: (1) they were part of the much broader and pervasive pictographic tradition; (2) within Indian cultures, artifacts that whites were later to categorize as maps were made and used in a wide range of indigenous contexts; and (3) maps made by natives in the course of the encounter stemmed from the pictographic tradition but had been adapted in the process.

Although Mallery's more than eight-hundred-page "Picture Writing of the American Indians" of 1893 had done much to promote an awareness of native North American pictography, its structure and emphasis had not placed maps firmly in the pictographic tradition.[15] Consciously, or otherwise, Mallery had not promoted the concept of map, and the range of maplike artifacts he mentioned was very narrow.

By 1970, awareness of the range of map types encountered was heavily biased toward those made for whites in the course of frontier encounters in the field. Although there was evidence scattered through more than two hundred years of literature that native North Americans had made maps for a wide range of indigenous purposes, there was no consolidated awareness of this.

Contrasts in style and content between maps made by natives for whites in the course of the encounter and the much smaller corpus of maps then known to have been made for indigenous purposes had rarely been commented on. The relatively few contrasts that had been made focused almost exclusively on differences between the physical media and, likewise, between the representational processes. That native peoples had been adept at drawing on paper in pencil or ink had occasionally been commented on because it was perceived to be alien. Traditional media and methods had been commented on because they were considered somewhat exotic. In contrast, aspects of geometry, modes of representation, and systems of information had been rarely mentioned.

Unperceived Aspects of Native Cartography circa 1970

In retrospect, as late as 1970, several important aspects of native maps appear to have gone unnoticed. Of these, the lack of concern for the provenance, descent, and survival states of extant maps was both surprising and serious. Observations had been made, and all too often conclusions reached, without any consideration of the extent to which the material evidence differed in form and content from the original. True originals were, as indeed they still are, very rare, especially indigenous originals, uninfluenced by whites and made in nonencounter contexts.

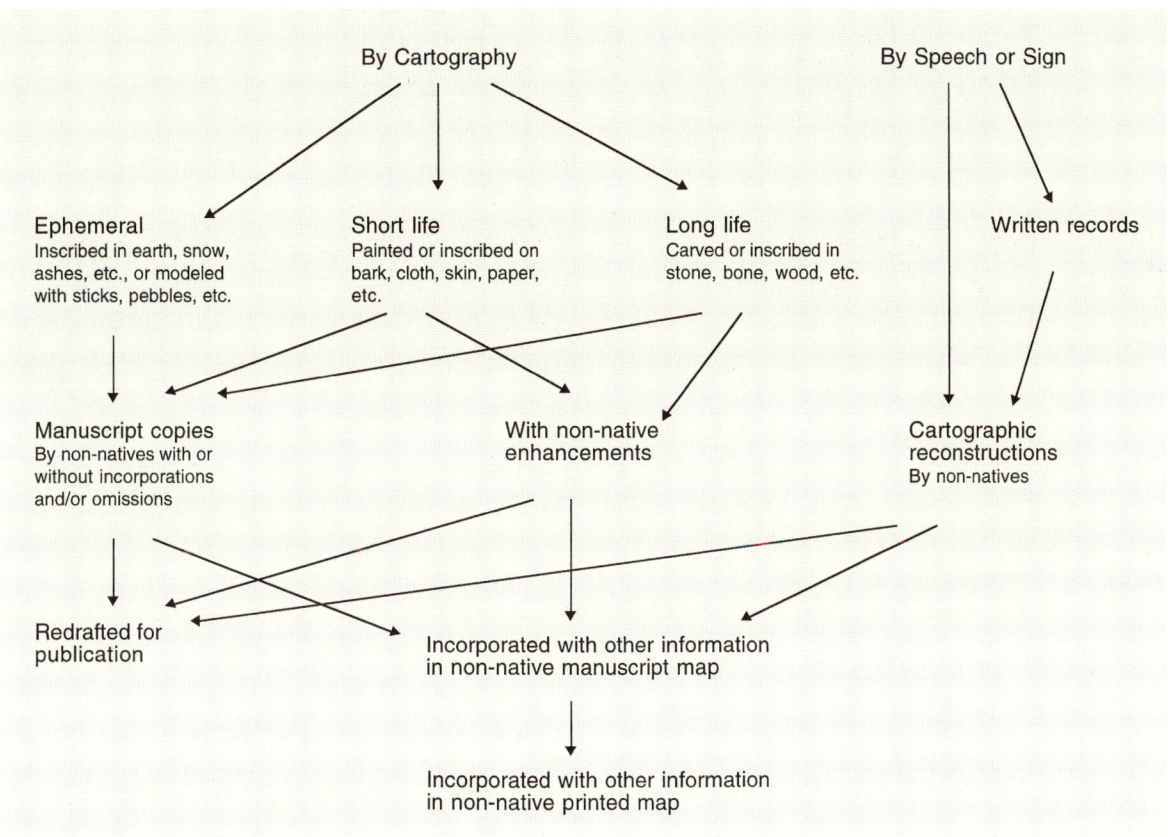

3.1 The descent and survival states of spatially arranged information originally communicated by Indians and Inuit. By the author.

Even now, there is no classification of maps by state.[16] Figure 3.1 is an indication of how complex a worthwhile classification will need to be. Both indigenously and in the course of encounters, ephemeral maps had almost certainly been the most common. Proof of that was, of course, and will always remain impossible. By definition, this category could only have survived as manuscript copies or as published redrafts.[17] Short-life and long-life maps differed only in the relative durability of the materials with which they were made. In 1970, relatively few of either were known to be extant, though unknown examples were scattered widely in public and private collections. Of those that were extant a high proportion had been enhanced or supplemented by the custodian or collector.[18] The highest proportion of native maps were known only as published redrafts.[19] Though not as numerous as the other states, some of the best known maps had survived as acknowledged incorporations in European or Euro-American maps. The difficulties with these were twofold: determining the precise limits of the incorporation, and establishing the original nature of the incorporated.[20] In retrospect, cartographic reconstructions of spatially arranged information that had originally been communicated by gestures or in speech are known to have existed. It is doubtful, however, if anyone even suspected this in 1970.[21]

Geometry was the second important aspect of Native Americans' maps that in retrospect appears not to have been noticed in the decades immediately before 1970.[22] When, toward the end of the fifteenth century, Indians first began to provide the Spaniards with directional and positional information, they unknowingly structured it in much the same way that the local maps and plans of late medieval Europe were still being shaped; according to experience or tradition and not plotted on mathematically derived graticules.[23] Soon after the Columbian encounter, however, European concepts of space and assumptions about maps had both begun to change. Very occasionally, perceptive Europeans had noted fundamental differences between their own and native concepts. In 1736, for example, Cadwallader Colden, the surveyor general of land of the Province of New York recognized some of these when Mohawk Indians complained "that they had not sold the quantity of land discrib'd in the Deeds of Purchase from them but a vastly smaller quantity according to other Boundaries describ'd . . . by s^d Indians." Colden, in his report to the lieutenant governor of the province, observed that in the deeds "the Boundaries of the Land said to be purchased are in several cases expressed by points or Degrees of the Compass [and] by English Measures [presumably yards, chains, furlongs, and miles] which are absolutely unknown to the Indians." In "order to prevent such Impositions or Deceits on the Indians of the future," Colden proposed that "all lands to be purchased of the Indians be survey'd in their presence before any Deed for the same be sign'd."[24]

Colden was one of a few exceptions. Most of the early evaluations of Indian maps consisted of favorable comparisons with European-style ones. For example, on the basis of experience in both the St. Lawrence and upper Mississippi valleys between 1683 and 1692, Louis Armand de Lom d'Arce, Baron de La Hontan had observed that the Indians:

> draw the most exact Maps imaginable of the Countries they're acquainted with, for there's nothing

wanting in them but the Longitude and Latitude of Places. They set down the True *North* according to the Pole Star; The Ports, Harbours, Rivers, Creeks and Coasts of the Lakes; the Roads, Mountains, Woods, Marshes, Meadow, Ec. counting the distances in Journeys and Half-journeys of the Warriers, and allowing to every Journey Five Leagues. These Chorographical Maps are drawn upon the Rhind of your Birch Tree; and when the old men hold a Council about War or Hunting, they're always sure to consult them.[25]

A few years before, Chrétien Le Clercq, a Recollect priest who had spent twelve years among the Micmac Indians in what is now northern New Brunswick, noted, "They have much ingenuity in drawing upon bark a kind of map which marks exactly all the rivers and streams of a country of which they wish to make a representation. They mark all the places thereon exactly and so well that they make use of them successfully, and an Indian who possesses one makes long voyages without going astray."[26]

Implicit or explicit in the above evaluations of native North Americans' maps were the European concepts of accuracy, orientation, linear scale, and utility in way-finding. La Hontan's observation even seemed to imply that the absence of meridians and parallels was a mere oversight. By the years preceding 1970, such evaluations had been replaced by assumptions, some of which were equally revealing. More than half the cards in the National Archives of Canada's catalogue of copies of "aboriginal maps" acquired from Richard Ruggles in 1951 indicated "No scale given" or "No scale." Even more revealing, of six Indian maps reproduced in 1970 in *Manitoba Historical Atlas*, four had computed linear scales, one was said to have four scales, and another to have a "Scale [that] varies over the map."[27] We now know that although it is possible to compute an average linear scale between any two points on a native map for which the ground referents can be established, that average varies—often enormously—among different pairs of points. Likewise, traditional native maps do not conserve direction; not even along particular axes. Yet, until very recently at least, interpretations have often assumed that they do.

Not constructed according to linear or angular metrics, the geometrical properties of Indian maps were the same as those studied by the branch of geometry called topology. These properties had always existed in nature. In the human world they were implicit in spatial behavior, in the spatial concepts embodied in ordinary spoken language, and in graphical representations. Unlike the Euclidean and projective geometries of Renaissance and post-Renaissance cartography, topological properties did not have to be invented or formally operationalized. Topology was the nineteenth-century branch of geometry that studied spatial relations unaffected by change of shape or size of figures.[28] It was, therefore, the appropriate branch of geometry with which to study the spatial properties of native maps. Yet, by 1970, more than one hundred years after it began to be formalized, topology appears never to have been mentioned in that context. Insofar as it was considered at all, the geometry of native maps was assumed to be less precise than but not fundamentally different from that of modern European- and Euro-American-style maps.

As late as 1970, the third aspect of native maps still unperceived by most white scholars was the manner in which elements were differentiated, grouped, and emphasized vis-à-vis each other.[29] It was still tacitly assumed by them that native systems of categorization, hierarchical organization, and classification were akin to their own. But this was not so. Indians quite frequently represented but did not distinguish between aspects of their terrestrial and cosmological worlds on the same map, something that Europeans had gradually ceased doing during and after the Renaissance. Each native map served a specific purpose. Hence, map content and magnitudinal ordering were determined by it. For example, a short portage around rapids could be emphasized on a map that omitted a long trail across the interfluve. Classifications of phenomena were not necessarily fixed and often differed fundamentally from culturally rigid European classifications. For example, not infrequently, Indian maps appeared to represent hydrological impossibilities; braided macro drainage systems, with streams connecting adjacent drainage catchments. In such cases, what had really been represented was an integrated route system that did not differentiate between canoe routes and land trails. This was as alien to the Euro-American intelligentsia as a late-eighteenth-century map would have been that represented but did not distinguish between canals and turnpike roads.

Overlooked Aspects of Native Cartography circa 1970

Whereas by 1970 those with an interest in native maps had a general awareness of the material, and to a much lesser extent behavioral, evidence, some of it had still to be revealed and some had been forgotten.

Even now it would be foolish to assume that examples of all types of indigenous maps have been encountered. As recently as 1985, the Lakota revealed star and earth maps that had been in their possession for at least three generations and perhaps much longer. Painted on hide, they came "as a big surprise" even to researchers who had a well-developed interest in Lakota knowledge of astronomy. "One hide was a star map, the other hide was an earth 'maka,' map [of] buttes and rivers and mountains even creeks clear out to Colorado Springs. Star map and earth map, they were really the same, because what's in the stars is on the earth and what's on the earth is in the stars." Another hide was said to combine "what is usually found on separate star and earth maps." According to its keeper "without proper instruction it wouldn't even be recognized as a star map."[30] In seeking the essence of native cartography, such revelations of the hitherto unencountered are far more important than the discovery of whole archives of pencil-and-paper maps.

Sometimes the encounters have been unexpected. The unpacking of a dream map by its Beaver Indian keeper in 1979 is a good example. A "thick layer of hide, pressed tightly together," it "was as large as the table top, and had been folded ... for many years." The keeper explained "that it was wrong to unpack a dream map except for a very special reason." The occasion was a public hearing at which "the Indians' needs had to be recognized." Natives and whites were encouraged to "look at the map." Indeed, those "who wanted to might even take photographs." None is known to exist today and this had been but a brief encounter with the "thousands of short, firm and variously coloured markings." The "intricate routes and meanings of a dream map" were not easy to follow and there "was no time to explain them all."³¹

On other occasions, renewed encounters with forgotten, ignored, and misinterpreted artifacts have been just as significant as new encounters. A good example of this is Morris Arnold's demonstration in chapter 8 that a more than 250-year-old painted Quapaw buffalo robe, for long overlooked in Paris, incorporates the indigenous roots of an Indian cartographic tradition. Arguably more significant was the convincing demonstration by Selwyn Dewdney that a few of the Southern Ojibwa birchbark scrolls, held for decades in scattered public and private collections, were maps of the Great Lakes–St. Lawrence river waterway. The idea was not new, but Dewdney was the first person able to examine eight pertinent scrolls and several potentially pertinent ones in places as far apart as New York, Ottawa, Calgary, Denver, London, and Edinburgh. Whether the function of the scrolls was to record the route via which the Southern Ojibwa believed their ancestors migrating over many generations had reached Leech Lake, Minnesota, or the route via which the midé religion had come to them is still not entirely clear, but Dewdney recognized the main features along it: from the Atlantic Ocean and eastern seaboard to Sault Ste. Marie, Lake Superior, the distinctive Minnesota Point creating Superior Bay and Duluth Harbor, the St. Louis River, and many minor features along the final stretch of the route to Leech Lake.³²

In 1970, the greatest barrier to progress in scholarship dominated by whites in understanding native cartography was their concept of map. It is unlikely that any of those involved would have been able to quote it, but most if not all probably assumed a definition very similar to that published nine years earlier by Sir Dudley Stamp. A map was "a representation of the earth's surface or a part of it, its physical and political features etc., or of the heavens, delineated on a flat surface of paper or other material, each point in the drawing corresponding to a geographical or celestial position according to a definite scale or projection."³³ With its emphasis on "representation," "corresponding," and delineation, rather than symbolic roles, its prerequisite of a flat surface, and the constraints imposed by "definite scale or projection," the definition precluded from consideration much that had been encountered and not necessarily forgotten. Very occasionally, there were glimpses of things that would later be considered pertinent. In 1951, for example, Delf Norona had suggested that "the most exotic [Indian] map is one tattooed on the human body" and tersely mentioned as an example "a symbolic chart of the Osage, preserved by tattooing on the bodies of their old men."³⁴ For the most part, however, the concept of map as a mathematically generated representation of the earth's surface, or sky, on a flat artifact, precluded from consideration several modes of spatial symbolism. In particular, consideration of space as symbolized in gesture, dance, built structures, and ceremonies was still to come. Under the conceptual constraints of 1970, Boone's chapter 5, Pearce's chapter 7, and Nabokov's chapter 11 could not or would not have been written, even though in each case much of the empirical material existed.

The changes that were to take place were in part cause and in larger part consequence of changes in the concept of map that by 1987 led to a "new definition . . . that [would be] neither too restrictive nor yet so general as to be meaningless . . . Maps are graphic representations that facilitate a spatial understanding of things, concepts, conditions, processes, or events in the human world."[35] Except for "graphic," which could be interpreted restrictively to exclude patterns of behavior, and "representations," which could be construed to exclude symbolizations, this looser definition of "map" is entirely appropriate for studies of native cartography of the types presented in part 2. Although Harley and Woodward's redefinition of the concept of map was not the trigger, it certainly reflected the loosening of underlying assumptions that had occurred gradually during the previous twenty years. Representation had been expanded to embrace symbolization. Their expression was no longer confined to artifacts. Above all, space was no longer restricted to that which could be generated by projective geometry and analyzed according to Euclidean principles. In many respects this had, of course, always been so, among ordinary folk in Western society as well as in traditional societies. What happened after 1970 was a growing awareness by cartographers, and others with professional interests in maps, of space as it had been experienced, represented, and symbolized by *Homo sapiens* for several millennia.

Notes

1. As defined in chapter 2, the first phase of scholarly encounters effectively ended in 1911 with the publication in English of a summary of Bruno Adler's Russian monograph of the previous year: H. de Hutorowicz, "Maps of Primitive Peoples," *Bulletin of the American Geographical Society* 43 (1911): 669–79. The long hiatus ended in 1970 with the publication of a work containing facsimiles or transcripts of five Indian or Indian-derived maps: John Warkentin and Richard I. Ruggles, *Manitoba Historical Atlas: A Selection of Facsimile Maps, Plans, and Sketches from 1612 to 1969* (Winnipeg: Historical and Scientific Society of Manitoba, 1970), figs. 24, 25, 32–34, each with facing texts.

2. Cottie A. Burland, "American Indian Map Makers," *Geographical Magazine* 20 (1947–48): 285–92; a popular piece, in a respected but popular periodical, by a Latin Americanist who used well-known published examples from several parts of the Americas to produce a "sequence of maps which leads from the simplest of direction signs to a good and accurate plan of a large extent of country." It was unashamedly developmental in tone. Leo Bagrow, "Eskimo Maps," *Imago Mundi* 5 (1948): 92 + two unpaged plates; a mere seven hundred words redescribing maps that had been collected and first described in the nineteenth century. Delf Norona, "Maps Drawn by Indians in the Virginias," *West Virginia Archeologist* (March 1950): 12–19; a chronologically arranged account of six maps, based largely on early published works and nineteenth-century histories.

3. William Kaye Lamb, *Report of the Public Archives for the Year 1951* (Ottawa: Edmond Cloutier, 1952), 27.

4. The original catalogue cards are in the National Map Collection, National Archives of Canada, Ottawa.

5. Warkentin and Ruggles, *Manitoba Historical Atlas*, mainly in part 1:5–132.

6. Frank G. Speck was perhaps the foremost among this small group with his work among the Naskapi and Montagnais of northeastern Quebec and Labrador.

7. In this volume, chapter 11 is the one drawing most heavily from the anthropological literature. Slightly less than 20 percent of its cited sources were published during the hiatus years of 1912 to 1969 inclusive. This proportion is much as might have been expected in the human and social sciences and suggests that cultural anthropologists

were making pertinent observations throughout the hiatus. Be that as it may, those with primary interests in cartography were either unaware of what was being published or failed to recognize its pertinence.

8. Maurice Metayer, ed. and trans., *I, Nuligak* (Toronto: P. Martin Associates, 1966), 115.

9. Diamond Jenness, "The Life of the Copper Eskimos," *Report of the Canadian Arctic Expedition 1913–18* (Ottawa: F. A. Acland, 1922), 12:229.

10. John Bockstoce, ed., *The Journal of Rochfort Maguire 1852–1854*, vol. 1, *Works Issued by the Hakluyt Society*, 2d ser. (London: Hakluyt Society, 1988), 169:234–35.

11. Robert J. Flaherty, *My Eskimo Friends* (London: Wm. Heinemann, 1924), 111.

12. Metayer, *I, Nuligak*, 115.

13. "Map Rock," Idaho, is probably the best example. Engraved on a large block of basalt near the north bank of the Snake River, opposite Givens Hot Springs, it was likened to a map by the early white settlers. In 1897, E. T. Perkins, Jr., of the United States Geological Survey, expressed the opinion that its "principal motif seems to be a mapping of the Snake River Valley"; letter to John Wesley Powell, Washington, D.C., January 14, 1897, MCB/1/1972, National Anthropological Archives, Smithsonian Institution, Washington, D.C. The name "Map rock (Petroglyphs)" appears at 43° 25'N, 116° 42'W on the Givens Hot Springs Quadrangle of the National Topographic Map Series 1:25,000, United States Geological Survey's 7.5 Minute Series (Topographic).

14. Formerly widely used by the Passamaquoddy and related tribes of the Northeast, among which they were known as *wikhegans*.

15. Garrick Mallery, "Picture Writing of the American Indians," in *Tenth Annual Report of the Bureau of American Ethnology 1888–89* (Washington: Government Printing Office, 1893).

16. As used here the word "state" combines two meanings: the physical condition in which a map exists; and any one of a sequence of developmental stages that began with an original artifact.

17. Occasionally two states of the same map have survived. In one example, very significant differences are apparent. In February 1854, a southern Paiute made a map of the lower Colorado River for Amiel W. Whipple. It was almost certainly traced on the ground. The manuscript in Whipple's notebook represents the Colorado river downstream from what was later to become Lake Mead. The dominant content is a series of enclosed irregular-shaped spaces, each delimiting the territory of a named tribe: Untitled sketch map, Lt. Amiel W. Whipple Notebook no. 20, February 21–25, 1854, Oklahoma Historical Society, Oklahoma City. An engraved version of the same map published two years later modifies the drainage pattern somewhat. More seriously, the enclosed tribal areas are replaced by clusters of tepee symbols; inappropriate in a region where the dominant dwelling types were either pitch-roof houses or domed bark, mat, thatch, hide homes. The tepee clusters are not congruous with the enclosed tribal areas of the earlier map. It is not known whether the draftsman-engraver was given instructions or license to modify, but the impression is that of a visually clearer, but less authentic map: Amiel W. Whipple, "Report of Explorations of a Railway Route near the Thirty-fifth Parallel," 33d Cong., 2d Sess., Sen. Exec. Doc. 78, *Report of Explorations and Surveys . . . for a Railroad from the Mississippi to the Pacific Ocean* (Washington, D.C., 1856), 3, pt. 3, pl. 3, p. 16.

18. A good example of an enhanced map on birchbark was included in "The Papers of Robert Bell, 1841–1917" when auctioned in Montreal on October 12, 1978. Now in the William L. Clements Library, Ann Arbor, Michigan, it is on two overlapping pieces, with line work and toponyms in pencil. The verso of the smaller piece carries the title "Indian Sketch of Country East of Lake [unclear but probably Wanapitei] (signed) R. Bell, 1890." The toponyms are almost certainly in the hand of T. Michaud of the Canadian Geological and Natural History Survey. Many such maps were made but few have survived.

19. A high proportion of these had been published in the *Annual Reports of the Bureau of American Ethnology*.

For the most part bold and clear, evidence has already been adduced (n. 17) to suggest that clarity, and perhaps conformity, took precedence over veracity.

20. By 1970, transcripts and one photograph had been reproduced of a particularly problematic example: the so-called Velasco manuscript map of circa 1610 of the northeast coast and adjacent interior. It had been published first in transcript form in Alexander Brown, *The Genesis of the United States* (London: William Heinemann, 1890), 1:456, and as a black and white photograph in Emerson D. White and Archibald Freeman, eds., *A Book of Old Maps Delineating American History* (Cambridge: Harvard University Press, 1926), 108. Neither, however, revealed that "all the blue is done by the relations of the Indians." This was not to be made clear until 1971 with the reproduction of the map in full color in William P. Cumming, Raleigh A. Skelton, and David B. Quinn, *The Discovery of North America* (London: Paul Elek, 1971), pl. 326. Even then, there was no indication whether the "relations of the Indians" were received directly or indirectly by the compiler or whether they originated in graphics, gesture, or speech. John Smith's well-known "Virginia Discouered and Discribed" of 1612 presented very similar problems: "In which Mappe observe this, that as far as you see the little Crosses on rivers, mountains, or other places have been discovered; the rest was had by information of the savages, and are set downe, according to their instructions" (reproduced in part in fig. 9.3 in this volume). John Smith and others, *A Map of Virginia with a Description of the Countrey* (Oxford: Joseph Barnes, 1612), 10.

21. David B. Quinn, *New American World: A Documentary History of North America to 1612*, 5 vols. (New York: Arno Press and Hector Bye, 1979), initiated my awareness of this by combining extracts from many accounts by early explorers with facsimiles of the maps that had incorporated their discoveries. An attempt to trace the sources of "all the blue . . . done by the relations of the Indians" on the "Velasco" map of circa 1610 revealed several examples. It represented the two lakes later named Champlain and George as rectangular, approximately equal in size, with their long axes oriented east-west, and together occupying much of what is now New England. In retrospect, all this was wrong, except that two lakes do exist at the head of the Richelieu River. In 1610, no European had seen either, but in 1603 Samuel de Champlain had been told of them by his Indian guides, who reported the upper lake to be "as large as" the lower one: Henry P. Biggar, ed., *The Works of Samuel de Champlain* (Toronto: Champlain Society, 1922), 1:143.

In a letter to me dated July 5, 1990, Kevin Kaufman explained how, in the course of editing *The Mapping of the Great Lakes in the Seventeenth Century* (Providence: John Carter Brown Library, 1989), the "more I worked with the material, the more I became convinced of a link between textual account and cartographic image. If the point appears slighted in my essay, it is due to my rather late awakening to the fact. . . . At some later stage, I'd like to examine this very subject in greater detail."

22. For a fuller treatment of this topic see G. Malcolm Lewis, "Metrics, Geometries, Signs, and Language: Sources of Cartographic Miscommunication between Native and Euro-American Cultures in North America," *Cartographica* 30, 1 (1993): 101–2.

23. For three early accounts see J. B. Harley, *Maps and the Columbian Encounter* (Milwaukee: Golda Meir Library, University of Wisconsin, 1990), 61–62.

24. Cadwallader Colden, "To the Honourable George Clarke Esqr. Lieutenant Governor of the Province of New York Ec . . . New York," November 3, 1736, *The Letters and Papers of Cadwallader Colden*, Collections of the New York Historical Society, vol. 51 (New York: Printed for the Society, 1918), 2:158–59.

25. Louis Armand de Lom d'Arce, Baron de La Hontan, *New Voyages to North America* (London: H. Bonwicke and others, 1703), 2:13–14.

26. Chrétien Le Clercq, *New Relation of Gaspesia, with the Customs and Religion of the Gaspesian Indians [1691]*, edited and translated by William F. Ganong, Champlain Society Publications no. 5 (Toronto: Champlain Society, 1910), 136.

27. Warkentin and Ruggles, *Manitoba Historical Atlas*, figs. 24, 25, 32–34, 55.

28. "Study of geometrical properties and spatial relations unaffected by continuous change of shape or size of figures." *The Concise Oxford Dictionary of Current English*, 6th ed. (Oxford, Clarendon Press, 1976). Sometimes known as "rubber-sheet geometry," to convey the idea of the properties of a drawing on such a medium that would be unaffected by differential stretching unless and until a rupture occurred.

29. For a somewhat fuller treatment of this topic see Lewis, "Metrics, Geometries, Signs, and Language," 103–4.

30. Ronald Goodman, *Lakota Star Knowledge: Studies in Lakota Stellar Theology* (Rosebud, S.D.: Sinte Gleska College, 1990), 16, 18.

31. Hugh Brody, *Maps and Dreams: Indians and the British Columbia Frontier* (London: Jill Norman and Hobhouse, 1981), 266–67.

32. Selwyn Dewdney, *The Sacred Scrolls of the Southern Ojibway* (Toronto: University of Toronto Press for the Glenbow-Alberta Institute, Calgary, 1975), 57–80, 183–84. See also fig. 3b in G. Malcolm Lewis, "The Indigenous Maps and Mapping of North American Indians," *Map Collector* 9 (December 1979): 27.

33. L. Dudley Stamp, ed., *A Glossary of Geographical Terms* (London: Longmans, 1961), 307.

34. Delf Norona, "Maps Drawn by North American Indians," *Bulletin of the Eastern States Archeological Federation* 10 (1951): 6.

35. J. B. Harley and David Woodward, eds., *Cartography in Prehistoric, Ancient, and Medieval Europe and the Mediterranean*, vol. 1 of *The History of Cartography* (Chicago: University of Chicago Press, 1987), xvi.

PART 2

The Ongoing Second Encounter

4

Recent and Current Encounters

G. MALCOLM LEWIS

Chapters 5–11 are contributions to the current encounter of scholars with Indian and Inuit maps. This chapter presents and illuminates them in the context of the considerable body of research and interpretation that has been published since the mid-twentieth-century hiatus in interest that ended circa 1970.

Recoveries from periods of intellectual stagnation are rather like the ends of economic recessions: easier to recognize in retrospect than at the time. Usually, recovery begins imperceptibly, proceeds slowly at first, and then accelerates. Scholarly interest in Indian and Inuit maps is currently beginning to accelerate. Almost thirty years after the recovery began, there is no indication of deceleration and certainly no prospect of recession.[1] I hope that by promoting awareness and stimulating active interest, this volume will help sustain the acceleration.

As in an economic recovery, quantitative growth in scholarly activity has been inextricably linked with diversification. Ways of seeking to know and understand (epistemologies), methods of finding out (methodologies), contexts in which maps have been studied, and the very concept of map itself have each and interrelatedly diversified. Issues raised and published findings have become more complex and, in some cases, controversial.

Posthiatus encounters have not been within any particular paradigm, stemmed from a particular subject discipline, or followed any particular leader. Indeed, there may be no specific causes, though the emergence of new epistemologies within academia in the postpositivist era, the renewed awareness of natives by non-natives, and native North Americans' increasing corporate awareness were and remain important.

The posthiatus literature is scattered and much of it is embedded as relatively minor components in larger works on wider topics. Some is in surprising, almost obscure, non-American publications.[2] Though a considerable literature exists for those prepared to search for it, until very recently, at least, there was no corpus. In these respects, the situation is not much different from that immediately preceding the mid-twentieth-century hiatus, but the rigor and perceptiveness of the published work, quality of the supporting graphics, and the bibliographic and cartobibliographic standards are each vastly improved. Whereas these improvements are heartening, the scatter and diversity are both regrettable and welcome: to be regretted because the critical takeoff point that marks the beginning of any vital body of interactive research may not have been reached; but at the same time welcome because diversity heralds multidisciplinism, potential in several different contexts, and an openness to new ideas.

Although classifiable in several different ways, it is useful to recognize three broad contexts in which significant achievements have been and continue to be made: historical, including exploration, archaeology, and the history of cartography; anthropological; and recent cartographic encounters on the part of native peoples themselves. As with any classification of human endeavor, these, of course, inevitably overlap.

Recent Scholarly Encounters in Historical Contexts

The process of geographical exploration is one important context in which the significance of Indian and Inuit maps has been recognized. In the posthiatus literature, John Allen was perhaps the first to recognize this fully. *Passage through the Garden: Lewis and Clark and the Image of the American Northwest* (1975) contained reproductions of transcripts of two Indian maps and (an important distinction) two other maps based on Indian information.[3] Allen pointed to the significance of these types of cartographic information in Lewis and Clark's strategic and route planning and in their formulation of geographical images of adjacent regions not explored by themselves and for which reliable non-native geographical information did not exist.[4] In 1987 James Ronda developed these themes in the same context in "'A Chart in His Way': Indian Cartography and the Lewis and Clark Expedition." In it he recognized "a whole battery of problems" faced by the two explorers when they "sought to interpret and then use Indian maps." Among these were the maps' "physical structure and expression," "the symbols Indians used to express map information," and "the different way of seeing the material world" that was reflected in "those lines, marks, and heaps of sand."[5]

In "Amerindian Maps: The Explorer as Translator" (1992), Barbara Belyea in part excluded Ronda from an otherwise blanket criticism of this type of interpretation. The essence of her case was that, unwittingly perhaps, they had adopted "the assumptions and standards of European cartography as universal measures of accuracy," translated "Amerindian maps into European terms," and defined "native convention in terms of absence or failure."[6] Belyea's case must be taken seriously and, if at all possible, acted on. She has developed it further in chapter 6 in this volume, concluding with both a condemnation and a challenge:

> The effort of explorers and imperial cartographers to translate native map images into the conventional terms of scientific mapmaking, to assimilate them, stripped them of their proper significance.... Instead of continuing to translate the native cartographic convention into our own, we must acknowledge that the gap between these conventions is essentially unbridgeable. The best we can do is to initiate a dialogue with native cultures as

they have survived, keeping in mind the dangers of upstreaming.... by considering such very different maps and cultures, we could learn more about our own.[7]

Whether we approve or not, past intercultural relationships have been in the nature of encounters. In these, the alien whites have almost always been the more dynamic in creating the contexts in which maps were made, thereby usually influencing the maps' nature and content, and almost always attempting one-way translation. It is surprising, therefore, that a cogent argument for dialogue to replace encounter did not originate until approximately twenty years after the hiatus. As so often in the history of cartography, the initiative came from a newcomer to the field.[8] The literature of the previous twenty years does, however, reveal less specific, but, nevertheless, perceptive antecedents.

In 1976, in a review of Spink and Moodie's "Eskimo Maps from the Canadian Eastern Arctic," almost all the maps in which had been made for white explorers, John Bockstoce had observed that though:

> their analysis, is in large part, a formal one, and to that end they have described accurately the physical characteristics of the maps . . . I believe that they have largely failed, or perhaps not gone far enough, in their search for the real essence of the maps: to probe for the Eskimo's own conception of their surrounding environment; to seek why certain spatial concepts—so foreign to Europeans—are employed by the Eskimos.[9]

As the curator of ethnology at the New Bedford Whaling Museum, Massachusetts, Bockstoce knew something of the ways and minds of the Eskimo. He must also have read Edmund Carpenter's *Eskimo Realities*, published three years earlier, with its many insights into Eskimo concepts of space, mapmaking styles, and world. These had included:

> Aivilik men are keen geographers when describing their immediate surroundings. But once they venture to tell of the outer world, geography gives way to cosmography.
>
> Once when I asked an old Iglulik hunter to draw a map of the islands north of Fox Basin (no simple task for they are most irregular), he rapidly and accurately sketched them in the air with his forefinger. But when urged to use paper and pencil, with which he was unfamiliar, he repeatedly declined and then, finally consenting, produced with difficulty an inferior map. He mentioned no names for most of the islands, though he did for salient points on their coastlines. In other words, he had no interest in land mass, only in geographical points.
>
> I asked several Aivilik to make sketches of this world.... The first, made by Karleaner, is limited to a series of dots, each representing a trading post and including all settlements known from experience and hearsay [but nothing else].
>
> Of course, what appeared to me as a monotonous land was, to the Aivilik, varied, filled with meaningful reference points.... By and large these are not actual objects or points, but relationships between, say, contours, type of snow, wind, salt air, ice crack.[10]

Bockstoce's and Carpenter's observations arose from experiences with Eskimos. In 1983, Frederick Turner published a book that had a global perspective. Subtitled *The Western Spirit against the Wilderness*, it drew attention to the fact that "Even 'primitive' peoples of the world have had maps, and the generic difference between these and the maps developed in the West is of interest." Using Micronesian stick charts from which to develop his argument, he compared the roles of what he cautiously referred to as "primitive" maps with those of the West. In common: "Their devices are, like our maps, abstractions from the physical territories. They are, again like our maps, attempts to 'control' those territories by representing them, since representation . . . is an attempt at mastery." Yet, relative to modern Western maps:

> both the level of abstraction and the impulse to control are minimal, and these maps . . . keep close to the realities of their worlds.
>
> In this one sees not so much the technological inferiority of these peoples as the possibility that their mythologies were still their basic technologies, providing them with spiritual as well as spatial orientation, answering the questions all humans pose of the uncharted spaces beyond experience. Spatially as well as technologically, their worlds remained closed, limited ones, and often these limits imposed severe hardships upon those within. But as Spengler observes, space terrifies because of its cognate sense of annihilation, and one thinks of certain comforts to be had in circumscribed worlds, comforts either forgotten or else remembered only in dreams, accidents, or passions by the people of the West.
>
> When the wider world of ours burst into the fragile perimeters, the little worlds of the primitives shattered as a Micronesian bamboo map might in a hurricane gust.[11]

Although not written with specific reference to attempts by whites to translate, understand, and use the maps of native North Americans, the last sentence poses serious questions for those who would follow Barbara Belyea's plea to replace primarily one-way encounter with dialogue. Have native cultures in North America been so "shattered" as to make dialogue at best misleading and at worst impossible? If the answer is yes, can one have meaningful dialogue with preshattered cultures on the basis of artifactual evidence and accounts written almost exclusively by whites? In chapter 6 Barbara Belyea has undoubtedly demonstrated that one can, but it will be a difficult road to follow. Theory and speculation must not be allowed to usurp cartographic evidence (incomplete and debased though it may be) however laudable the objective.

Archaeologists have shown less interest in native maps than have historians in general or historians of exploration in particular. This is somewhat surprising, because several hiatus publications had indicated their potential importance. The first of these was addressed to archaeologists but made no contribution to site archaeology.[12] The second did, however, reach a tentative conclusion that still awaits close examination. On the basis of a "preliminary study . . . of some hundred maps drawn by the aborigenes of North America," it suggested that

> after a definitive collection of data has been secured, tabulations of observable traits, critical evaluations, and re-appraisals of the entire subject will be in order. Perhaps then some useful contribution will be made to that branch of cultural anthropology [historical linguistics or the archaeology of language would now be the appropriate disciplines], which we term "Language," using the word in its widest sense, and of which in turn, "Cartography," though usually treated [as] a separate science, is really a sub-division.[13]

The "useful contribution" has never been made. This may have been because the idea was published somewhat obscurely and directed at the wrong audience. More likely, it was before its time; published before interdisciplinary walls came tumbling down. Even now, there is little evidence to suggest that space as expressed in spoken language is being considered together with space as represented in maps. Although there is a considerable and growing literature on spatial deixis ("A term used in linguistic theory to subsume those features of language which refer directly to the . . . locational characteristics of the situation within which an utterance takes place, whose meaning is thus relative to that situation."),[14] those interested in maps have remained unaware of this. Insofar as bridges have started to be built, the builders have begun from one side with little or no awareness of what exists on the other.[15]

In chapter 7, Margaret Pearce has made a seminal contribution to this bridge building. As with many new directions, it began by chance. Her initial interest was in Amerindian maps in general. A laudable decision to narrow the focus might have proved unproductive. An initial awareness of a few fairly well known native maps made in connection with early southern New England encounter land deeds led to a search for more. There were many deeds, but the hitherto known maps were found to be the exceptions. No additional ones were discovered. A less perceptive researcher might have become frustrated, but Pearce recognized that, although native maps were rare, native mapping was common but was expressed in spoken rather than graphical language. To the conventional historian of cartography, at least, her immediate conclusions will come as a shock verging on a contradiction: "To overlook written in favor of graphic (and thus 'more maplike') forms is to ignore a major part of property mapping in colonial southern New England." Those with interests shaped and constrained by graphical concepts of "map" may choose to ignore this and similar context-determined conclusions. But Pearce's plea for a "comprehensive inquiry . . . into the deed process as a whole" is followed by a much more challenging claim: "Such an inquiry would not only illustrate a distinct mode of colonial map communication, it would also greatly benefit our understanding of how the mapping of America has taken place."[16] That claim must be tested, and unless refuted, pursued. It is too challenging and important to be left dormant. It parallels conclusions I reached several years ago after reading a large sample of early-encounter literature in order to assess the influence of the cosmological ideas and geographical lore of native North Americans on early European exploration and geographical knowledge.[17] Far more spatially organized information was communicated by speech (often reinforced by gesture and inevitably distorted by translation) than by graphics.[18] Europeans occasionally transformed word maps into cartographics; more often they incorporated them into their own maps.[19]

We will never know what was in the archaeologist Delf Norona's mind when he wrote that a fuller knowledge of native maps would "perhaps" contribute to the understanding of language *sensu lato* and cartographic language *sensu stricto*. Beginning with an apparently minor thematic interest within a small region, Margaret Pearce has come close to asking perhaps the most fundamental question in the prehistory of cartography: which came first, word maps or graphical maps? Except in a short, eclectic, and somewhat theoretical essay by myself, this question was not even touched upon in the most authoritative and substantive work yet published on prehistoric cartography. In the essay I posited that "it is in the development of language in its broader sense that the origins of mapping are to be found."[20] To the best of my knowledge the statement has never been questioned.[21] Certainly it has not been tested. Indeed, at this stage it would be premature to attempt it, but when in the future the attempt is made, the evidence of Margaret Pearce's contribution will have to be taken into consideration.

Archaeologists have hardly begun to answer the question "How can we verify the possibility that some rock art panels contain maps?" Many glyphs and paintings have been said either to be maps or to incorporate them on the sole grounds that they appeared to have some of the general graphical characteristics of European-style maps; others because they appeared to be maps or plans of nearby features, terrains, drainage systems, or trail patterns. In the major posthiatus synthesis on North American rock art, Klaus Wellmann referred to a large number of these published claims and speculations, indexing them according to five categories: "Maps as Rock Art Design," "Game Trails," "Ground Plans of Houses and Lodges," "Battle Scenes," and "Astronomical Motifs."[22] Regionally organized, the treatment of these was inevitably fragmented. An evaluation of the case for terrestrial maps appeared in his chapter on the Great Plains where, in reviewing several interpretations published between 1950 and 1971, Wellmann noted,

> Here and there, meandering lines and other abstract elements on certain panels have suggested maps to experienced observers since the designs appeared to correspond rather closely with the features of nearby natural formations such as the contour of a mountain range or the course of a river.... In various parts of the country, including the Plains, Indians of the historic period not infrequently drew maps on material other than rocks, in order to commemorate important battles, expeditions, migrations, large hunts, horse stealing forays, or other notable occurrences.... It is possible that on occasion rocks were used for the same purpose, but reasonably convincing examples of this type are very infrequently encountered. It is perhaps pertinent to note here that the specifics of place, such as tipis, villages, hills, rivers and trees do not enter into the Plains Indian paintings on skin, cloth, and paper much before the 1870s, and that even thereafter "the actual number of instances of the adoption of such conventions remained small."[23]

Wellmann's evaluation of the case for maps in Great Plains rock art was inconclusive, verging on the skeptical. In retrospect it is revealing. No attempt was made to measure the extent to which specific "designs appeared to correspond with ... features." No mention was made of the considerable methodological problems that would have been faced by anyone attempting seriously to do so. Without reference to the date or cultural origins of specific examples of supposed maps in rock art, mention of "Indians of the historic period" having made maps implied that unknown ancestors or members of earlier cultures may also have done so. The implication was immediately countered with an equally dubious one. The "specifics of place" did not "enter into ... Plains Indian paintings ... much before the 1870s."[24]

Not much has changed since Wellmann wrote his unsatisfactory evaluation, but two developments within archaeology do merit mention as indicators that progress will be made in the near future. These have been in the contexts of rock-art dating techniques and archaeoastronomy.

Soon after my interest in Indian and Inuit maps first began, I wrote to a few friends and former associates requesting information about primary materials and secondary sources. Norman

Thrower kindly responded with, among other things, a copy of Robert F. Heizer's *Aboriginal California and Great Basin Cartography*.[25] It had been published twelve years before the end of what I was later to define as the hiatus and at the time I had no idea that the author was arguably the leading archaeologist of the California region. Nor had it occurred to me that rock art might contain evidence of pre-encounter cartography. I monitored Heizer's publications with interest. His early opinion on the possibility of maps in rock art had had the same ambivalence as Wellmann's. In a paper on the rock art of Nevada and eastern California coauthored with Martin A. Baumhoff, he had expressed the general "opinion that the petroglyphs of Nevada are not a form of communicative writing, nor are they maps," but conceded in a footnote, "Perhaps we should not make this denial too categorically," citing one example of a plausible cartographic interpretation of a petroglyph on the lower Colorado River.[26] Sixteen years later, in a much more detailed study of petroglyph sites in southern Nevada, Heizer, in collaboration with Thomas R. Hester, tentatively proposed that long ticked lines represented in plan diversion fences for the capture of game.[27]

The inconclusiveness of Heizer and his coauthors stemmed from two limitations in the evidence: the absence of reliable rock-art dating techniques with which to link panels to specific cultures, and the impossibility of finding ground referents for short-lived features such as drive fences made with brush.

In the late 1980s the cation-ratio varnish-dating technique was applied to a number of sites in southeastern Colorado.[28] One of the latest (youngest) panels coincided very approximately with the earliest European incursion into the general region of the southwestern Great Plains, when the culture and economy of the Indians was not dissimilar to that of later encounter times.[29]

> The scene appears to represent an animal drive. The humans with their outstretched arms are the herders who are driving the animals toward the net, which has been set across their pathway. The single animal with a possible arrow in it is located at the back of the drive scene, where it appears to have been shot while trying to escape. The possible dog is at the back of the group, where it may be in pursuit of the animal with the arrow or in a position to help with the drive. The bird, protecting the lower end of the net, may represent the power of birds in shaman lore. Fourteen of the animals are headed toward the net and ten are headed away. The latter includes the dog-like figure, but quite possibly the animals are shown in a surround where they are circling within a trap. Three animals at the top of the scene are outside the perimeter formed by the line of humans. All three are shown outside the confines of the net, where they may represent escape from the surround.

Although the pecked human and animal figures are all in profile, the description implies that the whole composition is a plan. This probability is further reinforced by the situation of the panel between

> standing blocks of sandstone, where hunters could have trapped animals by either lying in wait for them or by driving them into the rocks. Given the content of the panel, the latter explanation seems to be a good possibility. The historic use of the area between the large sandstone blocks as a corral is worth noting. It supports the assumption that the setting was ideal for trapping animals.

This very plausible hypothesis that the whole composition was the plan of an event (or recurring events) may even be testable:

The area may have held a symbolic meaning, with rocks that trap animals as the primary theme, or it may have been the location of an actual drive and surround. The latter possibility might be tested by excavating in the corral, but the tramping of historically corralled animals has eroded the ground area near the rock art.[30]

Because the referents are arranged and patterned in relatively small spaces, recognizing plans in rock art is likely to remain less speculative than detecting maps of extensive terrains, especially if the referents include man-made elements or interactive behavioral patterns. In both cases, however, detection rates and the degrees of certainty in making interpretations are likely to increase with the development and application of better absolute dating techniques.

Maps made by native North Americans have only occasionally been used as evidence in reconstructing former landscapes, environments, and events.[31] On the whole, historians, historical geographers, and others involved in reconstructing the past have opted for the supposed authenticity of evidence derived from early European and Euro-American maps in preference to the apparent incomprehensibility of that on maps made by Indians; often without recognizing that the former had been in part derived from the latter.[32] Conversely, reconstructions of early-encounter landscapes have sometimes led to a deeper understanding of Indian maps. Indeed, one recent study hypothesizes a prehistoric map of a pre-encounter political system. Having reconstructed the trade patterns in the lower Mississippi valley for the Mississippian period (A.D. 700–1540), the archaeologist Robert H. Lafferty III suggested that the "political structure of the Mississippian world may be represented by an engraved shell cup from Spiro [Oklahoma]."[33] Based on perceived similarities between the reconstructed pattern and the engraved pattern and between the engraving and the pattern of tribes and roads on the well-known manuscript copy of a 1737 Chickasaw map of Southeast North America (fig. 9.2), the hypothesis is plausible but ultimately unverifiable.[34]

The most significant single contribution by an archaeologist/ethnohistorian to an understanding of Indian maps was made by Gregory A. Waselkov. His "Indian Maps of the Colonial Southeast" is empirically the most scholarly of all publications exclusively concerned with maps made by native North Americans, and in its interpretations one of the most perceptive.[35] Perceptiveness was a quality revealed again a few years later in correspondence with me concerning the spatially fragmented pictograph painted on a tree as described and illustrated by William Bray in 1782 (fig. 1.1). Found some years before in the Ohio valley, it had been assumed to record events within that region. Then, Gregory Waselkov pointed out that the small but distinctively shaped plan of one of three forts was remarkably similar to that of the stockade of Fort Loudoun as reconstructed on the basis of archaeological investigation.[36] The other two forts had been identified by Bray's informant as Forts Detroit and Pitt.[37] If the third, unnamed one was indeed Fort Loudoun, then the whole painted pictograph was of warfare extending far to the south of the Ohio valley. Fort Loudoun existed from 1756 to 1760. If the pictographic plan had indeed been intended to represent it, then the painting was presumably concerned with events involving Indians in the French and Indian Wars. Further research could well strengthen the hypothesis. Meanwhile, it is based solely on a perceived visual likeness between a scaled plan obtained through archaeological investigation and a minute (10 x 6 mm) component of a line drawing at least twice removed from the original pictographic painting on the blazed tree.

In chapter 8 Morris S. Arnold has used methods similar to Gregory Waselkov's. His subject and objective are, however, different. Instead of a number of transcripts approximately contemporary with long lost artifacts, his subject is an original, remarkably well preserved and only slightly modified (by four written words, added at a later stage), supposedly Quapaw painted buffalo robe (fig. 8.1). Instead of increasing understanding of what have long been known as Indian maps, his objective has been to demonstrate that what at first appears singularly unmaplike is a map of an event in the real world. His evidence is partly architectural and in part ethnographic (figs. 8.2 and 8.3). Arnold's critical diagnostic evidence, unlike Gregory Waselkov's, is provided by dwellings and people presented in profile and arranged in a distinctive, one-dimensional, linear sequence, not by features represented in plan and distributed in two-dimensional space. Ultimately, the most important of Arnold's findings may be that an early-encounter Indian artifact, with some European content but little or no indication of European influence, incorporated the essentials of a line map. In aesthetically less rich and much simpler forms, such maps probably had long been used as the bases of messages. Left to inform friends, and occasionally to threaten enemies, they were used in virtually every part of North America. In the course of the encounter, these simpler forms of line map were also the bases of almost all pictographic aids made to assist white explorers, soldiers, trappers, traders, mineral seekers, and settlers. In European terms, they were not the equivalents of, say, the English county maps of the sixteenth to nineteenth centuries, but of road maps or itineraries, in a tradition extending from the Peutinger map (twelfth- or early thirteenth-century but derived from a fourth-century Roman archetype) to the road books of John Ogilby and later English mapmakers, which seventeenth- and eighteenth-century gentlemen took with them in their saddlebags and carriages.

Somewhat surprisingly, Indian maps have rarely been used in conducting searches for archaeological sites. The one apparently notable exception is atypical of Indian maps. Between 1906 and 1907 Sitting Rabbit, a Mandan Indian, painted a map of traditional Mandan and Hidatsa village locations along the upper Missouri River.[38] The map had been commissioned by Orin G. Libby, secretary of the State Historical Society of North Dakota.[39] Unfortunately, from the perspective of those later to be interested in Indian maps and mapmaking, Libby provided a detailed base map.[40] Hence, Sitting Rabbit's task was to plot rather than to map. From the point of view of later archaeologists this, however, was an advantage. Places and activities plotted by Sitting Rabbit could be related to sites on the ground; assuming of course that he had read the base data correctly. Almost seventy years later, an archaeologist, an anthropologist, and a linguist examined the utility of the painted map from their disciplinary perspectives. Their conclusion on its significance for archaeologists was that it is a native crosscheck on early Euro-American records. It also provides leads—such as the enigmatic "Earring House"—for features for which we should remain alert in the archaeological and historic record. The map should also be useful in providing leads for future archaeological surveys—keeping in mind that much of the Missouri River Valley in North Dakota has never been subjected to systematic pedestrian archaeological survey.[41]

In an earlier report, Raymond Wood examined the historical maps of the upper Knife-Heart region. The study was "not intended as an exercise in historical cartography, but simply as a guide to information contained on certain early maps of the upper Knife-Heart Region, especially as this information is relevant to the identification of historic and/or archeological sites."[42] In addition to that by Sitting Rabbit, Wood considered five other Indian maps. Four of these were part of a set sketched by Hidatsa informants for the ethnographer Gilbert L. Wilson in the early 1900s.[43] Wood's summary statement about these was factual rather than evaluative. Together, they depicted "the location of the stone on which the culture hero [Itsikamáhidic] purportedly rested his foot, and the several [former] Hidatsa villages."[44] Wood's concluding remarks did not discuss the complex issues to be overcome by archaeologists attempting to locate ground sites from the evidence of maps not based on surveys. In the introduction, however, he had offered the general observation that "the value of maps for historical [and, presumably, archaeological] detail tends to diminish rapidly as the area with which one is concerned becomes increasingly localized."[45]

Though potentially important aspects of the ongoing second encounter, the concerns of Raymond Wood and his colleagues were exceptional, perhaps even unique.[46] It was against the background of this near lacuna that I invited Gregory Waselkov to write a chapter for this volume "on the ways Indian maps helped—or if known [of] at the time, could have helped—archaeologists locate sites, discover routes, define territories etc." To this I added the critical observation that such an evaluation "should have been the culmination of your otherwise superb chapter in *Powhatan's Mantle*... but it was not there."[47] In accepting the invitation he observed that

> [the] problem I had while writing the piece for *Powhatan's Mantle*... was to relate the large-scale [presumably this should have been large-area or, in map-scale terms, small-scale] social views of the region found in the Indian maps to the rather site-specific data typically of interest to archaeologists. The Indian component of John Smith's map of Virginia has been used effectively by archaeologists at that level of analysis, but I don't know of many other comparable examples.
>
> Of course, I now realize that I was viewing the archaeological utility of those maps too literally. There are other ways that archaeologists could benefit from a consideration of the maps and contribute to their understanding.[48]

To this important general statement he added speculations about such things as field locations, trade networks in horses, slaves, and red pipestone, and the involvement of chiefs in long-distance travel to acquire both knowledge and the esoterica that helped to validate their higher social status in ranked societies.

Chapter 9 is the fruit of Gregory Waselkov's personal second encounter with Indian maps of the colonial Southeast. In "considering both the archaeological implications of these maps and some of the interpretative difficulties involved with their use by archaeologists," it is both insightful and a logical sequel to his chapter in *Powhatan's Mantle*.⁴⁹ By focusing his attention on a few maps from his twin regional perspectives as both archaeologist and ethnohistorian, he has amply justified his conclusion that "Native American maps clearly [do] have tremendous potential to inform us about past social landscapes."⁵⁰ By extending the interest from local sites to landscapes, territories, and spheres of influence he has demonstrated possibilities far beyond those envisaged by Raymond Wood and his colleagues in the late 1970s.

Historians of cartography were slow to recognize and examine the influence that native maps (and their geographical lore) exerted on European and Euro-American maps of North America. Serious work has been conducted almost exclusively during the still ongoing second encounter period. Indeed, little of interest appeared before 1986. In that year, I used a case study to develop ideas about what I called the "processes of assimilation": "interpreting information received from Indians . . . in the course of interrogating"; "mosaicing information received from several sources"; "transforming the mosaic according to some concept of scale"; "incorporating the transformation into an existing map of a wider area derived from a different source"; and "accommodating the original incorporation on other maps."⁵¹ The archival materials relating to French exploratory activities in the late 1720s around and beyond Lake Superior proved to be so complete that it led to two further studies, for the second of which fieldwork was undertaken.⁵² Visiting key sites and talking to Ojibwa Indians made me aware of two important principles: first, that whatever an Indian had represented boldly on a map would have been important in the context of that particular map but would not necessarily have been physically large, bold, or even visually conspicuous. Second, the map might not even have been physical.

More work with native people in the field is vital if, in the course of the second encounter, we are to reinterpret our earlier interpretations of the maps of their ancestors. Better still, they should be encouraged to research cartographic artifacts for themselves in the manner being pioneered by a group of Lakota, centered at Sinte Gleska College, Rosebud, South Dakota. They explain in cosmological terms why, by the principles of Euro-American cartography at least, a distinctive and very conspicuous natural feature (Devil's Tower, Wyoming) was mapped by a nineteenth-century ancestor as being far from its true geographical location. Things on earth mirrored patterns of stars in the sky. According to Lakota stellar theology, the heavens took precedence over mirror earth. Sometimes the latter deviated and had to be corrected; in this case by a matter of more than sixty miles.⁵³

Probably the most tangible outcome of my work on early French uses of Indian information (including maps) to map the trans–Lake Superior region was the explanation of why, during the second half of the eighteenth century, first French, then British, and eventually American printed maps perpetrated and then long perpetuated one of the grossest of all geographical errors: a great River of the West flowing from the heart of the continent, westward to the Pacific Ocean. The information given by Cree and other Indians was almost certainly valid according to their own frames of reference. The gross error arose from a combination of optimism and perhaps bias on the part of the French officials who had undertaken the various stages of assimilation. Louis De Vorsey, Jr., examined in depth the origins of an earlier gross geographical error. Maps of the lower Mississippi River after René Robert Cavelier, sieur de La Salle's successful descent and exploration from the Illinois River to the birdfoot delta in 1682 were, from our perspective, even more distorted and inaccurate than earlier conjectural maps. Rather than reaching the Gulf of Mexico in what is now Louisiana, the La Salle–based maps represented the River as swinging far to the west to enter the gulf in what is now southern Texas. The title of De Vorsey's study asked, "Product of Error or Deception?" Although making one reference to the well-known account of an Illinois warrior drawing a map of the river with charcoal, he concluded unambiguously that La Salle was "an opportunistic deceiver who, for a period, effected the greatest hoax in the history of North American exploration. It was a hoax that grossly distorted European maps of America for the final two decades of the seventeenth century." Interestingly, De Vorsey prefaced his forthright conclusion with the qualifying clause "from this reading of the evidence."[54] This raises fundamental issues that must never be overlooked in essaying second encounters with earlier frontier encounters. Is the available evidence complete? If not, is it biased? What was the background of the interpreter and from what perspective was the interpretation conducted? Were the causes really as different as our respective conclusions concerning the two gross geographical errors suggest?

The Lewis and Clark Expedition affords a less complex, better documented, but hitherto unanalyzed example of gross error on an American map arising from the assimilation of two Indian maps. On their return journey up the lower Columbia River in April 1806, the two explorers had missed for the second time the mouth of the Multnomah (later named Willamette) River, around which Portland, Oregon, was later to develop. It had been concealed from view by an island in the main river. Soon afterward it was represented for the explorers by two young Watlala men who drew it "on a [presumably rush] Mat with a coal" and, on the following day, by an old Indian man who drew "a sketch" and gave the names of "the nations resideing on it." Transcripts were made by William Clark, though with what understanding and veracity is not known.[55] What is certain is that, six years later, assimilation of the Indians' information led to an early-nineteenth-century error almost as great as those of the Mississippi and Great River of the West in the eighteenth. It was perpetrated on William Clark's highly original manuscript compilation: "Map of part of the Continent of North America."[56] It showed the Multnomah River rising just to the west of what was to become the site of Salt Lake City, rather than in the Oregon section of the Cascade Range, some five hundred miles west northwest across the Great Basin. We will probably never know what was in William Clark's mind when he made the decision to assimilate the Indians' scaleless information in this way: uncritical belief that this was the essence of their cartographic and other statements; unbiased evaluation of this and other evidence; an evaluation biased by the hope that there might be an easy river route from the soon to be opened up maritime world of what was to become known as the Pacific Northwest (thus bypassing Spanish control over the mineral-rich Southwest and offering a route to the fur-rich central Rocky Mountains); or a willful attempt to deceive.[57] Whatever the intention, the gross error was perpetuated on a number of important early-nineteenth-century maps.[58]

In chapter 10, Patricia Galloway has used manuscript sources in detecting how, at the beginning of the eighteenth century, the Delisle map house in Paris made use of Amerindian information and with what consequences. Like many who have developed an interest in the cartography of traditional peoples, she was drawn to it via a noncartographic topic: the ethnogenesis of the historical Choctaw tribe, whose peoples eventually occupied what is now east-central Mississippi. This led her first to seek evidence on European maps, next to reflect on the use of multiple symbols on these, and then to conclude tentatively that the technique may have been a direct consequence of the Delisles having used Indian information, including maps.

The operational problems faced by Patricia Galloway are almost inevitable for anyone beginning an investigation of this type when the appropriate archival data set has not survived or been discovered. In her case, the relatively recent breakup of a major corpus of Delisle papers (including maps) is unfortunate.[59] She has, however, supported the tentative conclusions based on map sources with other kinds of evidence: Indian mapping styles and conventions as evidenced in the considerable body of known Indian maps from the Southeast in general and noncartographic evidence such as pottery decoration, mortuary ritual, and origin myths. There is a principle to be enunciated here: convergent multiple evidence will always be more conclusive than that abstracted exclusively from Indian maps; especially so when the latter are one or more states removed from original artifacts.

Recent Scholarly Encounters in Anthropological Contexts

Anthropologists have shown less interest in maps than historians and archaeologists, and their work has sometimes been controversial.

Indians in several parts of the continent stored or repeatedly made maps that served as mnemonics of traditions and beliefs; often, though not necessarily, for use in rituals. Those who made and stored them were powerful individuals within their communities. In the early eighteenth century, for example, Iroquois wampum belts were "stored at Onondaga."[60] This was probably also the center at which were stored maps made on birchbark.[61] In 1820, somewhere near the White River in southern Indiana, a Delaware record keeper, in gratitude for being cured of an epidemic disease, presented a white physician with his most precious possession: a pictographic record on bark or wood slabs of his tribe's past.[62] In the mid-nineteenth century, an Ojibwa chief described how his and most adjacent peoples had "places in which they deposit the records which are said to have originated their worship." These were on slate, copper, lead, and birchbark. Stored in elaborate excavated structures located at "unsuspected spot[s]," their guardians were selected from among "the wisest and most venerable of the nation."[63] For some years after their relocation to Indian Territory in 1875, the Skidi band of the Pawnee practiced ceremonies, one of which involved a sky chart painted on buckskin. It had a hereditary keeper, whose wife or other female relative was responsible for its practical care, and its own priest charged with the responsibility for the ritual in which it was used and the knowledge it represented.[64] Between 1891 and his death in 1913, an Oglala Sioux made drawings in a large ledger. Most were records of historical events and some were maplike. On the artist's death, the ledger was inherited for keeping by his sister, with whose body it was interred in 1947.[65] The contents had been photographed on campus at the University of Nebraska some years before. In the late 1980s, in the small community of Hudson in northwest Ontario, I saw by chance a birchbark scroll that, until a few years before, had been preserved by an elderly Southern Ojibwa woman.[66] In recent years, these, similar, or related artifacts from the past have been interpreted or reinterpreted as maps; usually by anthropologists, sometimes involving native people, and almost always with an element of doubt verging in some cases on controversy.

Sometimes, repatriation of artifacts has been the cause of renewed encounters. On May 8, 1988, eleven wampum belts were returned to the Six Nations Iroquois Confederacy at the Onondaga Longhouse at Grand River, Ontario. They had been repatriated by the Heye Foundation, New York. George Heye had bought them for $2,000.00 in 1907.[67] One was a five nations belt: five lineally arranged, unlinked, paired red diamonds against a dark background. Of unknown date, it is believed to represent the five Iroquois tribes in the pre-Confederacy era, when they were frequently at war with each other.[68] Wampum belts were by their construction and shape unsuitable for the representation of all but the simplest and most stylized of topologically structured maps. This is what the war belt is believed to be: five identical red shapes, indicating five tribes at war; unconnected, indicating separateness and perhaps animosity toward each other; arranged lineally and spaced equidistantly, as indeed they approximately were in what is now upstate New York. In contrast, a now incomplete alliance belt originally showed five equally spaced, lineally arranged human figures in white on a purple background, holding hands with their elbows crooked.[69] The white figures symbolize the five tribes after they had entered into the Five Nations Confederacy. As in the war belt, the symbols are arranged lineally, much as in the real world. White indicates friendship and peace, in contrast to the red for the earlier animosity and war. The holding of hands indicates the formal linkages of the Confederacy, but the crooked arms signify that any of the tribes were free to leave it.

Once the League of the Iroquois had been established, specific individuals were designated to memorize its laws, constitution, and history. Wampum records were created as mnemonic aids. Not all incorporated cartographic principles, but the war and alliance belts certainly did: five nodes of equal importance, color coded to signify the most significant attribute, arranged lineally much as in reality, and either unconnected or connected to indicate the absence or existence of geopolitical links. Those who developed and interpreted such mnemonics certainly possessed several of the basic concepts of map, yet, outwardly at least, these belts were quite unlike the maps made by Indians for Europeans. In particular they incorporated no base data. This is the fundamental reason why their return to the custody of the confederacy chiefs "has not changed much. Most of the belts can't be read without [presumably "with" was intended here] certainty."[70] If Indian descendants cannot shed much light on the basis of oral traditions, anthropologists are unlikely to contribute very much, though ethnohistorians may still be able to do so.

The Delaware pictographic record on bark or wood given to a white physician in 1820 has recently and very controversially been interpreted as a map. Regrettably, the original artifact has not survived, but a transcript has. Known as the Wallam Olum, or Red Score, it consists of a creation myth, a deluge myth, and a history of the Delaware tribe. Much of the latter is a migration legend. Although not a map, several attempts have been made to interpret the route with reference to the real world. In 1885, Daniel G. Brinton interpreted it as beginning "probably at Labrador" then journeying south and west to "perhaps the St. Lawrence about the Thousand Isles" and so into the Delaware valley through parts of what were to become the states of New York, Ohio, and Indiana.[71] A mid-twentieth-century investigator traced the migration all the way from Asia.[72] At least one anthropologist then concluded that it was "difficult to accept the Wallam Olum as an authentic record of Delaware migrations," suggesting instead that it was "an account, by a despairing person . . . of a Golden Age which never was."[73]

Recently the argument for a geographical interpretation of the Wallam Olum migration legend has been renewed. Matching the sequence of sixty pictographic symbols (and translations of the Delaware words for them) with a sequence of ground referents, David McCutchen has made an apparently strong case for the migration having begun in Asia, somewhere near the border between China, Mongolia, and the former USSR. Its route is traced with apparently increasing precision via the head of the Sea of Okhotsk, Bering Strait, central Alaska, western Canada, the Columbia Plateau, Missouri River, and Ohio valley, to the Atlantic seaboard.[74]

McCutchen's interpretation has been subjected to serious criticisms on a number of grounds. Some relate to the authenticity of the pictographs and the supposedly Delaware text attached to them in the 1820s. In it unclear, for example, whether the latter came from a native speaker. The Wallam Olum originated at a time when the Delaware were undergoing a religious revival and themes from early traditions may have been reworked to conform to this. Notwithstanding personal and massive funding by Eli Lilly, a major program of research failed to prove that the text was Delaware either in origin or intent. The pictographic designs are not found in other examples of Delaware art. These and other criticism are concentrated in one review. It concludes that although "the Delawares have endorsed this book . . . much of it rings hollow. What appears to be precision and certainty is smoke and mirrors."[75] Such criticism is welcome and should be a recurring characteristic of the second encounter. In this particular case, however, it should not detract from the cartographic significance of the artifact. Unless the Wallam Olum is completely bogus or the associated text a gross fabrication, it is a linear sequence of places, conditions, and events as, according to tradition, they were experienced along the route. Reconstruction of that route will remain a matter for debate, but the original artifact was a graphical representation of unique nodes arranged in sequence. Unlike the nodes of the Iroquois war and alliance belts, each of the nodes of the Wallam Olum had distinguishing characteristics. Unlike the alliance belt, internodal linkages in the Wallam Olum were implied rather than represented. As with both belts, there was no network of base data, but many of its pictographs incorporated environmental information about nodes. Though not explicitly representing a route and apparently more abstract in its pictography, the Wallam Olum had one characteristic in common with the Quapaw map interpreted by Morris Arnold in chapter 8 and the *Mapa Sigüenza* discussed by Elizabeth Boone in chapter 5: distinctive nodes.

Like all Indians, by the eighteenth century the Delaware were perfectly capable of making what Europeans intuitively classified as maps. Though rich and diverse in content and graphical in form, the Wallam Olum was not at first classified in this way. Like the war and alliance belts, it probably represented an earlier stage in the development of mapping: halfway between an oral narrative and an itinerary map. If ever a phylogeny of maps becomes traceable (as yet, the evidence from prehistory is inadequate) this is the kind of anthropological evidence that will have to be taken into account.

The birchbark scroll I chanced to see at Hudson, Ontario, was almost certainly in the tradition of those described by the Ojibwa chief George Copway nearly 140 years before.[76] As a boy, the Canadian anthropologist Selwyn Dewdney had lived in Hudson, where his father was the Anglican missionary to the Ojibwa. In the early 1970s he made an inventory of, visited, and examined all known examples of Southern Ojibwa sacred scrolls, the results of which were published in a scholarly monograph. The scrolls were classified according to function. One class was made up of eight midé migration charts and seven possible ones.[77] Whereas the Wallam Olum supposedly recorded a tradition about the migration of a people, the midé migration charts recorded the tradition about the route via which the ancestors of the Southern Ojibwa or, more probably, their religion, reached what is now north-central Minnesota. As much as 106 inches long and made up of as many as six physically separate sections, the content of the best of these charts is remarkably detailed. They are made up of two graphically distinct but integrated sets of components: animals, birds, and mythical beings presented either in plan or profile, positioned purposefully in relation to an irregular and, at first sight, incomprehensible pattern of lines. It is the lines that are of particular interest here. Dewdney demonstrates conclusively that they are the base data of a map of an inland waterway system up which, according to tradition, the Southern Ojibwa received their religious system and beliefs. He further demonstrates that the base data is a map of the lake shores and major rivers of the Great Lakes–St. Lawrence drainage system: surprisingly detailed and eminently recognizable to the left (west)—especially Leech Lake, Minnesota, the center of the Southern Ojibwa's cultural world during the historical period; increasingly vague and disputable as one passes to the right (approximately east) through the system toward the Atlantic Ocean via regions further and further beyond the direct experience of the Southern Ojibwa in historical times. Even so, using Dewdney's interpretation, I have published elsewhere a cartographic correlation between one of the charts (Red Sky's, named after its last keeper before being transferred to the Glenbow-Alberta Institute, Calgary) and an outline of the Great Lakes–St. Lawrence drainage system based on a modern map.[78] This is a different order of mapping from that incorporated in the war and alliance belts or in the Wallam Olum; much closer in style and content to maps made by Indians throughout the continent for innumerable white explorers and to many of the birchbark messages left at strategic locations to inform their own peoples. Even so, it is fundamentally linear in structure.

Two late-nineteenth-century Plains artifacts have recently been the objects of second encounters: Amos Bad Heart Bull's drawing of the Black Hills, of which the original was interred with his sister but a photographic copy exists; and the Pawnee star chart that, like most of the known Southern Ojibwa midé migration charts, is in a museum. Both have been mentioned already, but separately and in somewhat different contexts.[79] In some respects, they are related, but it has required recent work by anthropologists and native people to reveal the links between what are superficially such different artifacts.

Amos Bad Heart Bull (1869–1913) was an Oglala Sioux. In 1890 or 1891 he obtained a large ledger in which, from time to time until his death, he made drawings. Some of these have maplike characteristics. One of the Black Hills, South Dakota, is particularly interesting (fig. 11.1).[80] The drainage network on the surrounding plains is so precise as to suggest that it was copied from a printed map. Representation of the meridians 103° and 104°W strengthens this probability. With the exception of the rivers rising in them, the Black Hills are represented quite differently: densely, abstractly, and several of the features totemically. The precise date of this ink and five-color crayon drawing is not known, but approximately a hundred years later it has been shown that Bad Heart Bull did not intend it to represent the Black Hills as a topographic reality but as it was according to Lakota theology, a consecrated enclosure:

> The constellations were the visible "scriptures" of the People at night; and the related land forms mirrored those stellar scriptures during the day. The stars were understood to be "The holy breath of the Great Spirit." ... The constellations expressed the basic symbols of the People, which in turn evoked their sacred stories and beliefs. Both night and day the Lakota lived between stories and symbols written in the sky and mirrored on the earth.[81]

Two constellations are mirrored by the Black Hills. The great circle of stars made up of Rigel, Sirius, Procyon, Pollux, Castor, Auriga B, Capella, and the Pleiades is mirrored by the red clay valley that encircles the entire Black Hills: the Race Track. All this has been explained by anthropologists working in conjunction with tribal elders and others. They also explained an apparent anomaly. Bad Heart Bull had represented pictographically the distinctive, free-standing, volcanic plug known now as Devil's Tower, but placed it well within the Black Hills and not, as it is, sixty miles away on the plains to the northwest. Topographically wrong, he was, however, theologically correct. The constellation correlated with Devil's Tower is located within the great circle of stars. On reading this explanation, I was ashamed of my own notes made many years before. I had accepted a published opinion that the tower was wrongly located and proceeded to question the authenticity of the drawing.[82] Indian maps are particularly likely to deceive whites when they appear to be most like maps made in the European tradition. Without the involvement of the Lakota people, much about Amos Bad Heart Bull's map would have remained confusing and opaque.

It is reassuring to know that the Lakota still possess hides with stars and topography painted on them. How long they have had them is not known, though according to an elderly informant,

> When our grandfathers came onto the reservation [presumably the Pine Ridge or Rosebud in or soon after 1878], they had three things: two hides and them sticks. One hide was a star map, the other hide was an earth, "maka," map—buttes and rivers and mountains even creeks clear out to Colorado Springs. Star map and earth map, they were really the same, because what's in the stars is on the earth, and what's on the earth is in the stars. Them sticks were used for time, for telling time...."[83]

Unfortunately, the transcript terminates before the elder's statement ended. It would have been interesting to know how and in what context the hide maps were used.

More is known about the Pawnee star chart than about the Lakota star and earth maps, though it has also been the object of recent reinterpretation. The archaeoastronomer Von Del Chamberlain has personally examined it and used polarized microscopy to identify the pigments. He supplements rather than takes issue with previous identifications of stars. Of more interest here are three of his concluding observations. First, "It is interesting to note that the star patterns that appear on the chart are generally reversed from their actual appearance in the sky.... [T]his is not surprising, since it is not uncommon for people to reverse patterns when they are transposing from overhead view to a flat sketch."[84] Presumably by "reversed" Chamberlain was referring to the phenomenon known to psychologists as lateral—or mirror—inversion.[85]

This may have some bearing on the second of Chamberlain's concluding observations: "When examining the Skidi star chart, one may be disappointed by the fact that there is no clear reflection of the village patterns described by Fletcher, nor is there a perfect rendering of all of the Skidi philosophy that we have speculated upon."[86] He was here referring to classic early work by the anthropologist Alice Fletcher. Writing at the beginning of the century, she had quoted an Indian informant: "The Skidi were organized by the stars; these powers above made them into families and villages, and taught them how to live and how to perform their ceremonies."[87] It is oversimplistic to consider the star chart simply as a mnemonic for use in conducting ceremonies, organizing lodges, positioning lodges within villages, and positioning villages vis-à-vis each other. A clearer understanding by future interpreters of the various kinds of inversion might clarify some of the issues relating to supposed correlations between the celestial world and the organization of Pawnee society on earth.

The last of Chamberlain's concluding observations suggests otherwise. The chart

> was an object that was of great spiritual meaning to a vanished priesthood; a marvelous painting, beautiful in its simplicity, on a canvas grown by nature as a covering for a living creature; and an unusually symbolic work of art, the full intended purpose of which will always elude us. We can continue to

speculate about it; no authority remains to interpret it for us.[88]

Without the help of appropriate native authorities it is indeed difficult for non-natives to interpret maps that are scaleless, lack the equivalent of keys, and do not contain written words or conform to the conventions of maps made in the European tradition. Yet the alternative option is fast disappearing, as the last native authorities and keepers die. One can only speculate about how many and which tribes had map-based belief systems of which we know nothing. There is certainly no reason to suspect that the Delaware, Southern Ojibwa, Lakota, and Pawnee were exceptional in having had one.

Occasionally, anthropologists have made latent contributions to the understanding of native maps. Because they are latent it is impossible to assess their number or evaluate their overall importance. Often their latency has had to be realized by others. In the early 1970s an attempt to reconstruct the subsistence strategy among the Weagamow Ojibwa of northern Ontario during the fish and hare period (1880–1920) led to an Indian making drawings of eight types of rapids. Each was in plan and incorporated a mix of essential characteristics. These included length, width, bed and flow conditions, canoeability, potential for fish, accessibility for predatory mammals, and suitability for constructing fish traps.[89]

The general appearance and adaptability of the Weagamow Ojibwa's plan pictographs are reminiscent of symbols used to represent rapids on transcripts of maps made by sub-Arctic Indians in the course of early encounters. One forwarded from Lake Superior in 1728 or 1729 to Charles, Marquis de Beauharnois de la Boische, in Quebec by Pierre Gaultier de Varennes et de la Vérendrye is a good example.[90] In the part of the transcript acknowledged to be based on a Cree map, La Vérendrye used keyed "marques" to distinguish between rapids, large portages, and small portages and to indicate the side of the river on which the portage was made. The "marques" share several characteristics in common with the drawings made by the Weagamow Ojibwa in the early 1970s. The Cree and Ojibwa were and remain culturally very similar. Falls, rapids, and associated portages were, and to a lesser extent are still, important to their way of life in a land of complex river systems and many lakes. It would not, therefore, have been surprising if they had had an adaptable pictography for representing these: pictographic plans, often, but not exclusively, for use on maps. Early-encounter transcribers may well have missed some of the graphical nuances when copying them, in much the same way that the cartographic significance of the Weagamow Ojibwa's drawings appears to have gone unnoticed for some twenty years.

Sometimes anthropologists have used native map evidence in attempting to establish principles. Scapulimancy is a good example of this. A form of divination, it involved the inducement on mammalian bones (usually the scapula) of random patterns of cracks and burns.[91] In North America the practice is best recorded among the Montagnais and Naskapi, but it is a circumpolar trait and, as such, has attracted considerable interest. Most anthropologists have claimed that, in part at least, the induced patterns were interpreted as maps. Divination was to some extent based on cartographic decoding of the oracle's message. Hunting was by far the commonest context in which this kind of divination was practiced. Not surprisingly, therefore, the ecological and social functions of scapulimancy have become matters for research and debate. Omar Moore has concluded that its use among the Naskapi randomized their hunting patterns and that as a consequence of this the caribou could not learn to anticipate the spatial strategies of the hunters.[92] Georg Henriksen thinks it probable that the same tribe practiced scapulimancy only in critical hunting situations, when it externalized the decision as to where to seek caribou. On failing to find them, good hunters could then explain the lack of success as being the fault of the shoulder blade. Consequently, it became easier for them to take the initiative in subsequent critical situations.[93]

Adrian Tanner reduced the emphasis on the map-reading aspect of scapulimancy. Working with the Mistassini Cree of central Quebec, he concluded that using environmental information, hunters were usually already aware of where new game was to be found. "The kind of information divination [including scapulimancy] gives is just that kind that cannot be known in advance from an examination of environmental signs. Divination fills in gaps in knowledge, which cannot be learned from the environment."[94] Tanner is somewhat vague as to whether filled gaps can be spatial. His own reports of specific scapulimancies, however, rather suggest otherwise. Although dominant opinion claims or assumes that scapulimancy in part involves a process akin to map interpreting, the matter is still one for research and debate.

Only very occasionally has a native map been used by an anthropologist in the way in which Gregory Waselkov has used them as sources of archaeological and ethnohistorical evidence. The one notable exception is a long, scholarly, and insightful paper by June Helm, a specialist in sub-Arctic anthropology.[95] Notwithstanding its many qualities, it appears to have been written almost as an aside. As volume editor of the *Subarctic* volume in the Smithsonian Institution's *Handbook of North American Indians* series, she had originally intended "to prepare a brief consideration" of a well-known eighteenth-century Chipewyan map "to be included in the chapter by [Beryl] Gillespie. But it became obvious that there was no way to keep it brief. It was the occasion of holding the Rupert's Land Research Centre Colloquium at Fort Churchill [in 1988] that spurred me to dig back in my files and work up the . . . paper."[96] Regrettably, June Helm's letter ends with the statement, "My work on Matonabbee's map was a one-time excursion into native maps and mapping."[97] Informed once-only contributions have unfortunately been common among recent and current encounters.

June Helm accepted a challenge laid down by the historian of exploration Glyndwr Williams, that Meatonabee's map "presents a fine puzzle to the investigator."[98] When "informed by ethnographic and ethnohistorical knowledge of the major water routes and markers followed by Chipewyan, Copper (Yellowknife), and Dogrib Indians past and present," she convincingly demonstrated first that "the features on the map, named and unnamed, can be identified" and second that the map can be "'deconstructed' to align those features with their actual topographic orientations and locations between Hudson Bay, the Churchill River, Great Slave Lake, and the Arctic coast from Coronation Gulf to Chantrey Inlet."[99] She revealed the remarkable achievement of the two Indians in making a map of what is still a little-known part of Canada that makes up approximately 10 percent of the nation's territory. Her deconstruction was distinctive in two features: the use of carefully constructed maps, and the ethnographic insights behind some of the interpretations. Four line maps illustrated the three-stage deconstruction from the original to a modern map. This process involved first "tilting," in order "better to conform to the cardinal orientations" of certain features, followed by the cutting, rescaling, and reorienting of waterways.[100] Effective and welcome though this was, the intuitions behind decisions to cut, shorten, and reorient were the more noteworthy. A few extracts illustrate their perceptiveness.

> The "puzzle" of the map endures only if the investigator tries to work his/her way inland from the map's coast of "Hudsons Bay," a coastline that extends north from Churchill for the entire length of the map to the mouth of the Coppermine River. The place to start in reading [the] map is where a widely travelled Chipewyan or Dogrib or Copper Indian would—from Great Slave Lake itself. The oversized depiction of the lake as the hub of the map suggests that the two Chipewyan leaders [the map's makers Meatonabee and Idotlyazee] saw it as just that. If one views the water routeways outwards to the south, east, and north from Great Slave Lake from the perspective of ethnographic and ethnohistorical knowledge of the major routes followed past and present by these Dene [speaking] peoples, the progressive truncations—or, in the case of the Coppermine River, relative elongation—and directional distortions of the routes as they proceed to the straight-line coast become evident. The routes are there, but as they proceed away from Great Slave Lake their scale and relative directions become progressively skewed.[101]

Farthest away from Great Slave Lake were the west coast of Hudson Bay and the Arctic coast of the mainland, the two represented together on the map as one, continuous, essentially straight line. Of this complex coastline, with its many estuaries, bays, headlands, and islands, "the Indians had at best only a very intermittent knowledge, observed or reported, in contrast to the topography of the Great Slave Lake hub."[102] It was the representation of an aspect of the Hudson Bay coast that presented June Helm with the "only severe problem in accounting for every feature on Matonabbee's map and matching each to a modern map":

> [It] occurs ... with the two entries titled [in the hand of Moses Norton, factor at Churchill from 1762 onward]... "Bowdens Inlet" and ... "Ye Grand Fish river."... To my mind, these can only be interpreted as duplications of, respectively ... "Little Head River" and ... "Sturgeon River" that lie sequentially on Matonabbee's map [to the] north of [them]. And I can only conclude that these duplications result from Moses Norton interposing into the Indians' original mapping his own firsthand knowledge of Chesterfield Inlet, which is the Bowden's Inlet of the map, and, as well, a "Grand Fish River" reported by earlier Chipewyan informants.[103]

Having detected the duplication, June Helm offered an explanation based mainly on linguistic evidence. Norton spoke English and Cree, but only a little Chipewyan. Meatonabee spoke Chipewyan, some Cree, but almost no English. Conversation between them would, therefore, have almost certainly been in Cree. Norton already knew of a Grand Fish River as reported by earlier Chipewyan informants. With an inadequate command of the Cree language, Meatonabee probably used the Cree for "sturgeon" (*numao*) to convey the concept of a big fish. Norton, whose mother was a Cree and whose duties at Churchill brought him into almost daily contact with Cree speakers, interpreted this literally and concluded that there were two rivers rather than one: a Big Fish and a Sturgeon. Notwithstanding a degree of speculation, this is a level of informed hypothesis formulation rarely equaled by interpreters of native maps. It was not, of course, a type of hypothesis that could have been formulated for a truly indigenous map artifact. Devoid of toponyms and inscriptions, linguistic evidence would not have been applicable.

June Helm observed that "the outline of Great Slave Lake is so proportionately realized as to be immediately recognizable. One marvels how their minds' eyes grasped the shape of this great body of water, some 300 mi long and encompassing over 10,000 mi^2."[104] Considerably more extensive than Massachusetts or Wales and shaped more complexly than either, this was indeed something to marvel about.

We are far short of being able to explain how, when necessary, some Indians could produce remarkably good plans of large and irregularly shaped lakes and islands, without a knowledge of absolute metrics, without employing the equivalents of trigonometric survey, and without having mechanisms from which to observe from above.[105] Most likely, the ability was based on experience of travel, probably as aggregated over several generations.

Such relatively well shaped maps pose questions very similar to those raised by some medieval European maps. Richard Gough's mid- or late-fourteenth-century map of Britain, for example, has an outline shape immediately identifiable with the surveyed shape of mainland Britain as now known. There is little understanding of how this was achieved, but, though not scaled accordingly, knowledge of distances along a few routes may have been taken into account.[106] Indians may have achieved something close to this in the course of canoe travel across and around lakes.

Inuit used "sky maps" to detect patterns of ice and water located beyond the horizon:

> A distant body of open water in the sea ice will often cast a dark shadow in the clouds above, producing a "water sky," while ice lying over the horizon often produces a soft white reflection in the air above called "ice blink." The term "sky map" refers to either phenomena or to the *pattern* [italics mine] of light and dark they create together in the sky. A very discerning eye can distinguish among several sorts of ice blink. Snow-covered land appears yellowish white in the sky. Field ice is a lucid white, tinged with yellow. Pack ice is pure white. Sea ice in embayments is a grayer white.[107]

It is unclear whether sub-Arctic Indians used this form of remote sensing. Even if they did, it is not known whether sky maps would have been an aid in making maps of extensive terrestrial and hydrological patterns. Within the range of physical sensing methods, however, the possibility must not be overlooked that such unusual ones might have been used.

Though almost incomprehensible to the Western scientific mind, many Indians had, and some continue to have, extrasensory experiences of terrain. How topographically detailed these were and to what extent they involved vertical perspective is not clear, but, for example:

> Like many northern hunting peoples, the Dunne-za or Beaver of northwest British Columbia used dreaming to know and to gain power in the world.

Individuals dreamed their hunting trails which they [then] followed in waking life. Strong dreamers could see sets of animal movements that would enable them to arrange collective hunts. They might also travel the trails of time, seeing moments of the past and future....

When Dunne-za dreamers heard about the Christian ideas of the afterlife, they began to make dream visits to heaven. There they saw relatives who had died, and became familiar with the route. If salvation depended on finding the route, then powerful dreamers could help others by teaching them how to recognize the way. So they made maps, marking the right trail, identifying the wrong ones.

Here were the important maps. Guides to the most elusive terrain, both descriptions of and metaphors for the ultimate journey. All other maps, perhaps, are fragments of this ultimate; signposts to where the route begins.[108]

"All other maps" presumably included the terrestrial, celestial, and, perhaps, cosmographical maps with which this book is almost exclusively concerned. Though important "signposts to where the [ultimate journey] route begins," the topographic exactitude with which terrestrial ones could be dreamed is unclear.

There were doubtless more powerful dreamers in the past. Whether either Meatonabee or his mapmaking associate Idotlyazee were included among these is not known, though, as Chipewyans, they were near neighbors of the ancestors of today's Dunne-za dreamers. For the time being, and perhaps forever, their remarkable shaping of Great Slave Lake (and similar achievements by some native people elsewhere in North America) remains both a "marvel" and part of Glyndwr Williams's "puzzle."

Like almost all who have written about native maps, mapmaking, and related intracultural activities, anthropologists have tended to adopt a Eurocentric perspective on them. Epistemologies, methodologies, and motivations have varied but very few non-natives have tried consciously to establish native perspectives on such topics as space, terrestrial patterns, orientation, navigation, maps, and mapmaking.[109] In chapter 11 cultural anthropologist Peter Nabokov announces in his title an intention to attempt precisely this. "Orientations from Their Side" brings together much of what is known about material representations of space in three native spheres: the domain of power, the realm of cosmology, and the conduct of life. The spatial representations are the media for five types of cartographic discourse, conducted within Amerindian and Inuit societies in the present, in the recent past, and perhaps in the more distant past. Most of this book is concerned with one only of the five: "depicting terrestrial or cosmological environments." With the exception of Elizabeth Boone, in chapter 5, very little attention is paid by the other authors to material representations of space in four equally important areas of discourse: "as a mode for cross-cultural argument, a mirror for collective self-expression, a rhetorical device for staking out social or diplomatic positions, or a visualization technique often used in conjunction with oratory or storytelling for the charting of proper behavior or spiritual development." Interestingly, Peter Nabokov finds only "glimpses" of these cartographic discourses "within American Indian traditions." It is in encounters with whites "over issues of land, political power, and cultural authority that they acquire higher profile" (chapter 11, p. 241) If this observation is valid, it raises questions about the nature, roles, and significance of cartography in pre-encounter Amerindian and Inuit societies. To what extent did the encounter process alter the forms, functions, and status of spatial representations that, until now, we have assumed were indigenous?

Nabokov has also introduced a corrective to this book's almost inevitable tendency to restrict the concept of map to spatial representations on essentially plane surfaces. Without organizing chapter 11 in terms of morphological structures and behavioral patterns, he draws his evidence from a variety of these: shrines, rituals, tipi paintings, architectural structures, sandpaintings, labyrinth designs, and artificial earth mounds, to name only a few. Diversification in awareness of forms, as well as of contexts, is essential if whites are to modify assumptions concerning the better-known, flatter varieties of Amerindian and Inuit maps, isolate and correct false interpretations of these, and attain a richer understanding of the role of space in native life and beliefs. Anthropologists with appropriate regional and cultural specialisms are the groups most likely to further diversification, and expand contexts.[110] From Elizabeth Boone's thorough survey of Aztec mapmaking in chapter 5, we learn the full extent of what more sensitive readings of Native American maps can reveal. She has convincingly shown that Mesoamerican cultures used artifactual maps in a variety of contexts and for multiple purposes, including the documentation of community histories and the military and legal defense of territory. It is hoped that each of these authors will continue to provide leadership, drawing further stimulus from research in traditional cultures elsewhere in the world.[111]

Recent Creative Cartographic Encounters by Native Peoples

Whereas non-natives working within the histories and anthropology have dominated recent critical and analytical encounters, native peoples have been making their own cartographic responses to the ongoing encounter. These have been in two contexts: mapmaking as a component of commercial art; and bioregional mapping, most of it in the course of mounting legal challenges against government control over their traditional territories. Of the two, the former is closer to traditional modes of mapmaking, but the latter reveals more about native knowledge of terrain, environments, and resources, and is a more direct response to change. Neither of these mapmaking contexts has generated the literature it deserves and much of what has been published is obscurely placed. Each merits a review beyond what is possible here.

The incorporation of maps in native commercial art made for white patrons had its origins in the late nineteenth century. The Alaska Commercial Company fostered the souvenir trade by supplying walrus ivory to Alaska coast Eskimo carvers and engravers and buying the finished products for their museum in San Francisco. In 1897 and 1898, demand for carvings and engravings was boosted by the Klondike and Nome gold rushes. The resulting art assumed several forms, but much of it consisted of pictorial engravings. The Eskimo had a long indigenous tradition of engraving on ivory and evinced a talent for copying illustrations and objects in two dimensions. "The first two decades of the twentieth century were probably the most bizarre in the history of Eskimo art in the transfer of subject matter from one medium to another."[112] Among the frequently transferred matter European-style maps ranked high. Nostalgic reminders to traders, miners, and, later in time, tourists, they were engraved on all manner of carved ivory objects. Of these, cribbage boards were perhaps the most common. Almost all the engraved maps were of coastlines; straightened somewhat and generalized as necessary to fit the gentle curve of the tusk or the more rectilinear forms of carved objects such as letter openers.[113]

Much more indigenous than the maps engraved by Alaska coast Eskimos were the celestial charts incorporated in Navajo textiles made for white collectors. As with the Alaska Eskimo's commercial engravings, the technique was an extension of a long craft tradition. Unlike the Eskimo's pictorial images, however, the designs were based on indigenous forms; particularly those incorporated in sandpaintings. Made in the course of religious healing ceremonies by or under the supervision of medicine men or women, these were made on the floors of hogans, using finely ground minerals and organic materials to create the required patterns. They were traditional, numerous, and complex. Toward the end of the nineteenth century, some of them began to be incorporated in textiles. At first, this was considered to be offensive to the Holy People and, hence, potentially dangerous. By circa 1920, however, they were beginning to be woven openly, in response to an ongoing economic encounter that was gradually weakening the traditional reliance on a resource-based economy and replacing it with one that was increasingly cash based. By 1917, Hosteen Klah, a respected medicine man, was beginning to weave sandpainting designs for sale.[114] His "Hailway: 'The Night Sky'" of circa 1920, woven in natural wool, incorporated a conventionally stylized Milky Way and several quite realistically represented constellations.[115] From the Euro-American perspective this came close to being a celestial chart; albeit fragmented, stylized, and selective. In the Navajo tradition from which it stemmed, it was "the small scale reconstruction of that part . . . of the universe . . . vis-a-vis which the 'patient,' [according to the medicine man's diagnosis, had] caused disorder or disharmony."[116]

Canadian Inuit developed forms of commercial art somewhat later in the twentieth century than the Alaska coast Eskimos. Although they were well known during the nineteenth and early twentieth centuries for the maps made on paper for Arctic explorers, whalers, and collectors on behalf of museums, no evidence of this was formally detected in their market art until 1988; though it was there for the perceptive viewer. In that year Robin McGrath published a study in which, having considered two somewhat unusually decorated late-nineteenth-century Inuit outline maps made for a whaler-collector, she proceeded to seek evidence of cartographic forms in prints made commercially for sale. Though soundly based, the search was not unbiased. She was apparently already aware that Inuit literature, for example, revealed "that even the most modern forms of composition—religious hymns, popular songs, radio plays, and science fiction—are firmly rooted in the old oral tradition, with themes and structures from the song duels and legends of ancient Inuit composers," and that "Inuit carvers and printmakers, seamstresses and designers all draw on the traditions of their ancestors to give meaning and integrity to their work."[117]

Perhaps not surprisingly, but nevertheless convincingly, Robin McGrath found considerable evidence for cartographic elements in commercial Inuit prints produced after the 1950s. The dominant characteristic of these was of events involving humans, animals, and human-built structures presented in profile against patterns of coasts, rivers, and trails in plan. By the 1980s there is some suggestion that the cartographic components were becoming more stylized, and thus less immediately recognizable. Robin McGrath offered an interesting interpretation of one of these: Pudlo Pudlat's "The Settlement from a Distance" (Cape Dorset, 1982). It shows houses in profile on land with relief and bounded by a coast. The sea beyond is confirmed by a ship in profile. Further out to sea are what could be islands with houses on them, but McGrath offered an "alternative interpretation ... that these are not islands [but] an artistic interpretation of the phenomenon [called] 'sky map.'"[118]

The ability to incorporate maps in two-dimensional art was not restricted to those who, by native ability or training, had become commercial artists. The Canadian photographer Marlene Creates spent 1988 in Labrador with elderly country people: Inuit, Naskapi Innu, and Euro-Canadians. For each of eighteen of these old people she produced an "assemblage" consisting of a photograph of the person, a story told in their own words, a "memory map" drawn at her request to show "how they remember their environment," a photograph of one of the landmarks on the map and, where appropriate, a photograph of an object from the landscape.[119] None of the old people were artists but, as a set, their memory maps, all drawn very simply in pencil on paper, were revealing. Nine represented coasts and buildings in plan; by contemporary Euro-American conventions, at least, four were very unmaplike or unclassifiable; and five represented coastlines in plan together with other features in profile. Quite clearly, slightly more than one quarter of these old people made maps according to the same principles as some more or less contemporary Inuit commercial printmakers. Stated another way, more than a third of those revealing an ability to make a drawing recognizable as a map, did so according to the principles of both contemporary native commercial artists and also many of those Arctic folk who, several generations before, had assisted Europeans in their search first for the Northwest Passage, later for Sir John Franklin, and last of all for whales.

During much the same period that some Inuit printmakers were incorporating cartographic content in their commercial art, others were making maps in the process of attempting to reclaim the rights to lands that they felt their ancestors had lost to whites. It was a renewal and it was hoped, a reversal of encounters lost by default many decades before. In 1973 and 1974, approximately 1,600 Inuit made maps as part of the Canadian government-financed Inuit Land Use and Occupancy Project. Each hunter compiled what was known within the project as a "map biography" to show the areas he had hunted, trapped, fished, and camped in during his adult life. Because the information recorded in the map biographies was to be integrated into a land-use study of the whole of the Northwest Territories (approximately 1,250,000 square miles; almost one-third of Canada by area), they were made according to standardized procedure. Hence, they were in no way indigenous in form, content, or style. Almost 150 people administered the work in thirty-three settlements. Topographic base maps were provided at scales of 1:500,000 or larger. Individuals were asked to identify land-use areas for each appropriate category of animals and plants from a list of thirty-three major types and many more subtypes. Felt-tipped pens with many colored inks were used to distinguish between these. Wherever possible, steps were taken to corroborate this data. In short, fairly high scientific field standards were imposed. For this reason alone the maps themselves hardly merit attention in the context of this volume, but several aspects of the mapmaking process do.[120] The published summary of these provides perhaps the best evidence we have with which to reconsider white assumptions about maps made by individual natives long dead. Factors including memory attrition, native perception and categorization, the communication of technical data across cultural boundaries, the language problem, and the concern for accuracy are each considered; albeit in a mid-1970s context and on the basis of a narrow (male Inuit hunters) but large (approximately 1,600) sample.

The attempt to obtain memory maps for the years before the local arrival of traders (pre-1912 to pre-1948, according to region) proved less successful than for later periods. "Much of the information obtained for this earliest period occurred in general statements, rather than specific, personal information, and resulted in large part from the need felt by informants, not to tell the literal truth . . . , but also to be seen to have told the truth through subsequent, independent corroboration."[121] Many of the maps made by Indians and Inuit in the frontier encounters between 1511 and 1925 (see chapter 1) were made by old men. Whereas the experience of the Inuit Land Use and Occupancy Project provides no evidence that the elderly drew inferior maps of coasts, rivers, and terrain, it does lead to doubts about the reliability of representations of events that occurred in their early lives.[122] The corpus of Inuit maps, and accounts thereof, obtained in the course of the several searches for the lost Sir John Franklin expedition should certainly be reexamined with this in mind.

Among the perceptual issues affecting the project's attempt to obtain map biographies the concept of hunting range raised the most serious problems. The aim was to "chart hunting ranges, in both their core and peripheral areas, and indeed to include any other terrain where a hunter has tried, successfully, or not, to find game [but] hunters describe[d] their ranges on paper in radically different ways." Some men indicated enormous areas, others comparatively tiny locales; the latter indicating either where kills of caribou had actually been made or the core area of the herd.[123] In the more than four hundred years of frontier encounters with Indian and Inuit maps after 1511, many were made in response to requests by whites. Corroboration was rarely possible or attempted. At the time, many significant perceptual differences must have gone undetected.

The project's field staff became aware that species that had low status and wide geographical range and that did not have year-after-year locales (ptarmigan, berries, and sculpin for example) tended to be minimized by hunters as they drew their maps.[124] This generalization tends to confirm a characteristic of many native maps; emphasis on the unique, rare, and occasional to the exclusion of the ubiquitous, undervalued, and frequent.

It was a basic requirement of the Inuit Land Use and Occupancy Project that all land-use categories be recorded on map biographies as circumscribed areas. Many hunters, however, were unable to

> perceive their hunting in terms of areas which could be so represented. There is no clear edge to where hunters look for caribou on the tundra—one year they go farther than other years. They do not think of the farthest they have been in each direction as the effective perimeters of a hunting range. Rather, these hunters tend to think of travel lines, routes across the land, around which, or alongside which, or at the end of which, they expect to find the game which they are consciously pursuing. Looking at a map a man can point to the routes—especially in difficult travelling country where much the same passes, valleys, and fiords would always be used—and indicate where along these routes he most often found a particular species. When asked to indicate a hunting range, then, many hunters liked to put in lines and loops, not circles [boundaries]. Since the finished maps had to be composed of circles, the hunters were urged to use circles. As a result they tended to mark inner hunting areas—the favourite spots, where kills had been made, the core areas—rather than outer perimeters. Thus the methods the study adopted tended to generate understatement of land use.[125]

Although made with reference to hunting territories only, this reluctance, verging on inability, to circumscribe areas (generalize regions) in part explains the extreme rarity of boundaries on Indian and Inuit maps. It also calls into question the intended meaning of these few boundaries that are represented, and whether or not they are interpretable. It could be that because they were so mobile, Inuit hunters had more difficulty in drawing boundaries (limits, regions) than more sedentary natives with limited territories, of which they had more intimate knowledge.

The vocabulary and syntax with which a question was asked was found to affect the cartographic response. The question of where natives had been "working at (i.e., killing) hare" elicited a map only of "locations where a hunter actually engaged with animals . . . [a] map of kills." In the case of bear hunting, that same type of response was an extremely serious error, since very large areas were hunted over but only small numbers of bears were actually killed. The question of where natives had been "seeking for" a given species resulted in a spectrum of interpretations between very restricted and all-embracing. The request to indicate "a place for finding," produced maps of favored hunting grounds.[126] Given the influence of the form of verbal requests on cartographic responses under the more or less controlled conditions of the project, it is debatable whether older maps made in response to unreported requests, and perhaps in a European language or possibly by gesture, can ever be used for purposes of historical reconstruction. What value, for example, can "Melikis Map of musk ox hunting . . ." inland from Wager and Repulse Bays in the years before 1898 have as a source of historical biogeographical data?[127]

Project workers in the field found a very high level of honesty among the Inuit hunters: "Respondents took enormous pains to be accurate

[and] it can safely be said that accuracy and honesty were in virtually every case beyond doubt." An observed consequence of this was that if, after consulting others, a hunter remained unsure as to where to place something on a map "he was inclined to leave it out."[128] Honesty and integrity would seem to have been an almost universal characteristic of native North Americans, at least in the earlier stages of encounters with whites.[129]

There is a very real possibility therefore that when viewed retrospectively by whites, what appear to have been inexplicable omissions from maps may have been negative expressions of truth, the silent equivalents of dotted lines in the *terrae semicognitae* on many European maps compiled during the Age of Exploration. From the perspective of Europeans and Euro-Americans, silences may yet emerge as the most troublesome problem in interpreting native maps.[130]

The kind of mapmaking done by male hunters in the course of the Inuit Land Use and Occupancy Project has since been done in other native societies. Doug Aberley, in a book subtitled *Mapping for Local Empowerment,* called it "bioregional mapping" and mentioned similar projects within North America; among the Inuit in Labrador in the mid-1970s, the Dunne-za (Beaver) of northeast British Columbia in the late 1970s, and later the Nisga'a of northwest British Columbia. In the 1980s, the Gitksan and Wetsuweten, near neighbors of the Nisga'a, produced an atlas of maps representing how they inhabited and stewarded their territory.[131] This or related forms of mapping for empowerment have been "exported" and used recently by members of traditional societies in other parts of the world.[132] Most, however, seem to have involved plotting data on the best available topographic maps. The Nisga'a actually "purchased state of the art Geographic Information System (GIS) computer software and are now digitizing satellite images of their territory to defend sovereignty, and aid in the stewardship of locally controlled forests, fisheries, and other resources."[133]

A recent review of past and present bioregional mapping by native peoples worldwide contains a digest of sixty-three projects. Twenty-six are North American; mostly initiated in the 1990s. In these, "basic mapping" (ephemeral mapmaking on the ground and map sketching), characteristic of some of the pre-1990s projects, has been superseded by the use of aerial photographs, satellite imagery, and a wide range of geographical information systems. Increasingly, the new techniques were being used by native specialists. Ironically, sophisticated nontraditional technology is now being used by members of traditional societies in the course of trying to conserve aspects of their traditional environments.[134]

In a paper published in the same year as the review, Robert Rundstrom considered some of the consequences of geographical information systems (GIS) for indigenous peoples in general, albeit mainly on the evidence of recent North American experience. He expressed skepticism "about motivations for inscribing indigenous geographical knowledge into GIS," concluding with the observation that "the institutions of assimilation that have silenced indigenous peoples throughout history have been those of the state, education, and religion. Clearly, GIS has been promoted by agents of the first two, with a missionary zeal exceeded only by agents of the third."[135] Though he gives cogent reasons for his skepticism, Rundstrom does not indicate an alternative future for "indigenous geographical knowledge." If it is not to be in the context of geographical information systems, will it develop independently in directions that cannot now even be guessed at, or will it be a modified revival of geographical lore and traditional cartography?

All this is far removed from perhaps the earliest example of bioregional mapping. In the early 1920s Frank G. Speck established the family hunting districts of Montagnais-Naskapi bands in the central and southeastern Labrador peninsula. "It was . . . upon the geographical knowledge retained in the memory of the Indians and their ability to demonstrate it on the inadequate charts only available at the time, that the demarcations of hunting grounds were based." On the evidence of the two examples reproduced by Speck, some at least of the Indians were drawing their own base data on which to plot their hunting grounds.[136]

Chapter 4 is by far the longest in this book. Its length reflects the recent growth of interest in Indian and Inuit maps and mapmaking and a diversification of contexts in which maps have been recognized and perceived to have significance. Chapters 5–11 are presented as independently prepared contributions to these developments.

In the course of this chapter an attempt has been made to place each of the seven contributions in the context to which it seems most appropriate, but, like categorization, contextualization is little more than an organizational convenience. Time alone will reveal the context or contexts in which each contribution proves to have been significant.

I hope each will be an important point of departure rather than a narrative of a journey that is coming to an end. It is I, in chapters 1–4, who have risked, but I hope not fallen foul of, the latter.

Notwithstanding the range of topics covered in chapters 5–11, we are aware of omissions. Some reflect oversights in planning and others are consequences of disappointments along the way. Among the most serious omissions are topics for which we knew of no potential authors.[137] Perhaps the most serious of all are the ones of which we still remain unaware. Failure to cover systematically any aspect of Inuit maps and mapmaking is particularly regrettable. The historical bias reflects the dominant thrust of recent work, but more contributions by cultural anthropologists would doubtless have resulted in new insights and suggested new lines of investigation. The absence of a contribution by a cognitive scientist verges on the unforgivable. In chapter 12 I will isolate other lacunae in the course of attempting to predict future developments. Readers must not misconstrue this apologia as in any way undervaluing what is included in part 2.

The Introduction stresses the freedom given to each of the seven authors of the chapters in part 2. It also admits to some "inevitable contradictions" between them. More important than the real or apparent contradictions, however, is the extent of their convergence. It is perhaps one of the best indications we have for a field that after haphazard infancy, followed by a long but restricted childhood and a somewhat aimless adolescence, is at last entering a vigorous and it is to be hoped integrated adulthood able to relate fully within the wider community of knowledge.

Notes

1. Major recent and forthcoming publications include: G. Malcolm Lewis, "Indian and Inuit Cartography," in David Woodward and G. Malcolm Lewis, eds., *Cartography in the Traditional African, American, Arctic, Australian, and Pacific Societies*, vol. 2, bk. 3 of *The History of Cartography* (Chicago: University of Chicago Press, forthcoming 1998); and Mark Warhus, *Another America: Native American Maps and the History of Our Land* (New York: St. Martin's Press, 1997).

2. Among the most surprising are: Hugh Brody, "Maps and Journeys," based on his experience with the Inuit and Dunne-za or Beaver Indians, in a publication on the history of the cartography of part of Scotland: Finlay Macleod, ed., *Togail Tir Marking Time: The Map of the Western Isles* (Stornoway, Scotland: Acair Ltd. and An Lanntair Gallery, 1989), 133–36; a synthesis of much that is known about Ojibwa maps in a Japanese series on geographical thought: Tetsuya Hisatake, "Indigenous Maps, Cosmology, and Spatial Recognition of the North American Indians: With Special Reference to the Ojibway around Lake Superior," in Hideki Nozawa, ed., *Cosmology, Epistemology, and the History of Geography: Japanese Contributions to the History of Geography* (Fukuoka: Institute of Geography, Kyushu University, 1986), 3:1–25; a chapter mainly on Indian and Inuit maps in an Australian publication produced for a distance learning course on nature and human nature: David Turnbull, *Maps Are Territories: Science Is an Atlas* (Geelong, Victoria: Deakin University, 1989), 19–27; D. Wayne Moodie, "The Role of the Indian in the European Exploration and Mapping of Canada," *Zeitschrift der Gesellschaft für Kanada-Studien* 14, 2 (1994): 79–94; and Michael T. Bravo, *The Accuracy of Ethnoscience: A Study of Inuit Cartography and Cross-Cultural Commensurability,* Manchester Papers in Social Anthropology, no. 2 (Manchester: Department of Social Anthropology, University of Manchester, 1996), 1–36.

3. John L. Allen, *Passage through the Garden: Lewis and Clark and the Image of the American Northwest* (Urbana: University of Illinois Press, 1975), figs. 6, 30, 39, 40.

4. Ibid. Allen claimed that the British concept of a continental divide, a central tenet in planning the transcontinental expedition, arose in part from Peter Fidler's 1801 copy of a Blackfoot's sketch map of the upper Missouri (fig. 6). A Minitari war chief's map of much of the upper Missouri was incorporated by William Clark in the cartographic compilation upon which in the spring of 1805 the strategy was based for proceeding beyond Fort Mandan (fig. 30). An Indian map of the Multnomah (now Willamette) River provided information abut a huge, unknown region that was not to be visited by the expedition (fig. 39). A sketch by Meriwether Lewis based on geographical lore offered by friendly Walla Wallas led to a decisive decision on a route to be followed on the return journey in 1806 (fig. 40).

5. James F. Ronda, "'A Chart in His Way': Indian Cartography and the Lewis and Clark Expedition," in Frederick C. Luebke, Frances W. Kaye, and Gary E. Moulton, eds., *Mapping the North American Plains* (Norman: University of Oklahoma Press with the Center for Great Plains Studies, University of Nebraska–Lincoln, 1987), 85, 87.

6. Barbara Belyea, "Amerindian Maps: The Explorer as Translator," *Journal of Historical Geography* 18, 3 (1992): 267. Belyea does not cite John Allen as an example, but authors whom she criticizes for not having "understood that mapping does not represent geographic knowledge in absolute terms" include (n. 3) June Helm, G. Malcolm Lewis, D. Wayne Moodie, David Pentland, and Richard Ruggles. Interestingly, the author Belyea does not cite may have been the first to introduce the concept of the translation of native maps by whites and certainly preceded her by seventeen years. In discussing the use made by William Clark of the Minitari map, Allen observed that "in spite of the difficulties that white men in the West nearly always had in translating Indian geographical information into their own frames of reference,

the contributions by the Minitaris were potentially valuable." Allen, *Passage through the Garden*, 214.

7. See chapter 6, pp. 149–50.

8. See Barbara Belyea's biographical entry in the Contributor's list, p. 265. Michael T. Bravo, another "newcomer," from social anthropology and bringing to bear a grasp of philosophy, has recently endorsed Belyea's position. The endorsement is the starting point for a case study of Inuit cartography and cross-cultural commensurability conceived and presented in the context of "the general question of how scientific practitioners pass judgment on other kinds of knowledge systems." Bravo, *The Accuracy of Ethnoscience,* 1–2.

9. John R. Bockstoce, Review of "Eskimo Maps from the Canadian Eastern Arctic," by John Spink and D. W. Moodie, *Terrae Incognitae* 7 (1976): 81–82.

10. Edmund Carpenter, *Eskimo Realities* (New York: Holt, Reinhart and Winston, 1973), 13, 14, 18, 21.

11. Frederick Turner, *Beyond Geography: The Western Spirit against the Wilderness* (New Brunswick, N.J.: Rutgers. University Press, 1983), 91–92.

12. Delf Norona, "Maps Drawn by Indians in the Virginias," *West Virginia Archeologist* (March 1950): 12–19.

13. Delf Norona, "Maps Drawn by North American Indians," *Bulletin of the Eastern States Archeological Federation* 10 (1951): 6.

14. David Crystal, *A Dictionary of Linguistics and Phonetics*, 3d ed. (Oxford: Basil Blackwell, 1991), 96.

15. Recent papers by James Kari of the Alaska Native Language Center, University of Alaska, Fairbanks, afford good examples of this one-sidedness. Although presenting and interpreting linguistic evidence cartographically, he seems unaware that the Athabaskan peoples whose languages he studies made maps. Yet his themes are eminently spatial, sometimes with a strong temporal component; his evidence is derived through fieldwork; and his analyses rigorously conducted. A recent publication on the distribution of "stream" and "mountain" arises from the observation that

> many of the systematic aspects of Athabaskan geography offer insights into what I regard to be the Athabaskan sense of territoriality in ancient times. The place names are "official" in that they are usually congruent and commonly known from the vantage points of different Athabaskan languages; i.e., for mutually known features there are multilingual pronunciations of the same name.

and concludes by suggesting that

> there are numerous systematic features of Athabaskan geography that convey a sense of the symbolic organization of the Athabaskan landscape. The place names appear in networks and constitute cognitive maps. The place names/signs have had several functions—to facilitate memorizability and way-finding, to mark travel corridors, to represent land use agreements, or to mark band boundaries. Of course, many other features of the languages are geographically grounded—dialect markers as well as lexical fields such as directionals, ethnonyms, and biota terms.

James Kari, "Names as Signs: The Distribution of 'Stream' and 'Mountain' in Alaskan Athabaskan Languages," in Eloise Jelinek, Sally Midgette, Keren Rice, and Leslie Saxon, eds., *Athabaskan Papers in Honor of Robert Young* (Albuquerque: University of New Mexico Press, 1996), 444. Although Athabaskan peoples had no means of writing on them, they did make maps. It would be interesting to know how the presence (including mode of representation and degree of emphasis), frequency, and absence of features confirmed or called into question these and others of Kari's conclusions.

16. See chapter 7, p. 182.

17. Almost exclusively the pre-1612 literature of discovery compiled in David B. Quinn, *New American World: A Documentary History of North America to 1612*, 5 vols. (New York: Arno Press and Hector Bye, 1979).

18. G. Malcolm Lewis, "Native North Americans' Cosmological Ideas and Geographical Awareness: Their Representation and Influence on Early European Exploration and Geographical Knowledge," in John Logan Allen, ed., *North American Exploration*, vol. 1, *A New World Disclosed* (Lincoln: University of Nebraska Press, 1997), 71–126.

19. G. Malcolm Lewis, "Words into Maps," a lecture given at the Thirteenth International Conference on the

History of Cartography, Chicago, June 1993. The so-called "Velasco" map of the northeast coast of North America and the adjacent interior drawn in 1611 is a good example, containing much content acknowledged to be "by the relations of the Indians." Untitled manuscript map, Legajo 2588, folio 22. Estado, Archivo General de Simancas, Simancas, Spain. It is not clear whether the "relations" were received directly or indirectly by the compiler or whether the information originated in graphics, gesture, or speech. Widely scattered, with the most distant elements almost five hundred miles apart, there were certainly two or more native sources and all may have been obtained indirectly. The northern Indian components may well have been constructed from geographical accounts by Indians as recorded in Samuel Champlain, *Des Sauvages* (1603).

20. G. Malcolm Lewis, "The Origins of Cartography," in J. B. Harley and David Woodward, eds., *The History of Cartography*, vol. 1, *Cartography in Prehistoric, Ancient, and Medieval Europe and the Mediterranean* (Chicago: University of Chicago Press, 1987), 51–52.

21. The origins of "mapmaking" (I now prefer this word to "mapping") probably has even deeper roots in the emergence of a concept or concepts of space, which in turn has been linked to the origin of language: B. Chairelli, "Spatial Coordination, Gestural Communication: The Origin of Language and Cerebral Lateralization in Man," *Mankind Quarterly* 29, 3 (1989): 195–210.

22. Klaus F. Wellmann, *A Survey of North American Indian Rock Art* (Graz, Austria: Akademische Druck-u. Verlaganstalt, 1979).

23. Ibid., 131. Recent reviews of evidence suggest that, though Plains Indian ledger art did not emerge much before the 1870s, the tradition of narrative scenes in buffalo hide robes, garments, and tipis goes back further: Janet Catherine Berlo, ed., *Plains Indian Drawings 1865–1935: Pages from a Visual History* (New York: Harry N. Abrams, 1996), 12.

24. Wellmann cites as his authority for this statement David S. Gebhard, *Indian Art of the Northern Plains* (Santa Barbara: Art Galleries, University of California at Santa Barbara, 1974), 13. Although southern Plains margin and not northern Plains, the mid-eighteenth-century Quapaw "Three Villages" painted robe described by Morris Arnold in chapter 8 does not conform to Gebhard's generalization.

25. Robert F. Heizer, *Aboriginal California and Great Basin Cartography*, University of California Archaeological Survey Reports, no. 41 (Berkeley: Archaeological Survey, Department of Anthropology, University of California, 1958).

26. Robert F. Heizer and Martin A. Baumhoff, *Prehistoric Rock Art of Nevada and Eastern California* (Berkeley: University of California Press, 1962), 279, and 394 n. 1 (app. B).

27. Robert F. Heizer and Thomas R. Hester, "Two Petroglyph Sites in Lincoln County, Nevada," in C. William Clewlow, Jr., ed., *Four Rock Art Studies*, Ballena Press Publications on North American Rock Art, no. 1 (Socorro, New Mexico: Ballena Press, 1978), 2–3, figs. 3a, 4b.

28. Lawrence L. Loendorf, "Cation-Ratio Varnish Dating and Petroglyph Chronology in Southeastern Colorado," *Antiquity* 65 (1991): 249 (site 5LA 5830).

29. The panel is dated 450 ± 75 B.P.; that is approximately 1550 A.D., not earlier than 1475 A.D. and not later than 1625 A.D. The expedition of Vásquez de Coronado almost certainly passed not too far to the south in 1541.

30. Lawrence L. Loendorf and David D. Kuehn, *1989 Rock Art Research: Pinon Canyon Maneuver Site, Southeastern Colorado*, Contribution no. 258 (Grand Forks, N.D.: Prepared for the National Park Service, Rocky Mountain Regional Office, Denver, Colorado, by the Department of Anthropology, University of North Dakota, 1991), 226, 228.

31. The most notable exception is the work of historical geographer Louis De Vorsey, Jr., whose studies of the Southeast have often revealed an awareness of Indian maps. Recent examples include "Silent Witnesses: Native American Maps," *Georgia Review* 46, 4 (Winter 1992):

709–26; and "Native American Maps and World Views in the Age of Encounter," *Map Collector* 58 (Spring 1992): 24–29. For a recent in-depth study of a probably Indian and certainly Indian-style map used in negotiations to purchase land, see G. Malcolm Lewis, "An Early Map on Skin of an Area Later to Become Indiana and Illinois," *British Library Journal* 22, 1 (1996): 66–87; also published in Karen Severud Cook, ed., *Images and Icons of the New World: Essays on American Cartography* (London: British Library, 1996), 66–87.

32. In chapter 10, p. 232–37, Patricia Galloway uses French maps to establish early-eighteenth-century Choctaw sites, but makes a case for the original information having been derived from Indian information.

33. Robert H. Lafferty III, "Prehistoric Exchange in the Lower Mississippi Valley," in Timothy G. Baugh and Jonathon E. Ericson, eds., *Prehistoric Exchange Systems in North America* (New York: Plenum, 1994), 201–5.

34. See figure 9.2 in this volume, "Nations amies et ennemies des Tchikachas," supplemented manuscript copy by Alexandre de Batz of a 1737 Chickasaw/Alabama map of Indian nations in Southeast North America, 1737.C13A, vol. 22, fol. 67, Archives des Colonies, Archives Nationales, Paris.

35. Gregory A. Waselkov, "Indian Maps of the Colonial Southeast," in Peter H. Wood, Gregory A. Waselkov, and M. Thomas Hatley, eds., *Powhatan's Mantle: Indians in the Colonial Southeast* (Lincoln: University of Nebraska Press, 1989), 292–343.

36. Gregory A. Waselkov to G. Malcolm Lewis, Mobile, Alabama, July 1, 1993, citing "Fort Loudoun on the Little Tennessee River: A Pictorial Restoration Based on an Archaeological Investigation," in Paul Kelley, *Historic Fort Loudoun* (Vonore, Tenn.: Fort Loudoun Association, 1958), 22. The pictograph to which the correspondence related was described, in part interpreted, and reproduced as a line drawing in William Bray, "Observations on the Indian Method of Picture Writing," *Archeologia* 8 (1782): 159–62. See fig. 1.1, above.

37. William Bray was an English antiquary, apparently with no direct experience of, or particular interest in North America.. By 1781, the fort at the confluence of the Monongahela with the Ohio was indeed Fort Pitt, but it had been built adjacent to an earlier French Fort, Fort De Quesne (1754–58). The pictographic plan is remarkably similar in shape and orientation vis-à-vis the rivers to that on a 1758 manuscript plan: "A Plan of Fort De Quesne," 1758, British Museum, London; reproduced in William P. Cumming et al., *The Exploration of North America 1630–1776* (London: Paul Elek, 1974), 66, fig. 94. Bray's interpretation of the third fort as Fort Detroit was probably an error. In the published line drawing it is square in plan with four square corner towers, and encloses two parallel rows of buildings, these apparently facing an open space at the center. In contrast, Gaspard Chaussegros de Lèry's 1749 plan of Fort Detroit represented it as markedly rectangular, with diamond-shaped corner towers and no central space. Michael Chartier de Lotbinière's 1749 plan of Fort Michilimackinac, however, shows it to have been square, with square corner towers, and a large central parade ground faced by trader's cabins. Modern transcripts of both plans are reproduced in R. Cole Harris, ed., *Historical Atlas of Canada* (Toronto: University of Toronto Press, 1987), 1: pl. 41.

38. Untitled painting on canvas by Sitting Rabbit of the Missouri River from Standing Rock Reservation upstream to the mouth of the Yellowstone River. The river is in eleven discontinuous but relatable sections, 7.07 m. x 0.46 m, State Historical Society of North Dakota, undated accession entry no. 679.

39. Sitting Rabbit, a native of the Fort Berthold Reservation, made at least three paintings for Libby, including the Missouri River map. Libby provided the canvas, certain drawing materials, and a modern base map of the river's course. Though it had been intended, no version of the painted map was ever published. The society's undated record entry describes it as an "archaeological map of Missouri River in North Dakota on canvas." It does not identify the painter, but from the Orin G. Libby Papers in the society's archives an incomplete account of the map can be reconstructed. Sitting Rabbit painted it between late 1905 and July 1906. Libby was dis-

satisfied with the map and apparently returned it to Sitting Rabbit for correcting. Some corrections were made. It is not clear how much Libby paid Sitting Rabbit, but $25.00 seems to have been expected. The last correspondence to mention the map was dated September 18, 1907. These and many other details concerning the circumstances of the map are included in Thomas D. Thiessen, W. Raymond Wood, and A. Wesley Jones, "The Sitting Rabbit 1907 Map of the Missouri River in North Dakota," *Plains Anthropologist* 24, no. 84.1 (May 1979): 145–67.

40. Among other promised essentials, Libby transmitted to Sitting Rabbit "a copy of the Missouri River Survey." The identity of this is nowhere specified but the "eleven individual segments of the Sitting Rabbit map . . . conform exactly with the boundaries of the individual maps of the sectional chart published for the War Department by the Missouri River Commission in 1892–1895 . . . this must be the one Libby provided—in spite of Sitting Rabbit's sometime liberal interpretation of it." Ibid., 147. The map referred to is *Map of the Missouri River from Its Mouth to Three Forks, Montana, in Eighty-four Sheets and Nine Index Sheets* (Sioux City: Missouri River Commission, 1892–95).

41. Thiessen, Wood, and Jones, "The Sitting Rabbit 1907 Map," 164.

42. W. Raymond Wood, *Notes on the Historical Cartography of the Upper Knife-Heart Region* (Columbia: American Archaeology Division, College of Arts and Science, University of Missouri–Columbia, for the National Park Service Midwest Region, Midwestern Archeological Center, Lincoln, Neb., 1978), 2–3.

43. According to ibid., 87–89, 105, the maps were sketched by Gilbert L. Wilson into his fieldnotes for 1908 (figs. 2 and 3), 1909 (fig. 7), and 1913 (fig. 54). These are on file at the American Museum of Natural History, New York, and the Minnesota Historical Society, St. Paul.

44. Wood, *Notes*, 88.

45. Ibid., 2.

46. For a very recent case study in the tradition of Wood et al., see Glen Fredlund, Linea Sundstrom, and Rebecca Armstrong, "Crazy Mule's Maps of the Upper Missouri, 1877–1880," *Plains Anthropologist* 41, no. 155 (1996): 5–27.

47. G. Malcolm Lewis to Gregory A. Waselkov, Sheffield, England, April 27, 1994.

48. Gregory A. Waselkov to G. Malcolm Lewis, Mobile, Alabama, April 28, 1994.

49. See chapter 9, p. 205.

50. See ibid., p. 218.

51. G. Malcolm Lewis, "Indicators of Unacknowledged Assimilations from Amerindian Maps on Euro-American Maps of North America: Some General Principles Arising from a Study of La Vérendrye's Composite Map, 1728–29," *Imago Mundi* 38 (1986): 14–15, developed fully on pp. 15–22.

52. G. Malcolm Lewis, "Misinterpretation of Amerindian Information as a Source of Error on Euro-American Maps," *Annals of the Association of American Geographers* 77 (1987): 542–63; and "La Grande Rivière et Fleuve de l'Ouest: The Realities and Reasons behind a Major Mistake in the Eighteenth Century Geography of North America," *Cartographica* 28, 1 (1991): 54–87.

53. Ronald Goodman, *Lak̇ota Star Knowledge: Studies in Lak̇ota Stellar Theology* (Rosebud, S.D.: Sinṫe Gleṡka College, 1990), 9.

54. Louis De Vorsey, Jr., "La Salle's Cartography of the Lower Mississippi: Product of Error or Deception?" *Geoscience and Man* 25 (June 30, 1988): 5–23, esp. 10, 21.

55. Gary E. Moulton, ed., *The Journals of the Lewis and Clark Expedition* (Lincoln: University of Nebraska Press, 1991), 7:56, 62, figs. 4, 6.

56. "A Map of part of the Continent of North America . . . By William Clark. Laid down by a Scale of 50 Miles to the Inch," manuscript, 29 x 51 in, William Robertson Coe Collection, Yale University, New Haven, Connecticut.

57. John L. Allen gave a completely different explanation. Placing the source of the Multnomah in the same general region as those of the Snake, Yellowstone, Bighorn, Platte, Arkansas, Rio Grande, and Colorado Rivers was necessary in order to conform to the theoretical concept of the continent's major drainage systems

flowing more or less radially from one height of land: John L. Allen, "Pyramidal Height-of-Land: A Persistent Myth in the Exploration of Western Anglo-America," *International Geography* 1 (1972): 396. This was restated by Allen with special reference to the Multnomah: "one momentous and erroneous feature of the imagination which marred Clark's otherwise superb geography." Allen, *Passage through the Garden*, 393–94.

58. "A Map of Lewis and Clark's Track, Across the Western Portion of North America from the Mississippi to the Pacific Ocean . . . Copied by Samuel Lewis from the Original Drawing of W$^{m.}$ Clark," was the mother of the printed maps containing this gross error; published in [Nicholas Biddle], *History of the Expedition under the Command of Captains Lewis and Clark*, 2 vols. (Philadelphia: Bradford and Inskeep, 1814).

59. See chapter 10, n. 31. Locating and then searching through the Delisle papers could reveal much about how, working in their drawing offices, European mapmakers utilized reports of Indian information written by whites on the basis of frontier encounters. There can be few comparable collections. Until someone locates and works diligently with one of them, cautious speculation is the best that can be hoped for.

60. *Council Fire: A Resource Guide* (Brantford, Ontario: Woodland Culture Centre, 1989), 5.

61. Joseph F. Lafitau, *Customs of the American Indians Compared with Customs of Primitive Times* [1724], edited and translated by William N. Fenton and Elizabeth L. Moor, Champlain Society Publication 49 (Toronto: Champlain Society, 1927), 2:130.

62. David McCutchen, trans., *The Red Record: The Wallam Olum; The Oldest Native North American History* (Garden City Park, N.Y.: Avery, 1993), 4–5.

63. George Copway, *The Traditional History and Characteristic Sketches of the Ojibway Nation* (London: Charles Gilpin, 1850), 131–33.

64. James R. Murie, *Ceremonies of the Pawnee*, edited by Douglas R. Parks (Lincoln: University of Nebraska Press, in cooperation with the American Indian Studies Research Institute, Indiana University, 1989), 33–37.

65. Amos Bad Heart Bull (drawings) and Helen H. Blish (text), *A Pictographic History of the Oglala Sioux* (Lincoln: University of Nebraska Press, 1967), 7–10. The absence of anything comparable to Bad Heart Bull's or Crazy Mule's maps among the more than 150 recently published Plains Indian ledger drawings could be a consequence of the principles of selection, but is probably evidence that few of the region's artists incorporated maps in post-1865 work done on paper. See Berlo, *Plains Indian Drawings*; and Fredlund, Sundstrom, and Armstrong, "Crazy Mule's Maps."

66. Although not known by me at the time, its existence had been announced some years before: Selwyn Dewdney, *The Sacred Scrolls of the Southern Ojibway* (Toronto: University of Toronto Press for the Glenbow-Alberta Institute, Calgary, 1975), 182.

67. *Council Fire*, foreword and p. 27.

68. Ibid., 14, including a photograph.

69. Ibid., 5–6, including a photograph.

70. Ibid., 29.

71. Daniel G. Brinton, *The Lenâpé and Their Legends* (Philadelphia: D. G. Brinton, 1885), 165.

72. Eli Lilly, "Tentative Speculations on the Chronology of the Wallam Olum and the Migration Route of the Lenape," *Proceedings of the Indiana Academy of Science* 54 (1944): 33–40.

73. William W. Newcomb, Jr., "The Wallam Olum of the Delaware Indians in Perspective," *Texas Journal of Science* 8, 1 (March 1955): 57–63.

74. This is a gross generalization of deduced and inferred sites as plotted on a series of maps; figs. 4.2–4.18 in McCutchen, *The Red Record*. For an even more rigorous evaluation and damning critique of McCutchen's and similar interpretations, see David M. Oestreicher, "Unmasking the *Walam Olum*: A Nineteenth-Century Hoax," *Bulletin of the Archaeological Society of New Jersey* 49 (1994): 1–44.

75. Jay Miller, Review of *The Red Record*, by David McCutchen, *American Indian Culture and Research Journal* 18, 1 (1994): 187–90, with quotation from p. 189. Oestreicher's more exhaustive critique does not refute the

idea that the Wallam Olum was derived either from a Delware migration legend relating to an origin within North America or perhaps from narrative accounts of long-distance hunting expeditions within the Midwest (Oestreicher, "Unmasking the *Walam Olum*," 31–32). Hence, theoretically at least, the Wallam Olum may still be interpretable spatially.

76. Copway, *The Traditional History*, 131–33.

77. Dewdney, *The Sacred Scrolls*, 57–80, 183–84.

78. G. Malcolm Lewis, "The Indigenous Maps and Mapping of North American Indians," *Map Collector* 9 (December 1979): 27, fig. 3b.

79. See chapter 2, p. 42; and chapter 3, p. 63.

80. Amos Bad Heart Bull, untitled map of the Black Hills, including the drainage on the surrounding plains; ink, pencil, and crayon on paper, probably approximately 31 x 36 cm, drawn between 1890 and 1913; the original was interred in 1947. There are several black-and-white reproductions, of which the earliest is perhaps the best: Bull and Blish, *A Pictographic History*, 289, no. 198.

81. Goodman, *Lakota Star Knowledge*, 9.

82. Bull and Blish, *A Pictographic History*, 289.

83. Goodman, *Lakota Star Knowledge*, 18.

84. Von Del Chamberlain, *When Stars Came Down to Earth: Cosmology of the Skidi Pawnee Indians of North America* (Los Altos, Calif.: Ballena Press for the Center of Archaeoastronomy, University of Maryland, College Park, 1982), 205.

85. Richard L. Gregory, ed., *The Oxford Companion to the Mind* (Oxford: Oxford University Press, 1987), 491–93.

86. Chamberlain, *When Stars Came Down*, 205.

87. Alice C. Fletcher, "Star Cult among the Pawnee: A Preliminary Report," *American Anthropologist*, n.s. 4 (1902): 732–33.

88. Chamberlain, *When Stars Came Down*, 205.

89. Edward S. Rogers and Mary B. Black, "Subsistence Strategy in the Fish and Hare Period, Northern Ontario: The Weagamow Ojibwa, 1880–1920," *Journal of Anthropological Research* 32 (1976), 1:8, fig. 2.

90. Untitled manuscript map of rivers and lakes to the west of Lake Superior, sometimes referred to by a bold legend that appears on the left-hand fifth of the map: "Carte Tracée par Les Cris." The original of 1728 or 1729 appears not to exist, but was photographed, probably by Louis C. Karpinski in Paris in 1926–27; usually referred to as La Vérendrye's map; NMC–24556, National Archives of Canada, Ottawa; reproduced in part in G. Malcolm Lewis, "La Grande Rivière et Fleuve de l'Ouest: The Realities and Reasons behind a Major Mistake in the Eighteenth-Century Geography of North America," *Cartographica* 28, 1 (Spring 1991): 60–62 and fig. 8.

91. Although this sentence is in the past tense, scapulimancy was practiced until modern times. See, for example, Alika Podolinsky Webber, "Divination Rites," *Beaver*, outfit 296 (Summer 1964): 40–41.

92. Omar K. Moore, "Divination: A New Perspective," *American Anthropologist* 59 (1975): 71.

93. Georg Henriksen, *Hunters in the Barrens: The Naskapi on the Edge of the White Man's World*, Institute of Social and Economic Research, Memorial University of Newfoundland, Social and Economic Studies, no. 12 (St. John's: Memorial University of Newfoundland, 1973), 49.

94. Adrian Tanner, *Bringing Home Animals: Religious Ideology and Mode of Production of the Mistassini Cree Hunters*, Institute of Social and Economic Research, Memorial University of Newfoundland, Social and Economic Studies, no. 23 (St. John's: Memorial University of Newfoundland, 1979), 133–35.

95. June Helm, "Matonabbee's Map," *Arctic Anthropology* 26, 2 (1989): 28–47.

96. Extract from a letter written by June Helm to G. Malcolm Lewis, Iowa City, April 10, 1990.

97. Ibid. The map was the supplemented manuscript copy on paper by Moses Norton of a 1767 Chipewyan map on skin of a large area west of Hudson Bay, with the inscription "An Explanation of a Draught Brought by Two Northern Indians Leaders Call[d] Meatonabee & Idotlyazee, of Y[e] Country to Y[e] Northward of Churchill River Viz[t] Hudsons Bay," 139.5 x 69.5 cm, manuscript map G2/27, Hudson's Bay Company Archives, Provincial Archives of Manitoba, Winnipeg. Helm reproduces this

somewhat inadequately in "Matonabbee's Map" (fig. 3), but also includes a modern redraft done by Richard I. Ruggles (ibid., fig. 4). The original skin map has not survived.

98. Glyndwr Williams, *The British Search for the Northwest Passage in the Eighteenth Century* (London: Longmans, 1963), 133.

99. Helm, "Matonabbee's Map," 28.

100. Ibid., figs. 4–7, with the deconstructing process illustrated in figs. 5 and 6.

101. Ibid., 32.

102. Ibid., 42.

103. Ibid.

104. Ibid., 32.

105. Lake Nipigon, Ontario, is only one sixth the area of Great Slave Lake and its shape is not quite as intricate. Nevertheless, the map of it made by Ojibwa Indians for the geologist Robert Bell in 1869 is another good example of this remarkable and as yet unexplained facility. "Sketch Plan of Lake Neepigon pronounced Am-Neepigon [by] Windigo (Chief) E. side of Neepigon + another Neepigon Indn. & corrected by seven other Indians and others," pencil and ink on paper, 53.0 x 65.5 cm, H2/410—(Lake) Nipigon—(1869), National Map Collection, National Archives of Canada, Ottawa.

106. P. D. A. Harvey, "Local and Regional Cartography in Medieval Europe," in J. B. Harley and David Woodward, eds., *Cartography in Prehistoric, Ancient, and Medieval Europe and the Mediterranean*, vol. 1, *The History of Cartography* (Chicago: University of Chicago Press, 1987), 496.

107. Barry Lopez, *Arctic Dreams* (New York: Scribner's, 1986), 291.

108. Hugh Brody, "Maps and Journeys," 135–36. The existence of not dissimilar maps is now beginning to be recognized in the Western world. Defining "psyche" as "the area of our experience that manifests itself in the world of feeling, memory and imagination (including fantasy and dreaming)," David Maclagan has demonstrated that like the physical world, it too has its maps. Beaver Indian dream maps seem to fall into his category of "subjective maps of the psyche," the key characteristics of which include: "the treatment of feelings, states of mind or psychic events (such as dreams) on the same footing as physical features or concrete facts; the use of a flexible framework or scale, rather than a uniform one; the adoption of a frankly idiosyncratic or personal perspective, i.e. subjective experience is the lens through which the convergence of phenomena is registered; [and] the inclusion of time (in the form of memory) and motion (in the form of narrative) in the map." David Maclagan, "Inner and Outer Space: Mapping the Psyche," in Gavin D. Flood, ed., *Mapping Invisible Worlds*, vol. 9 of *Cosmos: The Yearbook of the Traditional Cosmological Society* (Edinburgh: Edinburgh University Press, 1993), 155.

109. Though not with specific reference to the work of anthropologists, in chapter 6, pp. 140 and 150, Barbara Belyea deplores what she sees as "this persistently scientific, Eurocentric bias," and pleads for a "general recognition . . . that maps are purely conventional, that convention determines perception, and that this perception is culturally specific." Recognizing that "cartographic convention is only partially communicable from one culture to another" and that "'translation' is possible only in a superficial, limited sense" leads her to state a principle upon which to base a policy proposal: "Recognition and respect of difference is the necessary base and starting point for any intercultural dialogue. . . . Instead of continuing to translate the native cartographic convention into our own, we need to acknowledge that the gap between these conventions is essentially unbridgeable. The best we can do is to initiate a dialogue with Native cultures as they have survived." The proposal to "initiate dialogue" is very similar to "establish[ing] native perspectives." Initiating oral dialogues with extinct cultures is, of course, impossible. Doing so with members of changed cultures can lead to false perspectives and wrong conclusions, as, of course, has the traditional reliance on evidence mediated by whites in written reports and transcribed maps. Potentially, at least, cultural anthropologists are best qualified to evaluate together field, oral, written, and transcribed evidence.

110. Although pre–World War I field anthropologists provided much of the data later to be reworked by others, anthropology to date has not contributed as much as the several histories to our understanding of Amerindian and Inuit maps and mapmaking.

111. David Woodward and G. Malcolm Lewis, eds., *The History of Cartography*, vol. 2, bk. 3 (Chicago: University of Chicago Press, forthcoming), for example, contains sections by anthropologists on traditional cartography in Australia (Peter Sutton), Oceania (Ben Finney), Papua New Guinea (Eric K. Silverman), and South America (Neil L. Whitehead).

112. Dorothy Jean Ray, *Eskimo Art: Tradition and Innovation in North Alaska* (Seattle: University of Washington Press for the Henry Art Gallery, 1977), 43.

113. For example, the letter opener carved by Tony Pushruk of King Island, in Bering Strait, about 1958, the blade of which is engraved with a map of the more than fifty-mile-long mainland coast between Wales and Nome, straightened to fit its long, narrow form. Reproduced in Ray, *Eskimo Art*, 160, fig. 138.

114. Susan McGreevy, "Navajo Sandpainting Textiles at the Wheelwright Museum," *American Indian Art Magazine* 7 (1981), 1: esp. 53–57.

115. "Hailway: 'The Night Sky,'" woven by Hosteen Klah, ca. 1920, natural wool, vegetal and aniline dyes, 251.5 x 243.8 cm, cat. no. 44/14, Wheelwright Museum of the American Indian, Santa Fe, New Mexico; reproduced in McGreevy, "Navajo Sandpainting," 54, fig. 1. In addition to the Milky Way, the North Star and Big and Little Dippers are identified.

116. Ingrid van Dooren, "Navajo Hooghan and Navajo Cosmos," *Canadian Journal of Native Studies* 7, 2 (1987): 264.

117. Robin McGrath, "Maps as Metaphors: One Hundred Years of Inuit Cartography," *Inuit Art Quarterly* (Spring 1988): 7–8. Indian art with cartographic themes or styles is now beginning to be marketed by charities. For example, Tribale from Survival currently lists in its catalog "Past Journeys" a limited-edition original etching by the Hopi/Choctaw artist Dan Lomahaftewa. On seeing it I recognized subjectively some of its cartographic characteristics before seeing its confirmatory title elsewhere on the page.

118. Ibid., 9–10 and fig. 8.

119. Marlene Creates, *The Distance between Two Points Is Measured in Memories: Labrador 1988* (North Vancouver, B.C.: Presentation House Gallery, 1990), 16–51.

120. Milton M. R. Freeman, ed., *Report, Inuit Land Use and Occupancy Project* (Ottawa: Department of Indian and Northern Affairs, 1976), 2:49–52.

121. Ibid., 53.

122. Ibid., 52–53.

123. Ibid., 53–54.

124. Ibid., 54.

125. Ibid., 55.

126. Ibid.

127. Probably in the hand of George Comer, "Melikis Map of musk ox hunting different seasons," Aivilingmiut manuscript map of the west coast of Southampton Island, Roes Welcome Sound, and the Keewatin mainland between Depot Island and Repulse Bay; linework in pencil with names and legends in ink, 41.5 x 56.5 cm, ca. 1898, 60/2842/E, Department of Anthropology, American Museum of Natural History, New York.

128. Freeman, *Report, Inuit Land Use*, 55–56.

129. Although not specifically concerned with the moral qualities of honesty and truth, J. R. Miller made a pertinent generalization about "native-newcomer contact" (in the context of Indian-white relations in Canada) before 1700. Using words like "cooperative," and "tolerance," he concluded that "the ethical imperative to share made it difficult for Canada's native people to refuse the Europeans' demands for part of their fish, a share of furs they took, assistance in exploring and *mapping* the land and waterways, and, somewhat later, military aid. . . . After 1700 relations were modified, not radically transformed. It was still true that Europeans and Indians . . . cooperated [among other activities] in exploring and *mapping* the interior of the continent" (italics added). J. R. Miller, ed., *Sweet Promises: A Reader on Indian-White*

Relations in Canada (Toronto: University of Toronto Press, 1991), viii–x.

130. See chapter 3, p. 60, for another explanation of surprising silences. The concept of silences on maps was first introduced into cartography by J. B. Harley, "Maps, Knowledge, and Power," in Denis Cosgrove and Stephen Daniels, eds., *The Iconography of Landscape: Essays on the Symbolic Representation, Design, and Use of Past Environments* (Cambridge: Cambridge University Press, 1988), where silences were presented as "ideological filtering[s] . . . [enshrinements of] self-fulfilling prophesies about the geography of power" (pp. 290–92). The concept was soon incorporated into his epistemological writings based in part on the work of Michel Foucault and Jacques Derrida. Harley's use of it has been criticized by Barbara Belyea as a "complicity with the process of metaphorical reversal": "Images of Power: Derrida/Foucault/Harley," *Cartographica* 29, 2 (1992): 4.

131. Doug Aberley, *Boundaries of Home: Mapping for Local Empowerment*, New Catalyst Bioregional Series, no. 6 (Gabriola Island, B.C.: New Society Publishers, 1993), 14–16. The most accessible account of bioregional mapping by Indians is Hugh Brody, *Maps and Dreams: Indians and the British Columbia Frontier* (London: Jill Norman and Hobhouse, 1981), 146–77. This includes ten examples of maps, each with a brief text.

132. Recent examples from elsewhere in the Americas include: an early 1990s project in the tropical forest of northeast Honduras, mapping where and how the Garifuna, Pesch, Miskito, and Tawahka Sumu of the Mosquitia region lived, described by Derek Denniston in "Defending the Land with Maps," *World Watch* (January/February 1994): 27–31; and an approximately contemporary project in northern Argentina in which the Wichi of the semiarid Chaco compiled a map of 400,000 hectares of contested land to identify past and present village sites, their hunting, fishing, and gathering grounds, gardens, water sources, and places of religious and cultural significance, "The Land of Our Ancestors' Bones," *Survival International Newsletter* 33 (1994): 8–10. In Honduras topographic base maps appear not to have been used, but it is not clear whether they were used in Argentina. Both projects involved white consultants. In neither case is it stated what happened to the original Indian maps.

133. Aberley, *Boundaries of Home*, 15.

134. Peter Poole, *Indigenous Peoples, Mapping, and Biodiversity Conservation: An Analysis of Current Activities and Opportunities for Applying Geomatics Technologies* (Landover, Md.: Corporate Press, for Biodiversity Support Program, 1995), 6–9, 37–48, 71–73.

135. Robert A. Rundstrom, "GIS, Indigenous Peoples, and Epistemological Diversity," *Cartography and Geographic Information Systems* 22, 1 (1995): 56.

136. This statement is not intended to imply that the Indians made a distinction between base data and plotted data. Indeed, from their perspective, the two may not have been differentiated. Frank G. Speck, "Montagnais-Naskapi Bands and Family Hunting Districts of the Central and Southeastern Labrador Peninsula," *Proceedings of the American Philosophical Society* 85 (1942), 2:215, and figs. 1–2 on 217.

137. An attempt is made to anticipate some of these topics in chapter 12. Others include: mapmaking by Inuit and testing the opinion that their maps differ from those of north American Indians; and the need to explain why the Aztec maps described by Elizabeth Hill Boone in chapter 5 differed so markedly from those found in the first encounters with native North Americans, some of which were made and used only a few hundred miles to the north.

5

Maps of Territory, History, and Community in Aztec Mexico

Elizabeth Hill Boone

The Painters

It was Easter Sunday, 1519. Hernan Cortés and his group of Spanish adventurers had arrived only two days before at the natural port they named Veracruz, on Mexico's Gulf Coast; from here they would later begin their inland invasion of Mexico. That day two native lords, one a governor under the supreme Aztec ruler Moctezuma, came with gifts to greet the Spaniards; they also came with painters. As the *conquistador* Bernal Díaz del Castillo later recalled it, the governor "brought with him some clever painters such as they had in Mexico and ordered them to make pictures true to nature of the face and body of Cortés and all his captains, and of the soldiers, ships, sails and horses, and of Doña Marina and [the Spaniard] Aguilar [interpreters who had joined Cortés's forces earlier], even of the two greyhounds, and the cannon and cannon balls, and all of the army we had brought with us."[1] The painters also recorded the fine display of horsemanship and shooting that the Spaniards arranged to impress the Aztecs. Later that day, when the ceremonies and feasts were done, the Aztec governor hastily left on the overland journey to the imperial capital, Tenochtitlan, carrying the paintings for Moctezuma's inspection.[2]

The painters who recorded this first official meeting between the Spaniards and emissaries of Moctezuma functioned as reporters and news photographers combined, for they documented the larger event, its participants, and subplots, in paintings that could be read as text and viewed as pictures. They were *tlacuiloque*—the singular is *tlacuilo*—the Nahuatl word that means both painter and scribe.[3] The *tlacuiloque* did not preserve a series of words linked together in sentences and paragraphs, as do writers of alphabetic scripts. Instead, they recorded the facts of the meeting on that coastal beach by means of pictorial images—of varying degrees of abstraction—understood within an organized structure. Moctezuma, receiving the paintings, would have seen the arrangement of abstract and naturalistic images, understood the pictorial conventions and the hieroglyphs, and recognized meaning in their relative placement on the painted surface. The governor also told the story orally, but the painting was the medium of documentation. And Bernal Díaz del Castillo remembers that the likeness of Cortés was so accurate that Moctezuma recognized the uncanny resemblance with a member of his own court, whom he sent to Veracruz to greet the Spaniards.[4]

The occupation of *tlacuilo* was an honorable one in pre-Columbian Mexico. The sons of nobles, as well as promising commoners, were taught pictographic writing in the elite *calmecac* school. Some took it up as a profession, for the empire and its people relied on skilled painter/scribes. *Tlacuiloque* kept tax and tribute records, they documented land holdings, they served as court reporters, and they drew up battle plans in times of war.[5] Moctezuma's empire could not function without the plethora of these mundane records. On a higher level, painted books served the needs of historians, priests, diviners, astronomers, and rulers. The Franciscan friar Motolinía spoke of several genres of books on which the Aztecs relied, including histories, *tonalamatls* or divinatory almanacs, books of prophesies, and books of dreams.[6]

Physically, these books and documents were fashioned of long strips of animal hide or paper, which was primed on both sides with a white lime sizing, and the strips either rolled (as in a *tira*) or folded accordion-style (as in a screenfold). The paper was made by stripping, soaking, and then pounding thin the inner bark from the native fig tree.[7] For large-scale presentations, the scribes could also choose to paint on large panels of hide, paper, or cotton cloth, which could then be folded for carrying and storing. The Spaniards called the great cotton sheets *lienzos*, from their word for linen.

Although painted books pervaded almost all aspects of Aztec life, only a little over a dozen preconquest books have come down to us from all of Aztec Mexico.[8] From the vast realm of the empire, only five *tonalamatls* (divinatory almanacs) remain, and from the Mixteca, at the southeastern edge of the empire, eight histories have survived. Yet this tiny corpus is amplified by about 600 books, manuscripts, and documents painted in the native tradition after the conquest, when the need for painted documents continued to be felt among the Spanish as well as the indigenous populations. The corpus of early colonial pictorials is our critical key to the lost pre-Columbian tradition.

The Maps

Although no maps painted before the Spanish invasion have survived, there are probably a hundred maps, or documents with a cartographic component, that remain from the early colonial period. Almost all of them show a degree of European influence: either they incorporate European or colonial images, or their artists have adopted some European stylistic features, or they were painted on European paper. Many were painted to fulfill the documentary requirements of the new colonial administration; and they survived precisely because they entered administrative archives. But most of these cartographic paintings still retain vestiges of the pre-Columbian manuscript tradition. Some were executed fully in the indigenous painting style by artists who had not yet adopted European illusionism. Other maps show just how quickly the native artists incorporated new forms and painting techniques into their work. These colonial documents, coupled with the reports of the chroniclers, allow us to understand the pre-Columbian cartographic tradition.

We should understand first that the Aztecs did not think of maps as a separate genre of document, distinct from others. There was no Nahuatl word for "map." In all of Bernardino de Sahagún's bilingual *General History of the Things of New Spain*, there is no equivalent of "map" in the Nahuatl text. Sahagún mentions a map only once, when he describes a reconnaissance map painted of a town just prior to an attack on it; then he speaks of it simply as something painted.[9] Alonso de Molina's Nahuatl/Spanish dictionary of 1571 contains no entry for *mapa*. *Mapamundi o bola de cosmografía* ("world map or cosmographic globe") does appear, translated as *cemanauactli ymachiyo* ("the world, its image or model") and *tlalticpactli ycemittoca* ("the surface of the earth, the whole thing visible in one piece").[10] Francisco de Alvarado's Mixtec/Spanish dictionary of 1593 has an entry for *mapa;* there the short, simple Spanish word is rendered into the cumbersome *taniño nee cutu ñun ñayevui,* which can be read as "a reduced figure of the whole world."[11] None of the chroniclers, when they list the different genres of Aztec books, mention maps as a separate category. There are land records, to be sure, but they are not specified as maps. This suggests that the Aztecs did not distinguish maps from other kinds of writings.

We make such a distinction in Western culture, because we write with letters and words rather than with pictures and images, and we organize our writing in straight, unidirectional, lines rather than in a pattern that flows over a page. For us a map is distinctive in its pictorial quality and spatial organization, which set it apart from the alphabetic texts that compose our principal writing system. When the Spaniards introduced alphabetic writing in Mexico, the difference between a text and a map became plain, and the Nahuatl-writers adopted the Spanish loanword *mapa* when they referred to cartographic documents.[12] For the Aztecs and their neighbors prior to the Spanish conquest there was no such distinction between map presentations and "written" presentations.

Like us, the Aztecs relied on maps, or cartographic paintings, for three essential functions. They used maps to present routes or paths for travel, they used maps as the foundation on which to record past movement and action, and they used maps to explain how things are organized spatially. Maps excel over other kinds of texts at all these tasks, because geographic features and actions in space are best presented by their relative placement on a painted surface. Of course, none of these functions was exclusive of the others; all maps have an organizational component, and in the minds of the Aztec painters the subgenres blended freely. It can help us see the nature of Aztec maps, however, if we consider them from these three perspectives—as records of movements, future and past, and as organizers.[13]

Paths for Movement

One major reason for creating a map, in any culture, is to present paths for movement. When this movement is in the present or the future, such maps can be categorized as "way-finding." They are prescriptive, because they show how one is to go, perhaps also presenting optional routes. They specify the geographic or cultural features of an area that are likely to affect someone moving through that area. Like road maps and railway maps they are essentially guides for travel. The Aztecs used them effectively, as did the Spaniards during and following the conquest.

When Cortés was staying in Tenochtitlan, in 1519 before Aztec/Spanish relations had fully deteriorated, he asked Moctezuma if there were a river or cove on the Gulf Coast where his waiting ships might be safe. As Cortés later wrote to Charles V, Moctezuma "replied that he did not know, but would have them [the painters] make a map of all the coast for me with all its rivers and coves; . . . On the following day they brought me a cloth with all the coast painted on it, and there appeared a river which ran to the sea, and according to the representations was wider than all the others. This river seemed to pass through the mountains which we called Sanmin [San Martín], . . ."[14] The river was the Coatzacoalcos,[15] which Cortés's men easily found with the aid of this cloth map.

Three years after the fall of Mexico-Tenochtitlan, Cortés put his trust in another indigenous map, which proved equally valuable. Faced with a potential uprising in Honduras, Cortés decided to lead his men there himself. He chose to make the journey overland rather than sail around the coast, but first he marched to the large trading center of Xicalanco on the Gulf Coast. Then, according to his secretary and official chronicler, Francisco López de Gómara,

Cortés sent word to the lords of Tabasco and Xicalanco that he was there and that he wished to march by certain routes, and he requested them to lend him men who knew the coast and the back country. These lords at once sent him twelve of the most respectable persons of their towns, and sent him merchants also, with the safe-conducts they customarily carry. They, when they thoroughly understood Cortes' plan, painted him a canvas on which they depicted the whole route from Xicalanco to Naco and Nito, . . . and even as far as Nicaragua . . . This map was a remarkable thing, because it showed all the rivers and mountains that had to be crossed, and all the large towns and inns where they stopped when they attended the [market] fairs. . . . indeed, [Cortés] was astonished at their knowledge of such distant places."[16]

Neither of these maps has survived, of course, but a sixteenth-century map from Tuxpan (ancient Tochpan), along the Gulf Coast in northern Veracruz, gives an indication of what the lost maps must have been like (fig. 5.1). The Mapa Local of Tochpan is the earliest of a series of six maps from that area. Executed fully within the Aztec painting style and lacking any recognizably European elements, it is as close as we are likely to come to a pre-Columbian map. The large cloth panel presents two rivers (we know them to be the Tuxpan at the top and the Cazones at the bottom) flowing from left to right into the waters of the Gulf of Mexico, which borders the map's right edge. Although the map has no designated orientation, I have oriented it in figure 5.1 so that north is at the top and east is on the right, to agree with modern maps of the region. In keeping with the Aztec conventional treatment of water, the waters are characterized by stylized shells and disks appended to fingers of the currents. The gulf current swirls from bot-

MAPS IN AZTEC MEXICO

5.1 Mapa Local of Tochpan. This image also appears in José Melgarejo Vivanco, *Los lienzos de Tuxpan* (Mexico City: Editorial la Estampa Mexicana, 1970), first plate.

tom to top, and we can appreciate that the Aztec map painters understood the generally northward direction of the current along this coast.

A simple painting of a rabbit efficiently identifies and locates the provincial capital of Tochpan (the name means Place of the Rabbit). From here a road runs southward and bifurcates before it crosses the southern river. Black footprints designate the roads. Other towns along the route are identified (as is Tochpan) simply by the symbols of their name: a disk with many dots for Tlaltizapan (Place of Sand) located near where the road splits, a butterfly for Papalotlan (Place of the Butterfly) in the lower right corner. This is a local map, which covers the territory along about 50 miles of the Gulf Coast, from north of the Tuxpan River to south of the Cazones River. Within this area it locates the roads and the settled com-

5.2 Reconnaissance map, from Bernardino de Sahagún's *General History*. This image also appears in *Códice Florentino* (Mexico City: Archivo General de la Naciòn, 1979), vol. 2, bk. 8, fol. 33v. Courtesy of Archivo General de la Naciòn.

munities with respect to the basic topographic features. The map that led Cortés to Honduras was probably similar to this in its general features, although it was clearly much larger, extending from the Gulf Coast across the Peten and to the Caribbean.

Except for the two maps painted for Cortés, which proved crucial to the success of his ventures, way-finding maps are hardly mentioned by early colonial chroniclers. Motolinía omits them from his short list of Aztec books, and the other writers are equally silent.[17] The exception is the Franciscan friar Bernardino de Sahagún. In his pictorial encyclopedia of Aztec life, Sahagún explains how Aztec rulers conducted wars and how they relied on painted plans; he even illustrates one. Sahagún records that when war was waged, the ruler

first . . . commanded masters of the youths and seasoned warriors to scan the [enemy] city and to study all the roads—where [they were] difficult, where entry could be made through them. This done, the ruler first determined, by means of a painted [plan], how was placed the city which they were to destroy. Then the ruler noted all the roads—where [they were] difficult, and in what place entry could be made."[18]

The painting that accompanies this description (fig. 5.2) shows, on the right side, the route of the warriors in and out of the town on reconnaissance. In the lower left corner, the painted plan of attack is laid out before three high-ranking members of the military hierarchy. Wrapped in their cloaks and sitting on high-status reed mats, they all sport warrior topknots; the two seated on the far left also have (well above their heads) the pointed turquoise crowns of rulers. Speech scrolls curling from their mouths carry the tactical discussion.

Cortés's maps and this painted plan of attack all show the way for future travel; they are "way-finding" aides in this sense, and they work simply and directly. It is hard to say how widely and commonly they were used in pre-Columbian times, but if the Aztec armies used them to attack cities and towns, these armies must also have used them when they traveled through unfamiliar land. Long-distance merchants, *pochteca*, would have relied on them too.

Cartographic Histories

Maps that mark paths of travel are prescriptive in that they signal a route or routes to be taken. They look to the future in this respect by presenting the existing features of a place for the use of someone who will be in that place in the future. They suggest action and travel that has not yet occurred. With some amendments, however, such maps can also be made to look to the past and to be descriptive of that travel. The mere addition of a person or event can transform a way-finding map into an historical account. Most of the existing Aztec maps have some historical content.

A document becomes a history as soon as it records an event that has already happened. Usually these events are associated with people—they are actions carried out by individuals or they are events (earthquakes, for example) that affect people—who become the subject of the history.[19] Historical events, like all events, are wrapped in a temporal dimension. They occur at a specific time or over a certain period, usually with a beginning and end that can be variously defined. Too, and most important for this discussion, events will necessarily occur in a location. Each object, animate or not, has a physical and therefore a geographic presence, and each event has to happen *somewhere*. Every event, then, is attached with the other three elements: protagonist, date, and location.

Histories, like other stories, are built with these elements, although many, and probably most, histories emphasize one or more at the expense of others. For example, an annals history, which records events strictly according to the years in which they occurred, emphasizes the temporal element over place or person. A *res gestae*, or event-oriented, history follows events as they occur sequentially, irrespective of precise time. A cartographic history also sacrifices chronology to feature the events' spatial dimension.[20]

A map, in and of itself, establishes the element of location. It represents a physical area and then spatially positions geographic or cultural features on this area. A map becomes a cartographic history when an event is recorded on it, regardless of whether persons are also present, for usually the event itself implies protagonists, who may or may not be identified graphically.

In this respect, the Mapa Local of Tochpan, used as an example of a way-finding map, is equally a cartographic history (fig. 5.1). In the center of the map, near where the road bifurcates, the painter has drawn one of the Aztec conventions for war: a shield backed by a set of (usually four) spears. Suddenly the footprints are not there simply to designate the ribbons as roads; they are there to show movement from Tochpan to other locations. Tochpan has gone to war against its southern neighbors. Just to the right of Tochpan's toponym is the date 13 Flint, painted as a flint-knife accompanied by 13 small disks in a rectangular cartouche. This effectively sets the date (of the aggression, or perhaps of Tochpan's founding) to the year 13 Flint in the Aztec count.[21] Individuals, surely rulers of Tochpan and other towns, are identified by their name-signs attached usually to their heads. The map thus becomes a history of this coastal war, a war that involved the rulers of Tochpan and its neighbors. It is a brief and incomplete history, to be sure, for it omits most of the details, but all histories omit things.

Map-based historical accounts take location as the foundational and most important element of the story, into which people, the action, and the dates are inserted. Because these documents are spatially constructed, it is easy for them to record actions that occur in different places, either contemporaneously or sequentially. It is equally natural for cartographic histories to describe movement from one place to another. Above all, they make clear the spatial relation of one event or location to another. Aztec painters clearly understood these documentary strengths, for they relied on cartographic histories to record battles won and lost,[22] and they saw in them an ideal form for recording migrations, where the narrative is told through a people's wanderings. Moreover, the large panels of hide, bark paper, or cloth could often picture the whole of a story as a single conceptual statement.

The chroniclers do not describe cartographic histories in any detail. For map-based migration histories we have only the extant colonial documents to reveal the pre-Columbian tradition. The Mapa Sigüenza (fig. 5.3A), for example, presents the Aztec migration as a story line that moves across space. Painted on fine native paper with a delicate hand, it records a history that spans over a hundred years. The story begins in the upper right quadrant, where the Aztecs leave their mythical homeland of Aztlan, an island in the middle of a lake, the lake rendered as a watery square and the island as a green hill in its center (fig. 5.3B). Disguised as a bird, the tribal deity Huitzilopochtli perches atop a tall plant and commands a crowd of Aztecs to go forth; speech scrolls carry his words. The migratory path begins there as a narrow ribbon punctuated by footprints. Then as the leaders of the tribe actually set off on the journey, they are individually identified by name-glyphs attached to the tops of their heads—first five leaders and then below them ten more who are shown further along in the journey. From Aztlan the migration proceeds anonymously along the path marked by footprints, from site to site, each site identified by the hieroglyph or place sign of its name (fig. 5.3A). The footprints establish the direction of the path and, in the absence of walking figures, also provide the action. The reader knows to assume that all the Aztecs and their leaders are following this path. Small blue disks painted by each place sign indicate the number of years the Aztecs paused in each place. The migratory path leads around the upper right quadrant in a roughly counterclockwise direction to the lower right; then it doubles back and undulates along the bottom of the sheet, before it rises to the top of the sheet and crosses to the far upper left, where it doubles back again and drops down to Chapultepec, the Hill of the Grasshopper, situated prominently in the left side of the sheet.

With Chapultepec, the Aztecs have entered the lakeside around Lake Texcoco, and their history takes on more detail. Aztec leaders are once

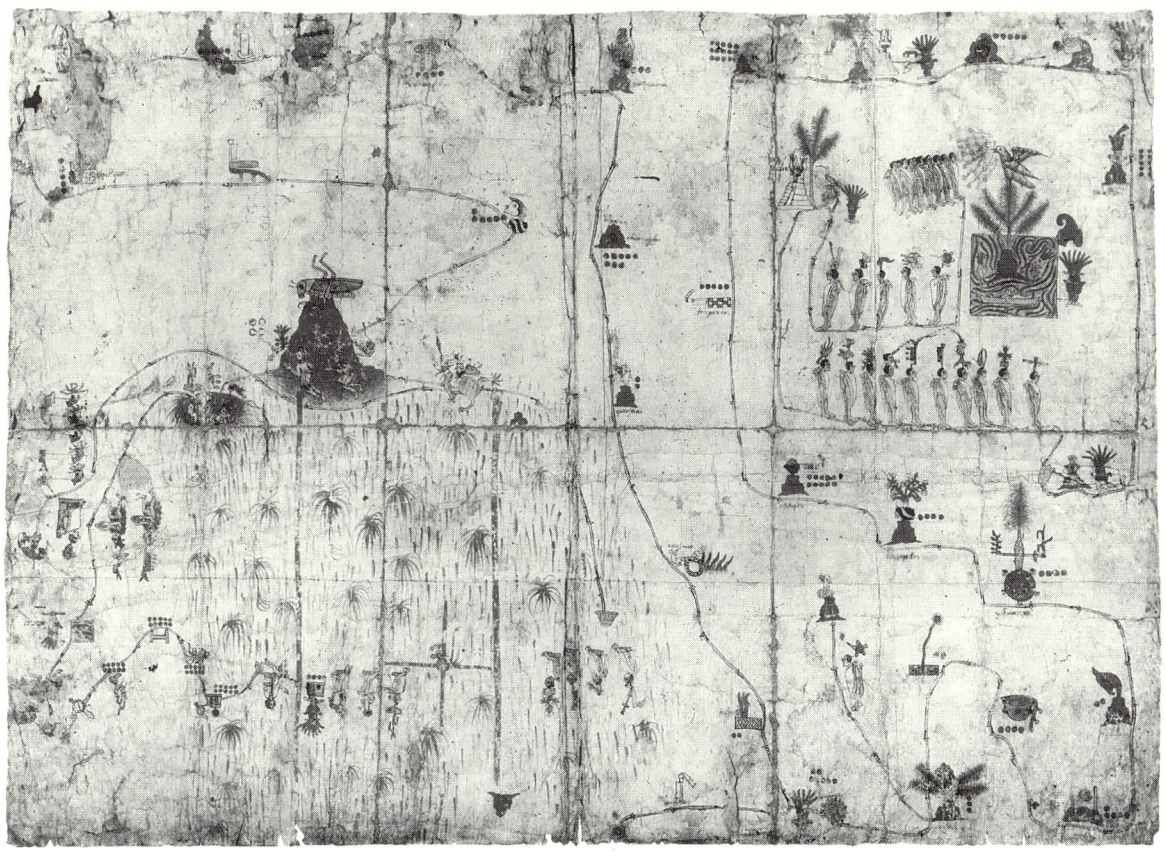

5.3A. Mapa Sigüenza. This image also appears in *El territorio mexicano*, pl. 3. Courtesy of Instituto Mexicano de Seguro Social.

again named; two of them perish at Chapultepec, their bodies covered with blood. The lake itself is described pictorially below Chapultepec as a marshy swamp, dotted with reeds and rushes and cut by straight blue canals. At this point, all the images are painted upside down, and one has to turn the sheet bottom up to read the story easily. The main branch of the migrating Aztecs proceeds to Culhuacan (the bent hill on the far left edge of fig. 5.3A; on the far right of inverted detail, 5.3C), where they have dealings with the local people. The Culhuacanos eventually evict the Aztecs, who must relocate again and again until they are finally pushed into the middle of the swamp. They finally settle at Tenochtitlan, depicted as a nopal cactus *(Opuntia phaeacantha)* on a rock below Chapultepec. While this is happening, a secondary branch of the Aztecs proceeds more directly from Chapultepec to Tlatelolco located to the right of Tenochtitlan.

If we consider the Mapa Sigüenza as a history, we see that in a very efficient way it gives us all the elements of the migration story. The protagonists are the Aztecs, led by tribal leaders who are named

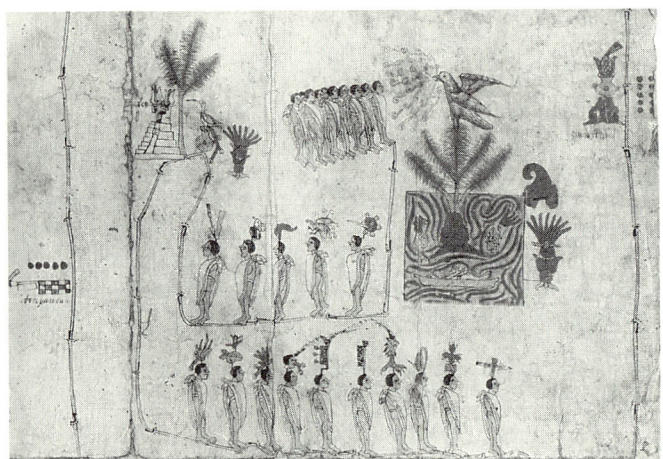

5.3B Mapa Sigüenza, detail of the Aztecs departure from Aztlan. Courtesy of Instituto Mexicano de Seguro Social.

5.3C Mapa Sigüenza, detail of the founding of Tenochtitlan; the leaders are pictured seated on either side of Tenochtitlan's place sign, here depicted as a small nopal cactus at the intersection of four canals. Courtesy of Instituto Mexicano de Seguro Social.

at the start of the story. The map format easily presents the changing locations of the story glyphically and pictorially. The path and the footprints provide the action, which is the migration, as well as the temporal dimension of sequence. Another temporal dimension, that of duration, is measured by the years the Aztecs stay in each place.

If we consider the Mapa Sigüenza not so much as a history but as a map, we can relate it to "way-finding" maps like Cortés had, for it records paths of movement, although in the past tense. In a straightforward manner, the map records the travel and events from Aztlan to Tenochtitlan.

Two kinds of space exist in this map, one sequential and the other geographic. On the right side of the sheet, and for most of the migration, the story line moves from place to place, but the places are not located within an actual geographic framework. Aztlan is not presented as being north or south of any place else, and the distance between the towns is ambiguous. The locations are related to each other sequentially rather than topographically. This part of the map is like a railroad map or timetable that lists in sequence the places the train will stop, but does not specify how they are geographically located with respect to each other. The Mapa Sigüenza gives us the sequence of the migration from one place to another, but it is not telling us about the terrain for the most part.

The map does give us terrain when the Aztecs arrive in the Valley of Mexico—when they reach Chapultepec. The lower left quadrant of the Mapa Sigüenza replicates the general geography of the valley (figs. 5.3A and C). Chapultepec is pictured above the swampy lake; in actuality it is located on the lake's western shore. If Chapultepec is in the west, Culhuacan (the bent hill on the far left edge) is correctly positioned to the south of the lake. Tenochtitlan is accurately located in the middle of the lake, north of Culhuacan and east of Chapultepec. Its sister city of Tlatelolco is just to the north, roughly where it lies geographically.

5.4 Cuauhtinchan Map 2. This image also appears in John B. Glass, *Catálogo de la colección de códices* (Mexico City: Instituto Nacional de Antropología e Historia, 1964), pl. 25.

Other cities, such as Chalco and Texcoco, are also sited geographically. The Mapa Sigüenza locates these places relationally to each other and gives the reader the topography of the lake, because territory and geography have finally become important for the migration history. The location of Tenochtitlan vis-à-vis the other lakeside cities is central to the Aztec story.

By using two separate spatial systems, one sequential and the other geographic, the history painter is differentiating between what is important in the two parts of the migration story. In the first half of the story, it was important to tell that the Aztecs left Aztlan and traveled from one place to another. The geographic location of Aztlan was unimportant to the painted story, as were the locations of the stops along the migration. What was necessary to the history was the sequence of the towns on the route. Once the Aztecs arrived in the valley of Mexico, however, geography became a real factor, because the geographic situation of the various lakeside towns was an integral part of the story about the Aztec rise to power. As Tenochtitlan grew to greatness in the years to come, its position in the middle of Lake Texcoco would also prove to be one of its chief advantages in commerce and in war. The history painter of the Mapa Sigüenza wanted his readers to understand just where Tenochtitlan was vis-à-vis its neighbors, so he positioned the second half of his migration story in real space.

Several other Aztec and Mixtec cartographic

histories use this combination of sequential and real space. Cuauhtinchan Map 2 (fig. 5.4), for example, traces the migration of the people of Cuauhtinchan from Chicomoztoc (the cave of origin) into the area of Cuauhtinchan, east of the city of Puebla, where they found their town and establish themselves territorially.[23] The events of the migration are detailed, as they are in the Mapa Sigüenza, along an undulating ribbon that moves sequentially from place to place in an ambiguous space. The people emerge from Chicomoztoc in the upper left corner and wind their way down the left side, then up, then down again, and then up (as if they were tracing a "W") until they pass through Cholula, roughly in the middle of the map. Leaving Cholula and going to the right, they enter the general area of Cuauhtinchan, where the features of the land are now fixed geographically. From this point onward, events are pictured as taking place in an actual landscape, and the focus is on the site of Cuauhtinchan, situated just to the right of the center of the map.

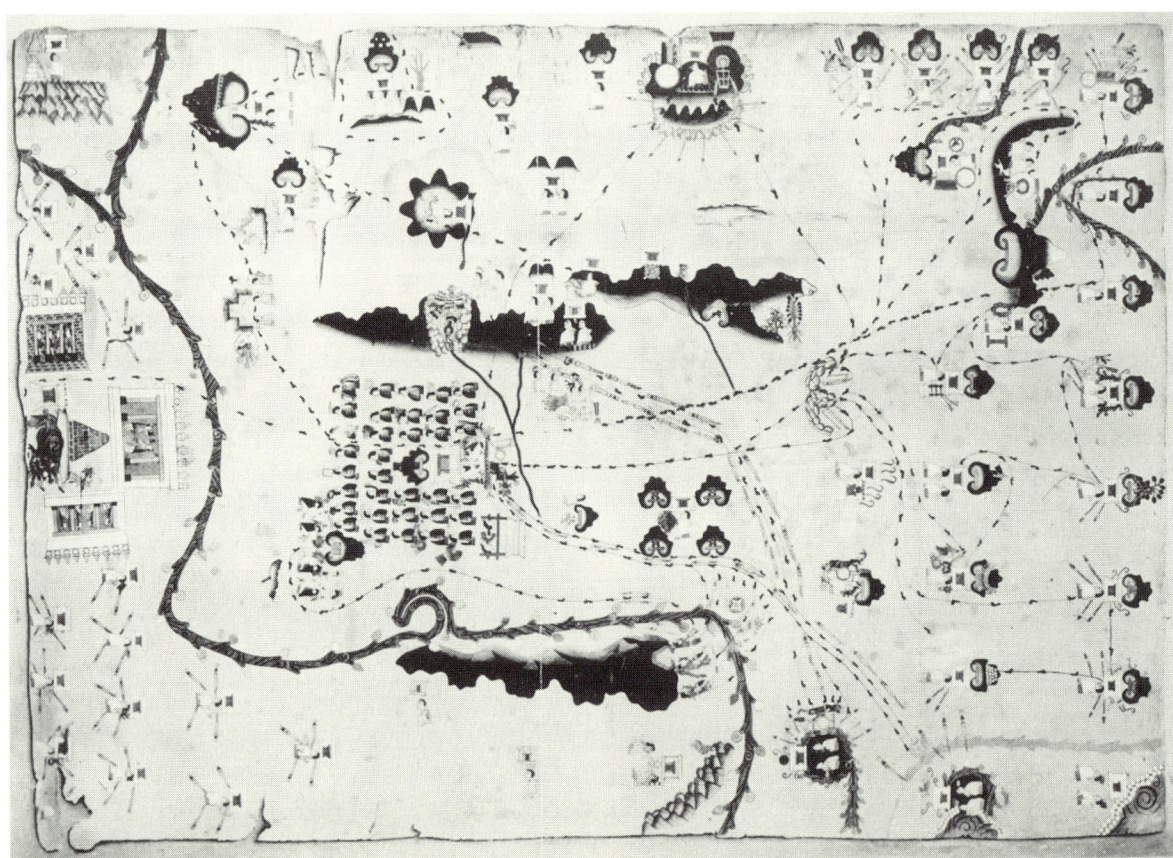

5.5 Cuauhtinchan Map 1. This image also appears in John B. Glass, *Catálogo de la colección de códices* (Mexico City: Instituto Nacional de Antropología e Historia, 1964), pl. 34.

Several *lienzo*s from Puebla and the Mixteca also juxtapose these different spatial presentations. The Lienzo of Tlapiltepec from the Coixtlahuaca valley also presents its story as occurring first in sequential space (for the migratory portion) and then in geographic space (once the people's territory has been reached).[24] The Lienzo of Zacatepec (see fig. 5.7) and, to a certain extent, the map of the *Relación geográfica* of Teozacoalco, both from the Mixteca, do also.[25]

Most Aztec cartographic histories define an identifiable territory and then place the history within that territory. The basic story these histories tell is that a people came into an area, laid claim to a defined territory, and established their dynasty. Cartography is the essential feature of these histories, even more so than in the Mapa Sigüenza. Such are the other Cuauhtinchan maps—maps 1 and 3 and those embedded within the Historia Tolteca-Chichimeca. The history told in Cuauhtinchan Map 1 (fig. 5.5) flows over a vast but well-defined territory, relieved by mountain ranges and rivers and punctuated by towns. Lines of footprints and arrows trace the routes of travel and military aggression. The multiplicity of the routes and the absence of date glyphs make it difficult for us to know the date of individual events; it is hard even for us to understand the sequence of the actions, which are made to appear contemporaneous. Temporal ambiguity is a feature of all the cartographic histories to one degree or another, because the history painter has chosen to stress relative location over chronology.

The cartographic structure also allows the painter to show how each episode fits with the others, for it enables him to outline the entire story at once on the map. Because each event is framed by the other events, the reader can see how one relates to the others and can judge their relative importance. This unity of expression is particularly noticeable in simpler presentations, such as the Cuauhtinchan maps included in the Historia Tolteca-Chichimeca (fig. 5.6). The story in this cartographic history roughly parallels the Mapa Sigüenza and Cuauhtinchan Maps 1 and 2 in depicting the migration and settling of people into the area (figs. 5.3A, 5.4, 5.5). Here the people enter the territory on the left, the area being framed by the place signs that mark the boundary. They defeat already established residents and settle in various sites; then they take possession of the land by walking its boundaries, this being represented by the footprints around the perimeter. In the center they found the city of Cuauhtinchan.

The Cuauhtinchan maps are migration histories like the Mapa Sigüenza in that they tell the story of a journey that culminates in the founding of a capital city, but their emphasis is on the establishment of territory. In this sense the Cuauhtinchan maps and, to a lesser extent, even the Mapa Sigüenza belong also to another category of map. They are presentations of the way land is organized, defined, and held, and they relate to those maps that function as community charters.

Community Charters

The political and social pattern of Late Postclassic Mexico was formed of many distinct and fairly autonomous community kingdoms, each ruled by a noble family. Although many had come under Aztec imperial domination and therefore sent tribute or provided services to the Aztec capitals, they all retained a degree of autonomy; they con-

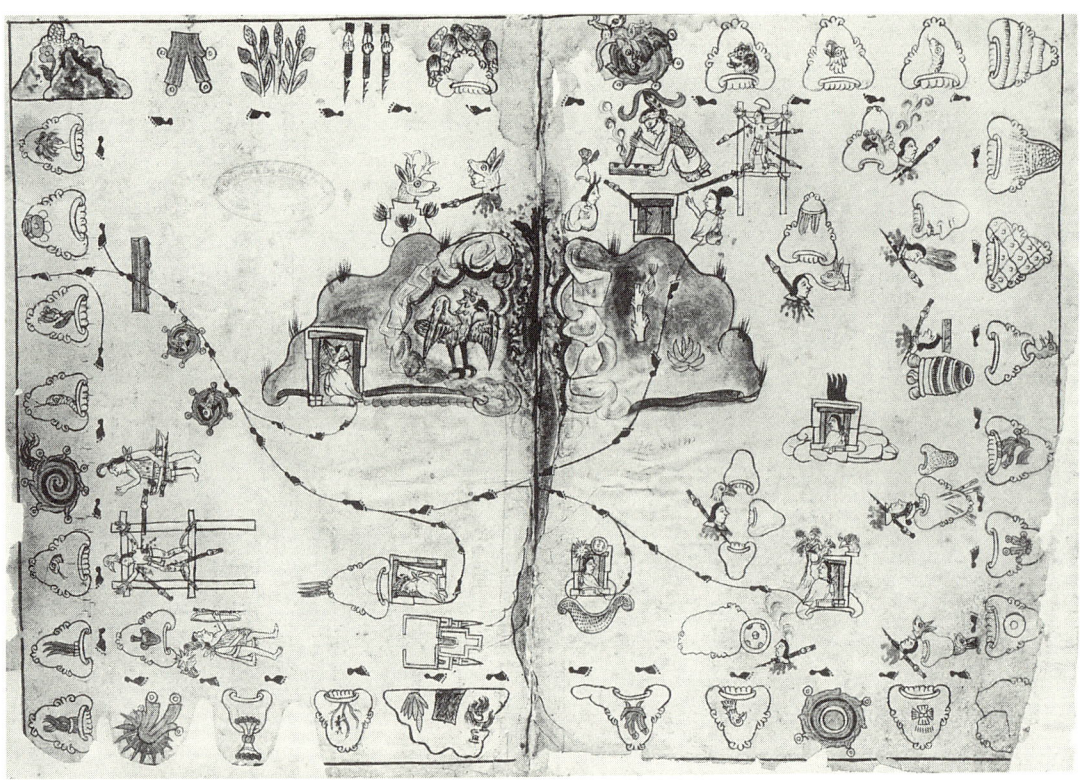

5.6 Map of Cuauhtinchan bound in with the Historia Tolteca-Chichimeca, fols. 32v–33r. This image also appears in Eugène Boban, *Documents pour servir à l'histoire du Mexique* (Paris: Leroux, 1891), atlas, no. 50.

trolled their own lands, received tribute from subject towns, and kept their own history. Documents that identified the people and their territory were kept in the archives of each community kingdom. Dozens of such documents have survived from the colonial period; often they are called *lienzo*s, because they are painted on large panels of cloth or paper. We might think of them as community charters for they are maps or diagrams of a community and its communal lands often also painted with historical and genealogical information about the ruling family. They served as land titles at the same time that they conveyed the community's identity, for they located the community in space and time.²⁶ All the surviving *lienzo*s date from after the Spanish conquest, but they recall pre-Columbian antecedents.

The *lienzo* from the small town of Zacatepec, in the coastal Mixtec region of far western Oaxaca, shows what kinds of information these town charters recorded and how it could be arranged and presented (fig. 5.7). On the rough surface of the great cotton sheet, the artist has outlined Zacatepec's territory as a "cartographic rectangle"—to use M. E. Smith's term—a rectangle on which are painted the place signs of Zacatepec's borders.²⁷ Such boundary markers can be hills, fields, or special topographic features like springs

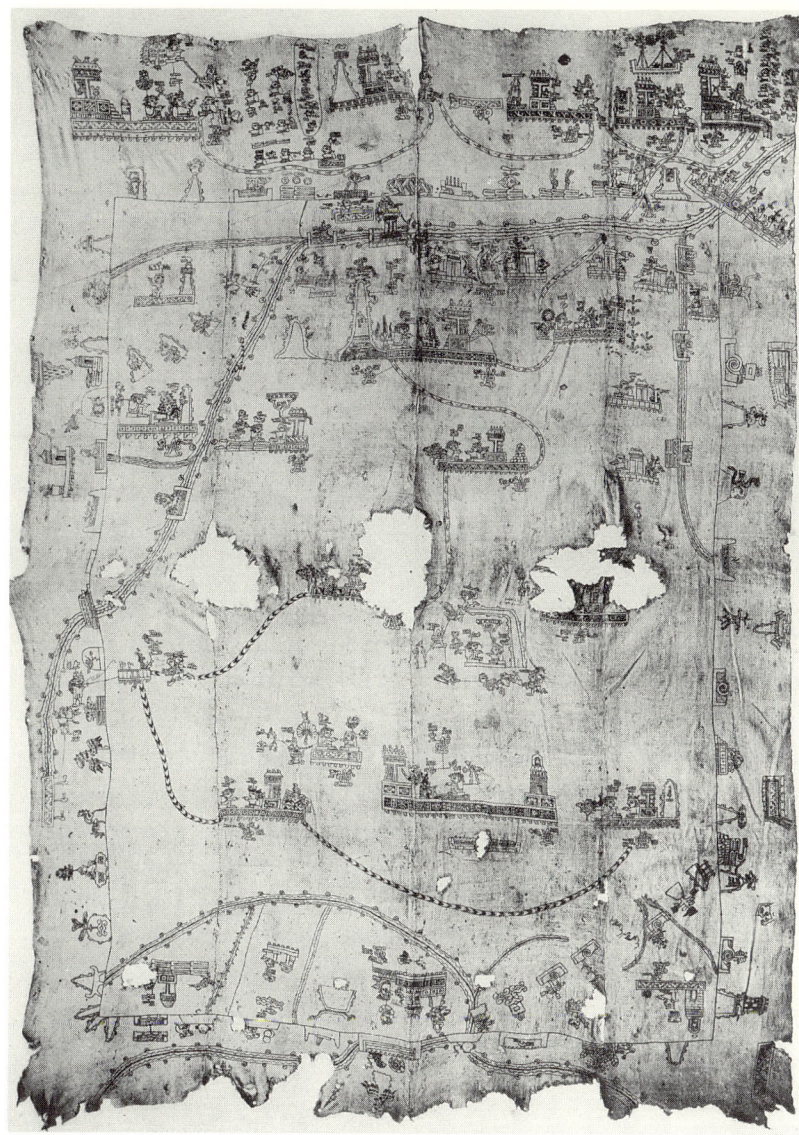

5.7 Lienzo de Zacatepec. This image also appears in Antonio Peñafiel, *Códice mixteco: Lienzo de Zacatepec* (Mexico City: Secretaría de Fomento, 1900), pl. 1.

or rocks. Between the borders, rivers cut across the land and subject towns are identified by their pictorial toponyms. Because some of the place signs have been identified, we know that many are arranged cartographically. In the center of the sheet toward the top, the town of Zacatepec is itself identified by the largest of all the place signs, which embraces both its Mixtec name (Hill of Seven Water) and its Nahuatl name (Zacate [grass] Hill).[28] The ruling couple, glyphically named according to their day signs, sit facing each other, their presence and attitude signaling the founding of the polity of Zacatepec. The *lienzo* thus establishes the geographic and political definition of the community.

Some *lienzo*s stop here, giving only the geography and rulers or founders. Many others, however, also add the polity's history and/or genealogy. Zacatepec's *lienzo* details the history of three generations of Zacatepec rulers, from their arrival into the area to their conquest and accommodation of neighboring peoples. The story begins in the upper left corner, outside the boundary of Zacatepec, where the founder of the local dynasty receives political authority from the Mixtec culture hero named 4 Wind. The founder's pilgrimage is shown across the top of the cloth to the far right, until he finally enters the boundary of Zacatepec, where he establishes his rule and is shown seated facing his wife; paths of footprints mark the progression. After this point, the dynastic founder is no longer an important character in the story, and the narrative moves to the left to the next generation of rulers, who are the couple seated on Zacatepec's place glyph. They also then drop from the narrative, which picks up the third-generation ruler and his conquests, which consolidate the Zacatepec domain. In this way, after defining Zacatepec's geographic extent, the *lienzo* presents its rulers and establishes their hereditary claim to rule by carrying the story back to the moment when the family patriarch first receives his authority. The Lienzo of Zacatepec was painted to identify this community kingdom, and the identity of this community was conceived in historical as well as geographical terms.

The maps of Cuauhtinchan also function in this same way to situate and define the polity and its people.[29] Cuauhtinchan Map 2 (fig. 5.4) may emphasize the long migration story, but it nonetheless presents Cuauhtinchan's territory geographically. Like the Lienzo of Zacatepec, it features, on its right half, a border of place signs that define the edges of the land in question; then it shows within this frame the actions of consolidation. Cuauhtinchan Map 1 and the map within the Historia Tolteca-Chichimeca (figs. 5.5 and 5.6) both emphasize Cuauhtinchan's territoriality and the battles fought to take control; they condense the journey to focus on the polity's establishment. All these documents define their community kingdoms geographically and set down the credentials of their rulers to govern, whether these credentials stem from a long migration, the receipt of emblems of office, or the conquest of neighboring peoples.

For the Aztecs and their neighbors, who wrote pictographically rather than alphabetically, such documents functioned as community charters and surveys. They document the town's founding and extent, and could be brought out as needed. Writing at the opening of the seventeenth century, the Texcocan chronicler Fernando de Alva Ixtlilxochitl recalled an old man from Huexotla who was very learned in matters of the land and who kept the land documents; Alva Ixtlilxochitl explained that when communities fell into dispute over territory, they would send representatives to this savant, who would settle the matter and who would demonstrate the origin of the territorial divisions.[30] After the conquest, such community documents continued to be brought out for this same purpose. The Spanish courts accepted *lienzo*s as valid land titles in the colonial period,[31] as do Mexican courts today. The Lienzo de Zacatepec itself was guarded in the municipal archive of Santa María Zacatepec until 1892, when townspeople took it and a second *lienzo* to Mexico City to submit as evidence in a suit petitioning that Zacatepec's ancestral lands be formally deeded over to the community. The suit was a success, but the *lienzo*s remained in the Secretariat of Agriculture and Development; the townspeople took home copies.[32] Nearly a hundred years later, the town of Tequixtepec was more cautious with its ancient *lienzo*s; in a boundary dispute with a neighbor in 1970, it decided to send photographs rather than the originals to Mexico City, and the town still has its paintings.[33]

The leaders of Tequixtepec saw their *lienzo*s as more than valid land titles: they recognized them as paintings that embody community identity. The *lienzo*s were, and presumably still are, kept in a case in the municipal building, where the Mexican flag is also kept.[34] Like the flag, the *lienzo*s testify to Tequixtepec's place in the world, not only to its political affiliation but also to its sense of itself socially, historically, and economically.[35] The flag symbolizes the community's link to the modern nation-state of Mexico, but the *lienzo*s speak to the community's ancient foundations and its Mixtec identity. They document how the people and land of Tequixtepec came to be who and where they are.

All these *lienzo*s and maps, these community charters, come from relatively small towns and villages. Although these communities were undoubtedly larger before the Spanish invasion, even then they were dwarfed by the provincial capitals and imperial metropolises. This raises the question: If most, if not all, of the independent community kingdoms in the Aztec and Mixtec worlds had documents, like these *lienzo*s, that laid out the town's identity, would the imperial capital of Tenochtitlan also have had one, and if so, what might it have looked like? A single document would certainly not have detailed all the relevant data about territory and history for the Aztec capital, as did Cuauhtinchan Map 2 for that town; there was too much information to include. In the two hundred years that followed its official founding, the island city of Tenochtitlan had grown from the insignificant site barely recognizable in the Mapa Sigüenza to the most powerful city in Mesoamerica. Its rulers dominated the Triple Alliance empire that controlled most of Mexico; luxury goods poured in from distant tribute provinces. On the island alone, the population swelled to 150,000, and another million people crowded the cities and towns along the lake shore and up the valley. The scale of the city and its territory was clearly too large to be framed by a border of place signs.

Within Tenochtitlan itself, the architectural and sculptural program proclaimed the city to be the center of the world, physically and metaphorically. Monuments like the Stone of the Sun proclaimed that the Fifth Sun, the present era of humankind, was the Aztec Sun and that all lands were Aztec lands; victory monoliths like the Stone of Tizoc announced that Aztec rulers had conquered all peoples between heaven and earth. The Templo Mayor stood at the center of the cosmos as the mountain of sustenance from which all abundance sprang. Each human sacrifice on the temple's steps virtually reenacted the victory of the Aztec patron god Huitzilopochtli over his enemies.[36]

Painted manuscripts also fixed Tenochtitlan centrally in time and space, and expressed community identity. The emphasis, however, was not on boundaries. Compared with their Zacatepec contemporaries, the Aztecs of Tenochtitlan were relatively unconcerned with precise geographic delineation. Their territory had long ago spilled over the borders of their community kingdom and had become inseparable with much of the empire's extent. There were contested borders and frontier zones, to be sure, but the empire lacked the same kind of fixed boundaries that concerned the people of Zacatepec. Rather, the Aztec empire was composed of a series of alliances, political and military obligations, and tribute arrangements owed by the provinces to the capital. Its edges were not an issue. Manuscript painters of the metropolis focused instead on locating the Aztecs and their capital city of Tenochtitlan at the center of the world.

This is seen clearly in the Codex Mendoza, where an idealized map presents both Tenochtitlan and the empire, conflating the two (fig. 5.8). The codex was painted to describe the Aztec empire to a European reader (probably Charles V);[37] it has three sections: an imperial history, a tribute roster, and an ethnography of Aztec daily life. Its history, and thus the manuscript itself, opens with the map of Tenochtitlan at the moment of its founding. A blue band of fifty-one year signs runs counterclockwise from the year 2 House (the date of the founding) in the upper left corner to 13 Reed at the upper center, framing most of the page. Within, the painter presents Tenochtitlan as an abstract geography, structurally similar to those in the maps and *lienzo*s. A cartographic rectangle, almost a square, is created by the border of blue waters that surround and define the island of Tenochtitlan. Running diagonally from corner to corner, the waters of canals cut the city into its four great quarters.[38] At the center is the place sign of Tenochtitlan—the nopal cactus growing from a rock—embellished here by the great eagle the Aztecs sighted and understood as the sign that they had finally arrived at the place of their destiny. Scattered around the place sign and interspersed among the reeds and grasses of the swampy island are the seated figures of the founding clan leaders, all of whom are named by the glyphs attached to their shoulders and heads. Like the leaders who appeared in the Mapa Sigüenza, these founders are the ancestry of Tenochtitlan. At the bottom of the page, scenes of two early military victories—over Culhuacan on the left and Tenayuca on the right—point to the victorious future. Belonging to a later time and to different places, these victories are appropriately painted outside the map.

In its general aspect, the foundation scene in the Codex Mendoza is very much like a *lienzo*. It establishes the territory and founders of the city of Tenochtitlan, just as the Lienzo of Zacatepec and the Cuauhtinchan maps do for those community kingdoms. First it describes the physical setting of the Aztec capital city, picturing the lake, the canals, and the swampy island. Then it presents and names the city founders and shows the initial conquests that helped to consolidate territory. It serves as a charter for the city, fixing Tenochtitlan geographically and outlining the historical events that were crucial in its early history.

Unlike the *lienzo*s, however, the Codex Mendoza assigns Tenochtitlan no named boundaries—nor did the foundation scene in the Mapa Sigüenza include boundaries. This, I believe, is a key to understanding how the Mendoza painting

MAPS IN AZTEC MEXICO   129

5.8 Founding of Tenochtitlan, Codex Mendoza (MS. Arch. Selden. A.1), fol. 2r. Courtesy of the Bodleian Library, Oxford.

represents the ideology of the empire and the place of Tenochtitlan within it. The absence of borders makes the Mendoza presentation more than simply a map of Tenochtitlan at its founding; it makes it a painting of Tenochtitlan as the center of the world. Without boundaries, the waters that surround the island capital metaphorically become the vast waters that surround the earth. The four quarters outlined by the canals then become the four world quarters of the cardinal directions,

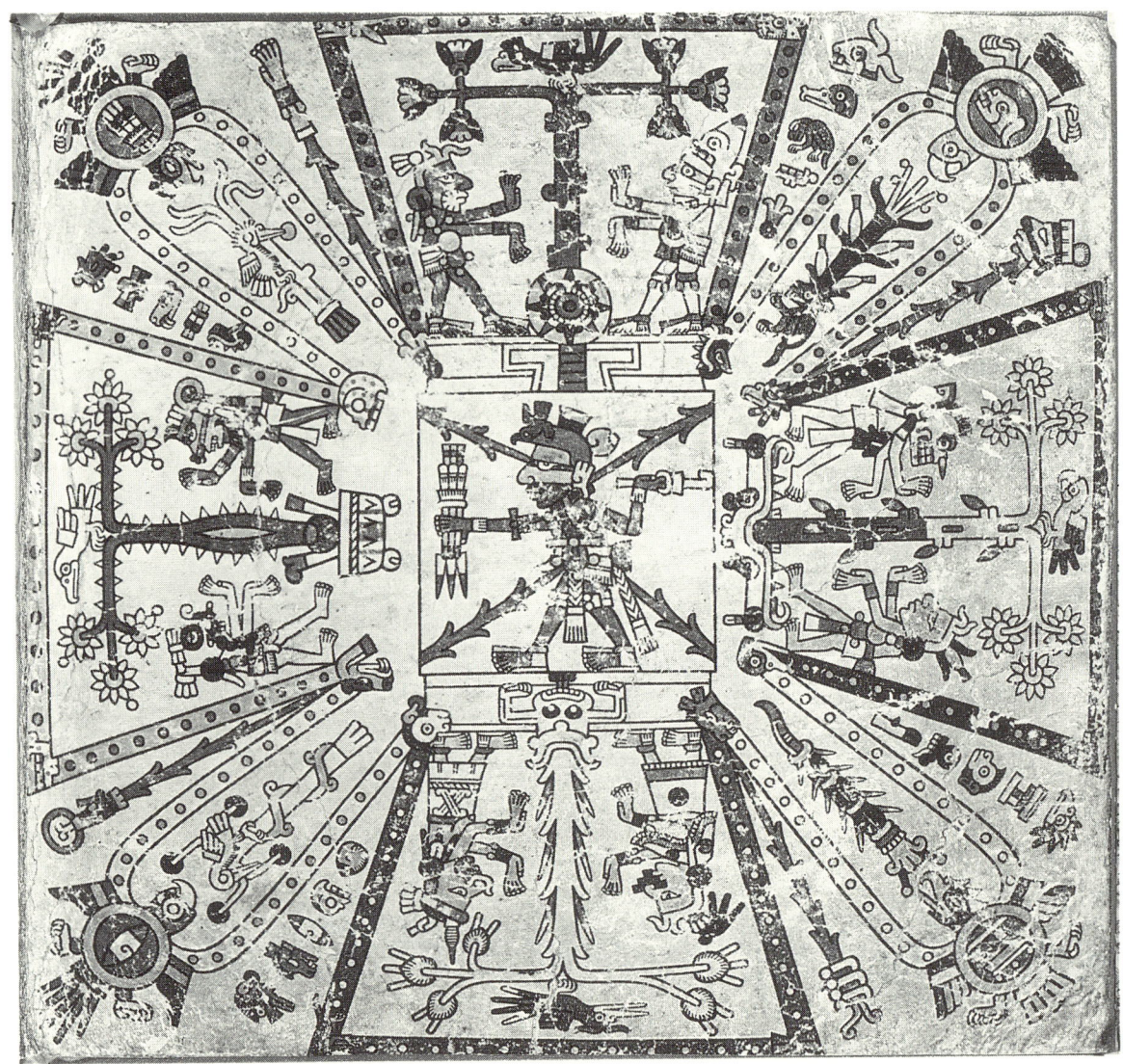

5.9 Codex Féjérváry-Mayer, p. 1. This image also appears in *Codex Féjérváry-Mayer: Colour Facsimile Edition of the Old Mexican Pictorial Manuscript* (Graz: Akademische Druck- u. Verlagsanstalt, 1971), 1. Courtesy Akademische Druck- u. Verlagsanstalt.

articulated just as they are in the ritual-divinatory codices, with Tenochtitlan solidly at the center. The Mendoza map compares easily with a diagram of time in the Codex Féjérváry-Mayer (fig. 5.9), where the 260 days of the ritual day-count are organized in a formée cross according to the four directions. Within the arms of the cross are the sacred tree, the bird, and the deities who preside over each world quarter.

In the Codex Mendoza (fig. 5.8), the community of Tenochtitlan is painted without borders because official imperial rhetoric held that its lands extended to the ends of the earth. These "community lands" of Tenochtitlan merged conceptually with the territory of the Aztec empire, which also had no borders. This painting in the Codex Mendoza preserves for us the Aztec view that their empire had no geographical limit. It was the center of the spatial world and controlled all the lands around it.

The Aztecs relied on maps to chart paths of movement, and they relied on them to organize space visually. Painted documents guided travelers and proposed routes into and out of battle. Practical, useful maps outlined property held and fields worked. Most important, however, maps encoded history and organized the community territorially. The cartograph became for the Aztecs a popular foundation on which to structure the telling of migration stories, where the narrative was a people's wandering across the land until they reached their destined home. Then the map form proved an efficient ground for recording territorial consolidation. These cartographic histories excelled at presenting action as it occurred in different locations. They emphasized the places where events happened rather than the date of these events or all the participants involved, which made them the choice for history painters who were recording stories about the land.

Historical information inserted into maps of community kingdoms transformed the maps into community charters. The autonomous kingdoms of Aztec Mexico thought of themselves as peoples with their own territory and their own history; community identity was inextricably tied to land and to the stories about how the people came to control that land. The large maps and *lienzo*s were an ideal document for presenting this kind of information. A typical *lienzo* or community map would feature the pictorial toponym of the town prominently near the center, along with the founder or ruling family; around the edges of the sheet would be outlined the community's borders, delineated as a series of place signs. These elements together effectively established the fact and extent of the polity. Many *lienzo*s and maps additionally presented the credentials of the rulers to rule or noted the rights of the people to occupy that territory. Some of these rights were based on the hardships of a migration or the conquest of neighboring peoples; others rested on a dynastic patriarch having received political authority through a series of rituals. The story of how this authority was gained was included within or beside the map of community territory, for the Aztecs and their neighbors conceived of their communities as being spread over the land because they were anchored to the past.

Notes

1. Bernal Díaz del Castillo, *The Discovery and Conquest of Mexico*, translated by A. P. Maudslay (New York: Farrar, Strauss and Cudahy, 1956), 72.

2. Ibid., 73.

3. Alonso de Molina, *Vocabulario en lengua castellana y mexicana y mexicana y castellana*, edited by Miguel León-Portilla (Mexico: Porrua, 1970), 2d pagination, 120.

4. Díaz del Castillo, *Discovery and Conquest,* 73–74.

5. For tax and tribute records, see, for example, Fernando de Alva Ixtlilxochitl, *Obras históricas*, edited by Edmundo O'Gorman (Mexico: Universidad Nacional Autónoma de México, 1975), 2:145; Hernan Cortés, *Her-*

nan Cortés: Letters from Mexico, edited and translated by Anthony Pagden (New Haven: Yale University Press, 1986), 109; Díaz del Castillo, *Discovery and Conquest,* 211; Alonso Zorita, *Life and Labor in Ancient Mexico: The Brief and Summary Relation of the Lords of New Spain*, edited and translated by Benjamin Keen (New Brunswick: Rutgers University Press, 1963), 110. For land records, see Alva Ixtlilxochitl, 1:286, 527; and Zorita, 110. For court records, see Bernardino de Sahagún, *Florentine Codex: General History of the Things of New Spain*, edited and translated by Arthur J. O. Anderson and Charles E. Dibble (Santa Fe: School of American Research and the University of Utah, 1959–81), bk. 8:42, 55; Motolinía, *Memoriales o libro de las cosas de Nueva España y de los naturales de ella*, edited by Edmundo O'Gorman (Mexico: Universidad Nacional Autónoma de México, 1971), 354. For battle plans, see Sahagún, bk. 8:51.

6. Motolinía, *Motolinía's History of the Indians of New Spain*, edited and translated by Francis Borgia Steck (Washington, D.C., Academy of American Franciscan History, 1951), 74–75; Motolinía, *Memoriales*, 5. For a review of the types of preconquest books, see Elizabeth Boone, "Pictorial Documents and Visual Thinking in Postconquest Mexico," in *Native Traditions in the Postconquest World*, edited by Elizabeth Boone and Tom Cummins (Washington, D.C.: Dumbarton Oaks, in press).

7. *Ficus petiolaris* is the most common.

8. The phrase "Aztec Mexico" here refers to the general area that came under the domination of the Triple Alliance empire as well as the areas just outside Aztec control, such as parts of the Mixteca. Specialists might object to my gathering different ethnicities under the "Aztec" label, but I believe these central Mexican peoples had more in common than not.

9. Sahagún, *Florentine Codex,* bk. 8:51.

10. Molina, 1st pagination, 82. Robert Haskett, who kindly helped me with the Nahautl translation, also pointed out that it is common for Molina to give two definitions that are different ways of saying the same thing.

11. Francisco de Alvarado, *Vocabulario en lengua mixteca*, edited and translated by Wigberto Jiménez Moreno (Mexico: Instituto Nacional Indigenista and Instituto Nacional de Antropología e Historia, 1962), 146r. John Monaghan kindly provided the translation from Mixtec to English.

12. I am grateful to Robert Haskett for pointing out the use of the loan-word *mapa*.

13. This article focuses on three kinds of maps from central Mexico; for a broader treatment of Mesoamerican mapping, including Maya maps, see Barbara Mundy, "Mesoamerican Cartography," in David Woodward and G. Malcolm Lewis, eds., *Cartography in the Traditional African, American, Arctic, Australian, and Pacific Societies*, vol. 2, bk. 3 of *The History of Cartography* (Chicago: University of Chicago Press, forthcoming 1998). For the maps painted to accompany the Relaciones Geográficas questionnaires, see, Mundy, *The Mapping of New Spain: Indigenous Cartography and the Maps of the Relaciones Geográficas* (Chicago: University of Chicago Press, 1996).

14. Cortés, *Letters from Mexico*, 94.

15. Ibid., 467 n. 49.

16. Francisco López de Gómara, *Cortés: The Life of the Conqueror by His Secretary*, edited and translated by Lesley Byrd Simpson (Berkeley: University of California Press, 1964), 345.

17. Motolinía, *Motolinía's History*, 74–75; *Memoriales,* 5.

18. Sahagún, *Florentine Codex,* bk. 8:51.

19. Most histories are stories of people, considered individually or collectively, although Stephen J. Gould's book *Wonderful Life: The Burgess Shale and the Nature of History* (New York, W. W. Norton, 1989) reminds us that fossils, as well as other animals and plants, can also be the protagonists in histories.

20. For the different kinds of Mexican pictorial histories see Elizabeth Boone, "The Aztec Pictorial History of the Codex Mendoza," in *The Codex Mendoza*, edited by Frances F. Berdan and Patricia R. Anawalt, 2 vols. (Berkeley: University of California Press, 1992), 1:35–54, 152–153; and Boone, "Aztec Pictorial Histories: Records without Words," in *Writing without Words: Alternative Literacies in Mesoamerica and the Andes*, edited by Eliza-

beth H. Boone and Walter G. Mignolo (Durham: Duke University Press, 1994), 50–76.

21. Mesoamerican years are identified by the combination of one of thirteen sequential numbers and one of four symbols; they total fifty-two before the cycle completes itself and begins again. The year 13 Flint is thus the equivalent of A.D. 1544, 1492, 1440, 1388, etc.

22. Díaz del Castillo, *Discovery and Conquest,* 157, recalls how the Tlaxcalans, once they had allied themselves with the Spaniards, brought out large hemp cloths on which were painted records of the battles they had fought with the Mexicans; he said the detail was sufficient to show "their manner of fighting."

23. See Bente Bittmann Simons, *Los mapas de Cuauhtinchan y la Historia Tolteca-Chichimeca* (Mexico: Instituto Nacional de Antropología e Historia, 1968); Keiko Yoneda, *Los mapas de Cuauhtinchan y la historia de cartográfica prehispánica* (Mexico: Fondo de Cultura Económica, 1991); Dana Leibsohn, "Primers for Memory: Cartographic Histories and Nahua Identity," in *Writing without Words*, 161–87.

24. See Alfonso Caso, "Los lienzos mixtecos de Ihuitlan y Antonio de León," in Ignacio Bernal et al., eds., *Homenaje a Pablo Martínez del Río* (Mexico: Instituto Nacional de Antropología e Historia, 1961), 237–74.

25. See Antonio Peñafiel, *Códice mixteco: Lienzo de Zacatepec* (Mexico: Oficina tipográfia de la Secretaría de Fomento, 1900); the *lienzo* has been thoroughly analyzed by Mary Elizabeth Smith, *Picture Writing from Ancient Southern Mexico: Mixtec Place Signs and Maps* (Norman: University of Oklahoma Press, 1973), 89–121.

26. A preliminary discussion of community charters appears in Elizabeth Boone, "Glorious Imperium: Understanding Land and Community in Moctezuma's Mexico," in *Moctezuma's Mexico: Visions of the Aztec World*, by Davíd Carrasco and Eduardo Matos Moctezuma (Niwot: University Press of Colorado, 1992), 159–173.

27. Smith, *Picture Writing,* 92. My observations are based on Smith's study of the *lienzo;* see Smith, 89–121.

28. Ibid., 96.

29. See also Leibsohn, "Primers for Memory."

30. Alva Ixtlilxochitl, *Obras históricas*, 1:286; see also 527.

31. Smith, *Picture Writing,* 170.

32. Peñafiel, *Códice mixteco,* 1–2; Smith, *Picture Writing,* 89.

33. Ross Parmenter, *Four Lienzos of the Coixtlahuaca Valley,* Studies in Pre-Columbian Art and Archaeology, no. 26 (Washington, D.C.: Dumbarton Oaks, 1982), 46–50.

34. Parmenter, *Four Lienzos,* 47.

35. Dana Leibsohn, in "Primers for Memory"; and "The Historia Tolteca-Chichimeca: Recollecting Identity in a Nahua Manuscript" (Ph.D. diss., UCLA, 1993), 1–4, 267–71, has pointed out how the Cuauhtinchan maps continue to function as anchors of Cuauhtinchan identity.

36. For the Stone of the Sun and the Tizoc Stone see Richard F. Townsend, *State and Cosmos in the Art of Tenochtitlan,* Studies in Pre-Columbian Art and Archaeology, no. 20 (Washington, D.C.: Dumbarton Oaks, 1979), esp. 43–49, 63–70; for the symbolism of the Templo Mayor see Eduardo Matos Moctezuma, "Symbolism of the Templo Mayor," in Elizabeth H. Boone, ed., *The Aztec Templo Mayor* (Washington, D.C.: Dumbarton Oaks, 1987), 185–209; Johanna Broda, "The Provenience of the Offerings: Tribute and *Cosmovisión*," in *The Aztec Templo Mayor*, 211–56; Johanna Broda, Davíd Carrasco, and Eduardo Matos Moctezuma, *The Great Temple of Tenochtitlan: Center and Periphery in the Aztec World* (Berkeley: University of California Press, 1987).

37. Evidence suggests that the codex was commissioned by the Viceroy Antonio de Mendoza; for its history see H. B. Nicholson, "The History of the Codex Mendoza," in Berdan and Anawalt, eds., *The Codex Mendoza*, 1:1–11.

38. Tenochtitlan was divided politically and socially into four quarters; see Edward E. Calnek, "The Internal Structure of Tenochtitlan," in Eric R. Wolf, ed., *The Valley of Mexico: Studies in Pre-Hispanic Ecology and Society* (Albuquerque: University of New Mexico Press, a School of American Research Book, 1976), 287–302.

6

Inland Journeys, Native Maps

Barbara Belyea

The familiar shapes of the continents on eighteenth-century European maps are deceptive; apart from Europe itself, little more than the coasts had been explored. Knowledge of the interior of non-European continents was limited to the reports of commercial or religious agents, who relied in turn on what the native inhabitants showed and told them.[1] Even into the first years of the nineteenth century, scientific mapping of North America beyond the Great Lakes and Hudson Bay was limited to fur trade surveys and information the traders gleaned from native maps. The problem with using Amerindian geographical knowledge lay in translating its descriptions and graphic signs into terms that were understood and accepted in the context of European scientific mapping. Although respectful of native cartography, far more than many historians of cartography are today, fur trade explorers stood outside native culture, reformulating and reconfiguring what they could of a foreign convention in terms of their own.

The rapid advance of French and British explorers into the North American continental interior can be traced on four exemplary maps: two by Jacques-Nicolas Bellin entitled *Carte de l'Amérique septentrionale,* published in 1743 and 1755; Andrew Graham's manuscript *Plan of Part of Hudson's-Bay & Rivers Communicating with York Fort & Severn,* drawn in 1774; and Aaron Arrowsmith's *Map Exhibiting all the New Discoveries in the Interior Parts of North America . . . ,* which appeared first in 1795 and in modified states for the next four decades. Each of these maps represents the extent of generalized European knowledge of the continental interior at the time it was drawn.[2]

On his 1743 map, Bellin reproduced the chain of lakes and portages from Lake Superior to Lake Winnipeg initially communicated to La Vérendrye by three Crees: Ochagach, Tacchigis, and La Marteblanche. The stages by which Parisian cartographers received this native knowledge has been well documented by G. Malcolm Lewis.[3]

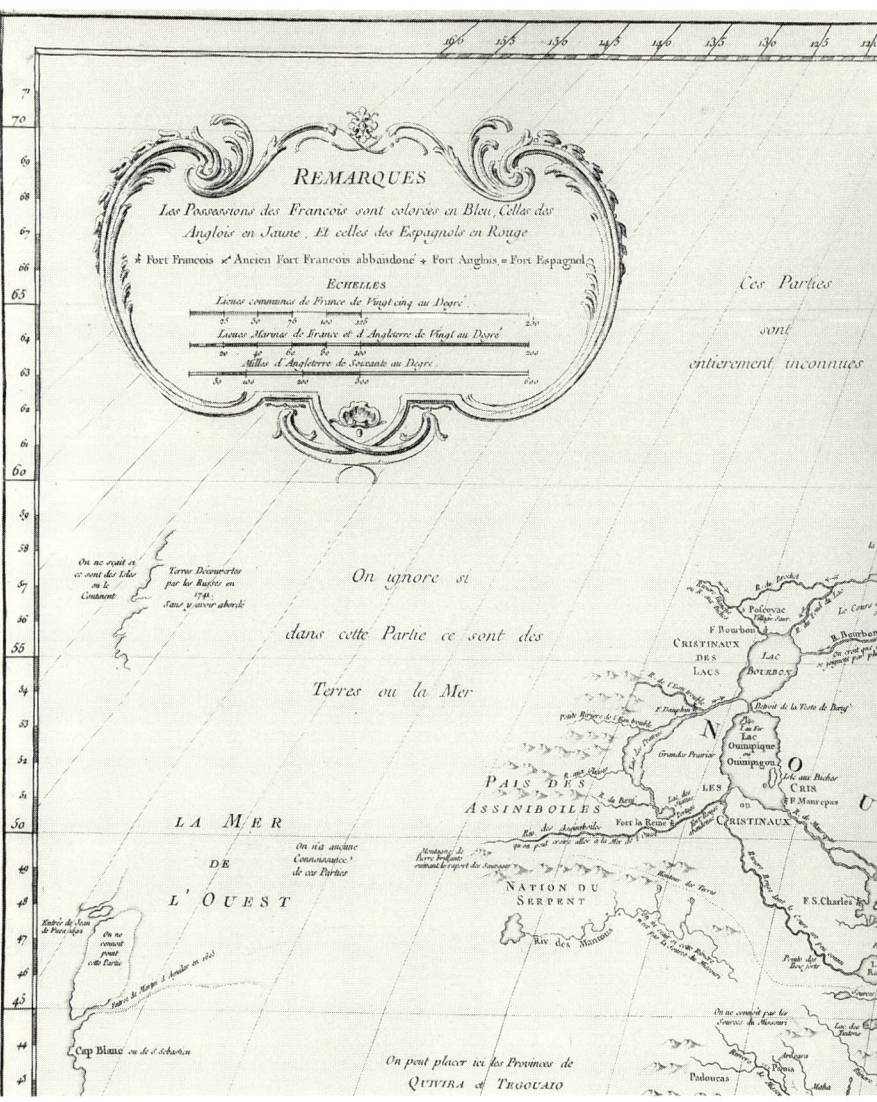

6.1 Detail from Jacques-Nicolas Bellin, *Carte de l'Amérique septentrionale* (Paris, 1755). North and west of the Great Lakes, French knowledge of the continental interior was limited to Amerindian reports, as indicated by the tribal names assigned to these frontier regions. Courtesy of the Newberry Library.

Bellin extended this river route across the North American continent, where only a few legendary details gleaned from Spanish reports of the previous century occupied the blank space west of Lake Superior and the Mississippi River. On his later map, Bellin reduced the chain of lakes so that it stretched only to the center of the continent. Otherwise the 1755 map still betrays complete ignorance of the region north and west of Lake Superior: repeatedly it notes that "Ces parties sont entierement inconnues," "On n'a aucune Connoissance de ces Parties," "On ignore si dans cette Partie ce sont des Terres ou la Mer." At the same time, La Vérendrye's composite of the Cree maps allowed Bellin and other French cartographers to blur the line of European penetration by including an intermediate zone—what Malcolm Lewis has called *terra semicognita*.[4] This gray zone must be deduced from stylistic changes of form and nomenclature. Bordering on the blank space of the 1755 map are the names of several tribes ("Les Cris ou Cristinaux," "Nation du Serpent," "Nation des Monsonis") and the name of a country ("Pais des Assiniboiles") also identified by tribe. Until Europeans surveyed the land itself, tribes and *pais* denoted regions still unexplored and known only by native report. Bellin's maps frame and present the unknown continental interior as what is not yet known.

On Bellin's 1755 map (fig. 6.1), both native-derived knowledge and admissions of European ignorance are artfully subordinated to the map's scientific pretensions. The French cartographer took care to balance the lower right cartouche (fig. 6.2), which gives his credentials, with another cartouche in the upper left corner. Placed in the blankest part of the map, where uncertainty reigns as to whether there is land or sea, the second cartouche explains the color washes demarcating European colonies and provides multiple European measurements of distance. The map reader's eye moves from the colored colonies, across the disturbing blank of unknown space, to the reassuring explanations of the second cartouche. These assertions of imperial and scientific authority impose on the still unmapped territory a promise of rapid discovery and subjugation.[5]

Graham's *Plan* of 1774 (fig. 6.3) reveals how little was generally known to the British of the Hudson Bay drainage even after a century of trade on the coast and twenty years of travel to the

6.2 Title cartouche, Jacques-Nicolas Bellin, *Carte de l'Amérique septentrionale* (Paris, 1755). Royal patronage and the cartographer's scientific credentials lent authority to Bellin's image of North America. Courtesy of the Newberry Library.

western edge of the prairies. In 1754–55 Anthony Henday went inland with a party of Crees in order to promote trade with other native groups on the plains. His map and original journal have both been lost, and Henday does not appear to have kept a journal during his second trip in 1759–60. Other winterers left no better record until Samuel Hearne traveled from Churchill to the Coppermine River in 1770–72. Graham's map shows the river routes inland from York and Severn forts to Lake Winnipeg, and the Saskatchewan River up to the fork. Lake Winnipeg, which extends almost due north/south on modern maps, inclines west-northwest/east-southeast on Graham's sketch. The salient features of Graham's map are limited to a few named groups of hills, but rivers, lakes, and portages are clearly marked. All of these characteristics reveal Graham's debt to Amerindian mapping. Yet his *Plan* shows the homes of various native groups as separate terri-

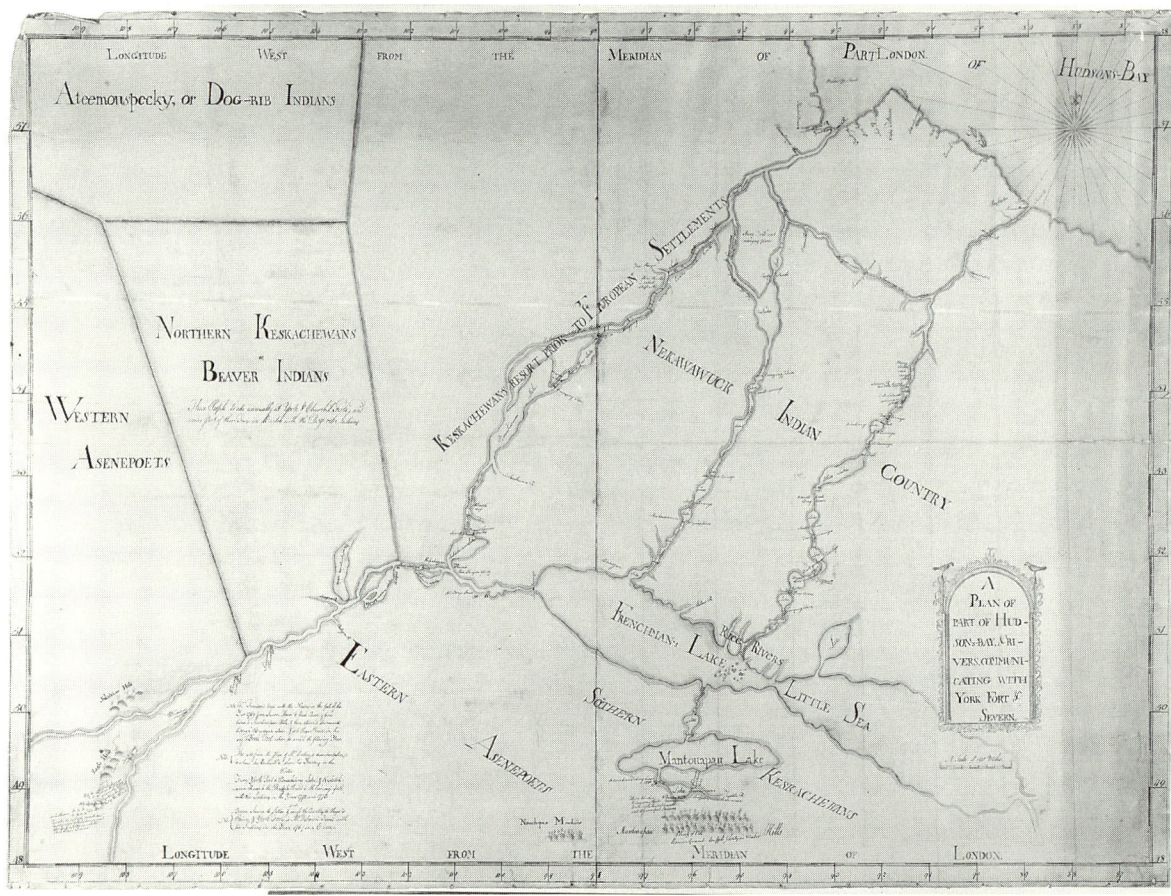

6.3 Andrew Graham, *A Plan of Part of Hudson's Bay and Rivers, Communicating with York Fort & Severn* (1774). Graham's manuscript shows the "countries" of western tribes as clearly demarcated territories comparable to European colonies along the St. Lawrence River and the Atlantic coast. Courtesy of Hudson's Bay Company Archives, Provincial Archives of Manitoba, HBCA G.2/17 (N4607).

tories enclosed in borders, very like the European colonies of Bellin's map. Another border frames the sketch, marked off in degrees of latitude and of longitude from London.[6]

The original 1795 Aaron Arrowsmith's *Map Exhibiting all the New Discoveries* includes the river routes to Lake Winnipeg as they appeared on earlier maps. The upper Saskatchewan and lower Missouri rivers are partially drawn in, as well as rivers and lakes leading to Lake Athabaska. Along the western edge of the prairies is a long ridge of "Stony Mountains"; at their southern limit, a note marks Peter Fidler's journey with a Peigan band in 1792–93. Arrowsmith added a pasted-on slip showing Hearne's route from

Churchill to the Arctic Ocean when the explorer's narrative was published in 1796. On successive states of the map, the features that progressively filled its blank interior indicate the sources of Arrowsmith's knowledge: reports of HBC winterers, notably Hearne and Fidler, who traveled with native bands. Fidler, who kept detailed journals and was a far better surveyor than Hearne, was the last of the HBC employees to accompany natives on their seasonal circuit; later he wintered at inland posts along the Saskatchewan River. Subsequent states of the Arrowsmith map, beginning with the first issue dated 1802, extended the Rocky Mountains south and added rivers of the Missouri watershed. All of these additions were derived from maps and descriptions sent by Fidler to the HBC London Committee.[7]

What distinguishes the HBC winterers from other North American explorers is that for four decades they were followers, not leaders, of Amerindian groups, living for months and sometimes years according to customs, values, and ideas radically different from those of their own culture.[8] These native groups had their own cartographic conception of where they lived: for them the continental interior was obviously not a blank space anticipating the Europeans' full knowledge. Henday's second trip inland is mapped on a "Draught of the Northern Parts of Hudson's Bay," drawn in 1760 by Moses Norton, the Métis governor at Churchill. Norton's map, "laid down on Indn. Information," includes not only the Saskatchewan River where Henday apparently wintered, but also Lake Athabaska, Great Slave Lake, and the Arctic coast.[9] Norton's sketch thus shows many more features of northern North America than Bellin's roughly contemporary 1755 map, which declares that "Ces parties sont entierement inconnues." Graham's *Plan*, drawn after Hearne had returned from the Coppermine River but with no indication of his route, resembles Bellin's in recording only the names of tribes that inhabit this otherwise little-known region: "Northern Keskachewans or Beaver Indians," "Western Asenepoets," "Ateemouspecky, or Dog-rib Indians." In contrast, Norton's map depicts the region in some detail, according to the native cartographic convention. This system of mapping is so different from the European system of spatial coordinates that Norton's map puzzled even June Helm, a leading anthropologist, who finally explained its unfamiliar design by locating its features on a modern topographical sheet.[10]

Helm is not alone in her response to a design which purports to be a map, but which looks so little like the maps produced by our own culture. At least she labored over Norton's "Draught." The treatment accorded such maps in more than one respected history of cartography is to relegate them to the "primitive" category and to focus on the development of European cartography from its roots in Greek theory to its status as a scientific record based on empirical observation.[11] This persistently scientific, Eurocentric bias is seen in Christian Jacob's choice of an anecdote in *L'Empire des cartes*. A Huron asks a missionary to draw a map of the world and to explain its features. The direction of inquiry in this story is intriguing, since in most records of exploration it is Europeans who ask natives to draw maps. Posing as the "maître," the priest gives a "leçon de géographie" to the Huron who has no idea of the wider world beyond his own forest: "la carte est médiation entre le monde et le Huron, entre le Père et le

Sauvage: elle synthétise, elle enferme sur l'espace d'une écorce l'immensité du monde réel."[12] The interest of this anecdote lies not so much in the crude partiality of Jacob's terms "maître" and "Sauvage" as in the conception of cartographic space it conveys—the *mise en abyme* of drawing the world in little, of confining its extent to a piece of bark. The map is a scaled-down reproduction of the earth's dimensions; features must therefore be drawn "to scale," each one occupying a precise point on the representative surface. By this definition of the map—since Jacob considers his anecdote to be a "récit exemplaire et édifiant"—cartographic space is a space extending uniformly in all directions, which therefore allows exact location of points and correlation with the earth's surface.

Pierre Clastres called anthropology's scientific claims to objectivity and universality an exercise in ethnocentric ideology,[13] and the same could be said for claims made by certain historians of cartography. Although the multivolume history edited by J. B. Harley and David Woodward will go far to modify this focus of cartography, one series cannot be expected to transform the axioms of an entire field of study.[14] What needs general recognition is that maps are purely conventional, that convention determines perception, and that this perception is culturally specific. Moreover, cartographic convention is only partially communicable from one culture to another, and from one period to another: "translation" is possible only in a superficial, limited sense.[15]

Appeal to scientific cartography as a standard by which native map images are to be understood therefore guarantees that they will be *mis*understood. The procedure of translating the terms of one cartographic convention into those of another may serve the purpose of rough, practical route-finding, and certainly it was for such a purpose that explorers solicited Amerindian maps. But this process has almost always privileged one convention: the explorers' own, and now the historians' own. Translation obviates the need to recognize native maps for what they are: graphic forms which represent a worldview that is utterly different from that produced by European scientific cartography.[16] Traditional modes of conserving and communicating these maps make historical analysis of the graphic designs which survive very problematic. Amerindian maps were copied as simple sketches, without the descriptions and toponyms that would have been communicated orally. The graphic form would also have been occasional; normally these maps were stored in the memories of elders and leaders, and were drawn in response to a specific situation or to questions. A parallel might be drawn with the memorial culture of Inuit north of Churchill, as revealed in a comment made to Robert A. Rundstrom: "During fieldwork in 1989, one Inuk elder told me that he had drawn detailed maps of Hiquligjuaq from memory, but he smiled and said that long ago he had thrown them away. It was the act of making them that was important, the recapitulation of environmental features, not the material objects themselves."[17] To recognize the distinctiveness of Amerindian maps and to understand their role in the scientific mapping of North America, we must resist the temptation to translate their signs into ours, and accept that these maps constitute a complete and valid cartographic convention without recourse to "accuracy" or explanations in scientific terms. Native maps are not crude attempts to render geometric space, inferior to contemporary European maps such as Bellin's or Arrowsmith's, or to modern topographical surveys.

This kind of mapping is not restricted to "primitive" cultures. Not all maps, not even all European maps, conform to the Ptolemaic system of spatial coordinates on which modern topographical surveys and GIS mapping are based. The linear principle is operative in the ephemeral sketch maps we all make to indicate a route from A to B; it is also the principle underlying the guides to transport systems such as the London Underground.[18] Users of transport maps are unconcerned that junctions are not "actually" at right angles, or that the distances between stops and choice points are not "really" equal. Such concerns are not part of these maps' information or function. Grid maps operate by locating positions along axes of latitude and longitude. Amerindian maps rely not on fixed positions in space but on a pattern of interconnected lines.[19] Spacing and directions of north, south, east, and west are simplified, even ignored, since the key to reading the map is not to locate points in space but to trace a continuous path from one geographic feature to another. Intersection rather than spacing determines the cartographic design. We need to look again at the defining characteristics of native cartography, and instead of reading them as distorted spatial indicators, see them as consistently derived from a principle of linear coherence. Then native maps will begin to look less crude and to make more sense.

Of course, it is extremely hazardous to generalize about the hundreds of very distinct indigenous cultures of North America. A cautionary example is James Isham, mid-eighteenth-century factor of York Fort, who sent several of the HBC winterers inland. Isham generalized about the route-finding abilities of "Indians," and produced contradictory statements. At one point, he noted in his *Observations on Hudson's Bay* that "the four corner's of the world they Stile, and observe, But have no knowledge of all the Various points of the Compass, Neither have they any call for such, they seldom or ever going out of Lakes, Rivers & creeks, which they can see a cross from Land to Land." He also remarked that "the Natives are Seldom at a loss in their travelling, no part being Difficult for them to find, tho never was in some parts of their Country in their Lives,—yet will steer by the sun, or moon."[20] Obviously the technique of lining up "Lakes, Rivers & creeks" was appropriate to native groups who lived in the Canadian Shield forests, while astronomical referents became more important when crossing relatively featureless terrain such as parkland or prairie.

Nevertheless, native maps have surprisingly constant characteristics, even though they were drawn in places very distant from each other and over a period of centuries. Indeed, this consistency strongly argues for consideration of Amerindian cartography as a fully developed convention: these maps are highly stylized, highly standardized geographical indicators. They are easily recognized by their depiction of geographic features: round lakes, rivers drawn as straight or curved (not wavy) lines, slashes across the river lines to indicate portages, dots to show campsites and hunting areas, commemorative signs for raids and battles.[21] The most striking and important characteristic of native maps is that their network of lines is unframed, hence independent of a spatial grid or ground. In contrast, the frame of a traditional European map determines its form and function as a spatial construct; as Jacob remarks, "Le cadre, en délimitant l'espace de la représentation, est l'une des conditions de perception et de compréhension de la carte."[22] Stories of native cartographers who found too limiting the sheets of paper given them to draw on are evidence of the absence of a framing element in their conception of maps.[23]

Since native maps, like all maps, direct and organize perceptions of the regions they depict, the geographical features themselves must be separately defined and recognized. Native cartographers did not see rivers or hills in the way we have been taught to see them; there is thus no measure of "accuracy" or basis of comparison between ways of seeing. There is no direct, unmediated perception of the world which can confirm or correct cartographic images. The empirical argument that we align the pattern of the map with the pattern on our retinas and can thus determine cartographic error is simplistic. The notion of error, like that of scale, is conventional and culture-specific. We do not see with the naked eye; instead, we see what we are trained to look for. The map's connection with landforms out there is arbitrary, tenuous, and culturally imposed. When the map image is constructed according to a convention unfamiliar to the user, the landform represented on the map is as mysterious as the signs used to represent it. After much effort, we may be able to recognize places along the Arctic coast that are marked on Norton's "Draught"—that is, identify the map marks with those places as they are known in our culture—but we cannot begin to know the significance of those places as they relate to each other in the Dene worldview recorded on the map.

Eighteenth-century European cartographers tried, and failed, to absorb native data into their own maps. Either the native map signs were simply juxtaposed to their own surveys, or they were radically reconfigured to appear like the signs of surveyed regions, with all the risks of misunderstanding attendant on this transformation.

We have seen that Bellin's maps of 1743 and 1755 depict a route from Lake Superior to Lake Winnipeg ultimately derived from Amerindian maps. Malcolm Lewis has traced the process of "unacknowledged assimilations" and "French misinterpretation of . . . Indian cartographic information." Certainly, as it passed through intermediary hands to Parisian cartographers, La Vérendrye's transcript of native maps was redesigned and reinterpreted. At every stage, the Cree designs were read as one would read a foreign language, trying to recognize familiar forms in an unfamiliar context. The difficulty lay precisely in the fact that La Vérendrye and the French mapmakers were not dealing with "*in*formation" but with a foreign set of conventions which required them to *de*form the map images and *re*form them according to their own cartographic understanding. On his *Carte physique des terreins les plus élevés de la partie occidentale du Canada* (fig. 6.4), Philippe Buache acknowledged this difficulty by choosing to copy an intermediary form of the map still recognizably native in style, as well as to present his further interpretation of it.[24] In contrast, Bellin chose to "assimilate" the native convention, that is, to dissimulate native forms—to redraw the circular lakes and leaf-vein rivers as irregular

INLAND JOURNEYS, NATIVE MAPS

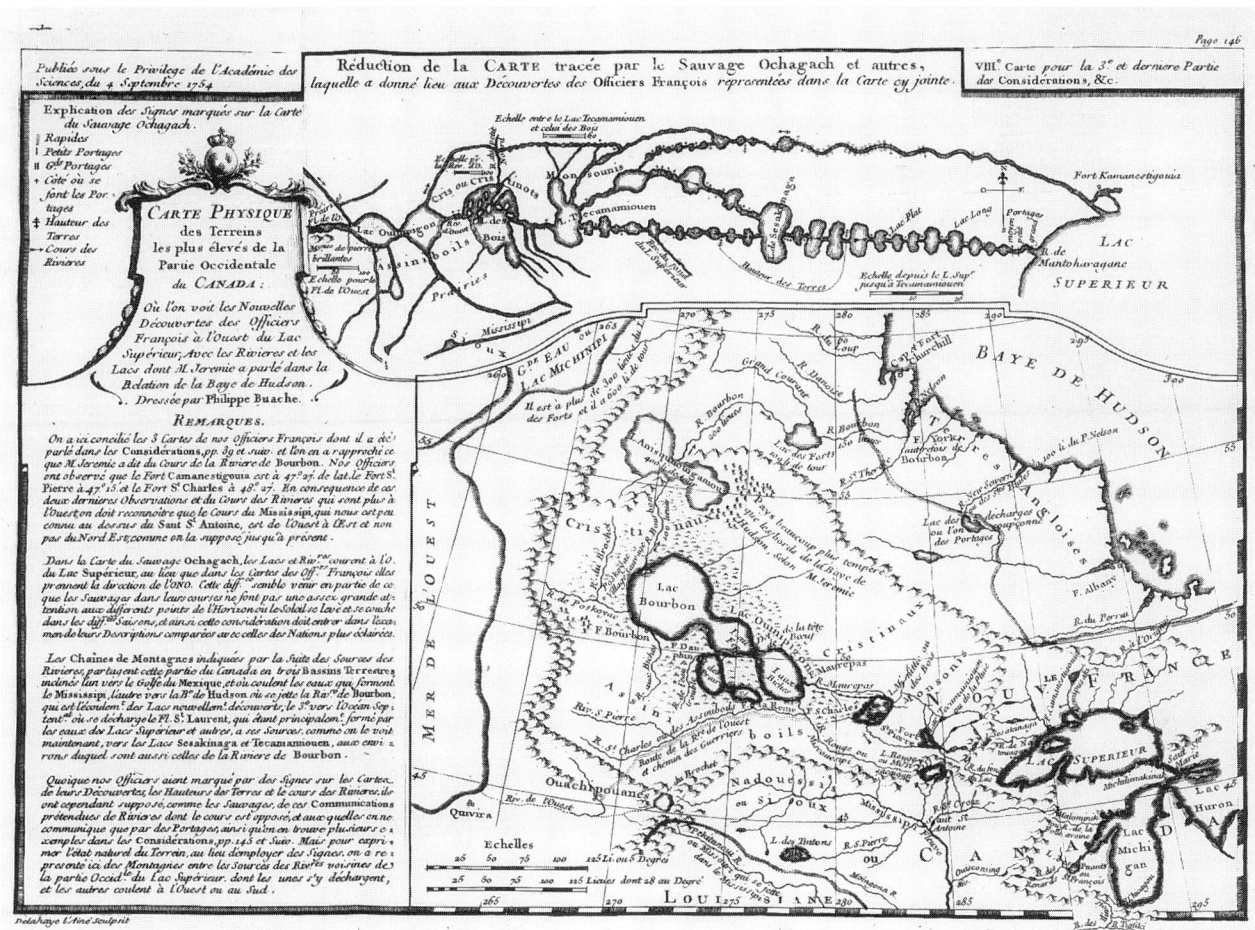

6.4 Philippe Buache, "Carte physique des terrains les plus élevés de la partie occidentale du Canada," from *Considérations géographiques et physiques sur les nouvelles découvertes au bord de La Grand Mer* (Paris, 1753). By reproducing a map noticeably indebted to native sketches as well as incorporating its "information" into his own scientific convention, Buache acknowledged the validity and distinctiveness of Amerindian cartography. Courtesy of the Newberry Library.

forms more akin to typical rivers and lakes on European maps. In this way he tried to mask the line between regions already explored by Europeans and those known only by report. The result is similar to Graham's *Plan*: however naturalized the forms, the succession of features is still identifiable as a native canoe route.

The "assimilation" of Amerindian map images on three issues of Arrowsmith's *Map Exhibiting all the New Discoveries* is a far more radical attempt to translate the native map convention. Over a period of seven years, Fidler's reports had supplied Arrowsmith with surveys and native information, which allowed the London cartographer to add features to his maps. In 1801 and 1802 Fidler requested maps from several native visitors to his post on the South Saskatchewan River: these were three "Blackfoot" (Siksika) leaders named Ackomokki, Kioocus, and Ackoweeak, as well as an unnamed "Fall Indian" (Atsina). The native cartographers produced five sketches of the upper Saskatchewan and upper Missouri rivers.[25] What is remarkable about these maps is their very similar style and repeated signs: the same selection of geographic features is shown on all of the maps; all are oriented west; all show the Rocky Mountains as one or more straight lines; all show the rivers as straight or slightly curved lines joined to the line of mountains; the three that mark campsites show them as small circles. Fidler's copies of these maps can be set beside the third issue dated 1796 and the two issues dated 1802 in an attempt to see how Fidler and Arrowsmith read the native designs. The process is very difficult to reconstruct, since the dating of Arrowsmith's map issues can be misleading and Fidler's own map of 1802 is not extant.[26] The first of two maps drawn by Ackomokki provided Fidler with the most detailed image of the Missouri watershed, the region that interested him most. Ackomokki's map was copied onto a separate sheet, which may have been one of the maps (the one he described as "taken solely from Indian information") sent by the HBC governor to Arrowsmith late in the fall of 1802.

Fidler's awareness of the native mapping convention, so different from European scientific cartography, made him wary of transferring the native designs directly onto his own map. On the third issue of Arrowsmith's 1796 map, the rivers of the Missouri watershed appear as separate, truncated streams emerging from the east side of the Rocky Mountains. The number of streams and their relation to prominent peaks conform closely with the first Ackomokki map. On both 1802 issues of the Arrowsmith map (fig. 6.5), these streams are linked by dotted lines into a tributary system. Since the conjectural linking first appears on the first 1802 state, and for only some of the streams shown, the pattern of the Missouri as it takes shape on the Arrowsmith map most probably owes more to Fidler's or even Arrowsmith's imagination than it does to native "information." Arrowsmith's dotted lines, the sign of conjecture in scientific mapping, are the graphic sign of this invention.[27] As they appear on the two 1802 issues of the *Map Exhibiting all the New Discoveries*, the rivers of the Missouri watershed are reconfigured to make them look like rivers on European maps. The Blackfoot Missouri is a symmetrical trellis of equally strong curved lines converging, far downstream, on a central line. The native maps may well depend on a pattern which would have conferred spiritual significance on the region and made its features easier to hold in memory: geographical features were named as body parts (nose, backbone) of Napi, the Creator.[28] All the features are shown as relative to each other rather than spatially located. The rivers join the mountains in a connected design that is purely terrestrial, without the sky referents on which the European grid map depends. In contrast, the Arrowsmith Missouri is a pattern of tributaries which begin as small streams and join into progressively larger rivers, shown by progressively heavier lines. Fidler and Arrowsmith had to *reimagine* the Missouri of the Blackfoot maps to fit the scientific conception of rivers, and to demarcate the region as a grid, not a body.

INLAND JOURNEYS, NATIVE MAPS

6.5 Detail from Aaron Arrowsmith, *Map Exhibiting All the New Discoveries in the Interior Parts of North America*... [second of the two 1802 states] (London, 1802). Arrowsmith's depiction of the hypothetical courses of the Missouri watershed was based not on scientific surveys, but on an Amerindian map forwarded to him from Peter Fidler. Courtesy of the Edward E. Ayer Collection, the Newberry Library.

We cannot know exactly how much the trader himself adapted the native image of the Missouri River to his own cartographic convention, and how much of this process was left to Arrowsmith. Fidler's appreciation of native cartography is evident by the number of native maps he requested, copied, and kept. Both he and William Clark seem to have suspected that there was more to native mapping than they could immediately understand or make use of. It is tempting to think that more or less prolonged exposure to native ways of living and seeing would have enabled a willing and open-minded explorer to understand something more of this culture than is possible for us. Apparently the HBC winterers enjoyed their years inland and tried hard to win acceptance from the groups they accompanied. But the daily records they were instructed to keep must have worked against their adaptation to native cultures more than the habits of daily living allowed it. In the best empirical tradition, the HBC employees observed their positions and plotted their routes on paper lined with a grid for the purpose; in their journals, they were equally observant of weather, soil, vegetation, and native customs, all noted according to prescribed categories of usefulness and exoticism. The duty of keeping such records ensured that the winterers saw native life, and the landscape, with European eyes.[29]

To this point we have assumed that the conventional nature of cartography is analogous to that of verbal language. This assumption seems reasonable enough, especially for cartography of the eighteenth century, when language was thought to be a verbalization of mental images. But Enlightenment savants were only too aware that language could misrepresent even as it codified and communicated. They tried to look beyond language to a form of expression which was immediate, visual, and hence (they thought) closer to its object. Gesture was proposed as better able to satisfy criteria of naturalness and direct communication. In his *Essai sur l'origine des langues,* Jean-Jacques Rousseau, for example, argued that gesture is "plus facile et dépend moins des conventions" than speech.[30]

This speculation was applied to North American native cultures during the following century. Observers appreciated native sign language as direct and spontaneous, hence an unambiguous alternative to verbal "dialects." Sign language was thought to be more natural than speech, because it was used when verbal communication failed. Here the advocates of Rousseau's preference for gestures erred by oversimplification: although sign language reached across verbal barriers, it was, like mapping, and no less than verbal language, a highly stylized, highly standardized convention of prescribed elements. Nevertheless, the observations encouraged by this line of reasoning are of interest. Garrick Mallery's commentary on native languages, published in 1881, claimed that sign language and pictographs were "representative of ideas without the intervention of sounds," to which he added, "This will be the more apparent if the motions expressing the most prominent feature, attribute, or function of an object are made, or supposed to be made, so as to leave a luminous track impressible to the eye."[31] The "luminous track" of the pictographic design recalled the movements of the rock painter as he drew in imitation of the sign-speaker. Rundstrom suggests that the "astonishing precision" of Inuit maps drawn for explorers of the eastern Arctic may relate to an Inuit tradition of "kinesic or gestural mimicry."[32] Given the gulf of cultural difference between Amerindians and Inuit, we should guard against making too much of this tradition if it were not paralleled in Amerindian culture. A Spokane woman whom Jack Nisbet met told him the story of how the bat got its wings:

> Bat . . . gathered some broken arrow pieces [and] poked them into his blanket as struts, then held the blanket over his shoulders and leaped out of the clouds. For a long time Bat floated gently underneath the blanket parachute on his way back to earth. . . . As she finished her story, Selena seemed

to shrink deep into her cushion. Moving her hands slowly, she lifted a black, loose-knit afghan that hung behind her... For a moment she held it over her head, her small arms crooked like newfound wings.[33]

For us, this propensity to enact ideas and events, to give them bodily as well as or instead of verbal or graphic expression, is highly unusual. Rundstrom also remarks that a major difference between these cultures and ours is the difference between incorporation and inscription as modes of retention and valorization.[34] Clastres and Michel Foucault define torture as (re)inscribing social law on the body of a transgressor; Claude Lévi-Strauss and Jacques Derrida see the gesture of a Nambikwara chief as the imitation of writing: for these writers, incorporation is defined in terms of inscription.[35] Denis Wood places the two modes on an evolutionary scale; he distinguishes between "mapping" (ephemeral gestures) and culturally more advanced "mapmaking" (graphic artifacts):

> I don't have to *explain* Native Americans' sketching of river systems for European explorers (we might say it comes with the species); but because I insist on the existence of a relatively sophisticated sign system on which to erect map signing, I posit as a requirement for the emergence of a mapmaking society [a culture that is] at least poised on the verge of literacy.[36]

But if we follow Rundstrom's lead and apply Mallery's observations to Amerindian cartography, we may well conclude that a closer analogy can be made between maps and gesture than between maps and language/writing. When we are asked for directions, few of us can resist pointing and waving our arms, or tracing the traveler's route in the air above the surface of his map.[37]

This gesture becomes part of the map, a feature of its reception. An anecdote in Hugh Brody's *Maps and Dreams* furnishes an example of such gesture in native map use. Brody describes how Joseph Patsah participated in a land-use study of northern British Columbia by indicating Quarry, his hunting area, on a Canadian government topographical sheet. Although the map was constructed according to scientific convention, Patsah read it in a way that is suggestive of Mallery's rock painter.

> He stood by the table, and looked at the map, and located himself by identifying the streams and the trails that he used.... He had many things to say about his way of life... His presence and manner caused the map to fade into the background.... At times, as he talked, Joseph made use of the map in the most general, even abstract way: with no more than a slight flutter of a gesture he established Quarry as his.... Mapping as such was now completed... Joseph turned to me and said, "We'll go to Quarry.... Good country. You'll see."[38]

Brody's description could not be more unlike Jacob's anecdote of the Huron and the missionary. The topo map does not reveal the world to the "sauvage" Joseph Patsah; instead it is a prompt for the old man's memories and a link with continuing experience. "We'll go to Quarry," he says to the researcher, "You'll see." For Patsah, the area in question is a place where he has lived and hunted, rather than a space that is prescribed, circumscribed, and possessed. He lays claim to it not by framing it within a boundary line on the map, but by a flutter of his hand in the air, a "luminous track" which confirms his physical association with the land in the form of a gesture. Brody's account may provide valuable hints by which we can try to understand, however imperfectly, how

Amerindian cartographers recorded their experience of the land they inhabited and how they communicated it—graphically, verbally, gesturally—to the traders who wintered with them and lived for a time as they did.

At the same time, we need to see Joseph Patsah's use of the topo sheet as a concession to scientific cartography—a complex response whose traditional and adaptive elements cannot ultimately be distinguished. A third anecdote, told to a journalist by Matthew Coon Come, grand chief of the Quebec Crees, indicates strenuous resistance to European-style maps. When Coon Come unfolded a topo map during a hunting trip with his father Daniel, he provoked a strong reaction in the older man. "The first thing my Dad did was tear that map into tiny little pieces," the chief recalls. "He said I was committing the white man's mistake, making plans for the land without ever setting foot on it, without ever getting a feel for it."[39] Whatever knowledge is valuable cannot be prescribed on the topo sheet; it must be gained by long personal experience. Patsah also insists, though more gently, that the scientific map is inadequate: to know Quarry, one must go there.

Traditionally maps stored in memory, drawn from memory, had this guarantee of experience behind them. Certain Amerindian maps cover a considerable part of the continent—most of the Arctic coast on Norton's "Draught," and the vast area from the Saskatchewan and Missouri watersheds to the Pacific Ocean on Ackomokki's 1801 map. The extent of such maps may well be greater than the country personally known to each cartographer, although we should keep in mind the distance Hearne walked during his two years inland, and the fact that the features shown on these native maps have nothing of the vagueness and fantasy of contemporary European maps drawn from rumor and report. The conformity of Amerindian maps indicates a cultural memory at work. The Missouri of the Blackfoot maps is a common image of that river, not invented on the spot by each mapmaker, but repeated from time to time and from cartographer to cartographer.

When Fidler asked his four native visitors to map the country south of his post, he transcribed the graphic designs they made for him but kept no record of their conversation. What questions did Fidler ask which were answered by drawing the maps? What comments were offered as they were drawn? What gestures were made which, like Patsah's, may well have "caused the map to fade into the background"? Unfortunately, missing the dialogue is missing the point. All we have left are the graphic transcripts, without explanations, memories, or associations. The maps are deprived of the "luminous track"—the ephemeral structure of their significance.[40] Fidler considered the maps useful for indicating certain features, but scientifically unacceptable: the maps were not drawn to scale, that is, the features were not located as points on a uniform cartographic space. "On some occasions," he conceded, "they [the native maps] are of much use, especially as they show that such & such rivers & other remarkable places are, tho' they [the native cartographers] are utterly unacquainted with any proportion in drawing them."[41] Each mapping convention is considered inadequate by the other culture: Fidler and others disregarded the natives' memorial emphasis, just as Daniel Coon Come refused the documentary resources of the scientific map.[42]

There can be no resolution of this impasse, no better translation from one cartographic convention to the other than what Fidler managed. When he copied the Blackfoot maps, Fidler conserved part of the process by which they were communicated to him. Simultaneously he began a process of alienation. Fidler realized that the natives he conferred with could not understand the European system of cartographic signs. What Fidler failed to recognize was that his own grasp of native mapping was also limited.[43] He selected and recorded only what could be more or less integrated into his own system, and quietly made his own surveys when he could. At a meeting with Kootenays near the Oldman River, Fidler recorded that "after smoking a Pipe with the Cottonahaws [Kootenays], I went away privately up the river to examine and measure the altitude of the Mountain in this place—had the Indians have known, they would some of them have accompanied me & hindered me from making the remarks I wished." That evening Fidler smoked again with the visitors, a solemn occasion for the Kootenays, though all in a day's work for the trader. Since he and the Kootenays communicated in sign language, in theory they should have been able to establish a direct, spontaneous, and unambiguous understanding. Fidler's account reveals nothing of the sort. Despite his easy tolerance and mastery of outward form, his sense of his own cultural ascendancy limited his perception.

> I was according to the customs I had seen amongst other Indians smoking away at my ease, but after the 4th wiff [the Cottonahaw chief] took the pipe from me and made me to understand by signs that 4 was the number upon extraordinary occasions . . . [and] that I should light it & make the same ceremonies with the Pipe as we did in our own country . . . I made several curious motions with it that they could not comprehend—or myself either; however, as I kept my gravity (tho' with great difficulty) during the ceremony, . . . these people appeared to be highly pleased at my dexterity with the Pipe.[44]

Although he satisfied the natives' expectations, participated in their way of life, and shared their experiences, Fidler's remarks underline the distance between their cultures and his own. He was intrigued enough with native cultures to learn several languages, respectful enough of native cartography to collect at least twenty maps. But he remained an outsider whose job was to exploit the elements in these cultures that would be useful to his own.

In describing native maps as "primitive" and lacking in features characteristic of European topographical surveys, historians of cartography have been less self-aware than Fidler. We—since I count myself among them—have overlooked the crucial fact that, like Fidler, we are looking at a convention from the outside, centuries later, and with less if any lived experience of it. We can never understand more than partially a cartographic convention that conceptualizes and organizes the world so differently from the way ours does. And we should be aware of the dangers of anthropological "upstreaming," by which cultural traits from a later period are assigned to an earlier period: North American native societies have been too disrupted and subverted to make such a technique reliable.[45] All we can tell for sure is that while Amerindian maps lack many of the characteristics we assume that maps must have, they apparently satisfied the cartographic needs of many indigenous North American cultures for hundreds of years. The effort of explorers and imperial cartographers to translate native map images into the conventional terms of scientific mapping, to "assimilate" them, stripped them of their proper significance.

Recognition and respect of difference is the necessary base and starting point for any intercultural dialogue. As Gilbert Vaudet observes,

> C'est le moment de l'épreuve: je veux dire le moment où le savoir se retire.... Là, un monde doit céder pour qu'en advienne un autre.... Là, dans l'entre-deux, une chance surgit qui est celle du contact et—par-delà ce que cela comporte de désarroi, d'opacité, de silence—celle d'un échange peut-être. Cela, la meilleure ethnologie le contient, et un tremblement de la parole en elle garde la trace de cette approche.[46]

Instead of continuing to translate the native cartographic convention into our own, we need to acknowledge that the gap between these conventions is essentially unbridgeable. The best we can do is to initiate a dialogue with native cultures as they have survived, keeping in mind the dangers of upstreaming. The instant of hesitation "dans l'entre-deux" could be salutary: by considering such very different maps and cultures, we could learn more about our own.

Notes

An earlier version of this chapter was published in *Cartographica* 33, 2 (Summer 1996). I would like to thank Theodore Binnema for allowing me to read the prepublication typescript of his paper on the "Blackfoot" maps, and Edward H. Dahl for advice on the states of Arrowsmith's *Map Exhibiting all the New Discoveries* mentioned in the text.

1. Peter Hulme and Ludmilla Jordanova, eds., *The Enlightenment and Its Shadows* (London: Routledge, 1990), 7.

2. National Archives of Canada: Jacques-Nicolas Bellin, *Carte de l'Amérique septentrionale,* 1743 (NMC 13814); Jacques-Nicolas Bellin, *Carte de l'Amérique septentrionale,* 1755 (NMC 21057); and Aaron Arrowsmith, *A Map Exhibiting all the New Discoveries in the Interior Parts of North America, Inscribed by Permission To the Honorable Governor and Company of Adventurers of England Trading into Hudson's Bay, In Testimony of Their liberal Communications,* 1795 (NMC 97818), third of three issues dated 1796 (NMC 17396), and the second of two issues dated 1802 (NMC 19687). For Aaron Arrowsmith, *A Map Exhibiting all the New Discoveries . . . ,* 1802 first issue, see Warren Heckrotte, "Aaron Arrowsmith's Map of North America and the Lewis and Clark Expedition," *Map Collector* 39 (1987): 16–20. Hudson's Bay Company Archives, Provincial Archives of Manitoba: Andrew Graham, *A Plan of Part of Hudson's-Bay & Rivers, Communicating with York Fort & Severn,* 1774 (HBCA G.2/17), reproduced in Richard Ruggles, ed., *A Country So Interesting: The Hudson's Bay Company and Two Centuries of Mapping, 1670–1870* (Kingston and Montreal: McGill-Queen's University Press, 1991), pl. 6. An earlier sketch by Andrew Graham, *A Plan of Part of Hudson's-Bay & Rivers Communicating with the Principal Settlements,* 1772 (HBCA G.2/15), reproduced in Andrew Graham's *Observations on Hudson's Bay,* edited by Glyndwr Williams (London: Hudson's Bay Record Society, 1969), endpaper, extends only as far west as Squaw Rapids on the Saskatchewan River.

3. See notes 4 and 11 below.

4. G. Malcolm Lewis, "Indian Maps: Their Place in the History of Plains Cartography," in Frederick C. Luebke, Frances W. Kaye, and Gary E. Moulton, eds., *Mapping the North American Plains* (Norman: University of Oklahoma Press with the Center for Great Plains Studies, University of Nebraska–Lincoln, 1987), 78.

5. See the "correction" of native geographical knowledge, in Matthew H. Edney, "The Patronage of Science and the Creation of Imperial Space: The British Mapping of India, 1799–1843," *Cartographica* 30, 1 (1993): 63: "Eighteenth-century Europe was rooted in an *esprit géométrique,* the desire to order both nature and human

society through laws and formulations which possessed the simplicity and universality of geometric theorems.... The advocation of science, that is, rational thought, was tantamount to the advocation of the European state and its stable social hierarchies.... The British ... had no qualms about using geographical information from native sources, but they discarded it as soon as even the sparsest survey had been completed."

6. See note 1 above and cf. Glyndwr Williams's evaluation of Graham's very similar *Plan of Part of Hudson's-Bay* dated 1772, in Andrew Graham's *Observations on Hudson's Bay,* verso of endpaper: "As a work of cartography the map has little to commend it, and reveals how limited was the company's knowledge of the interior a century after the establishment of the first Bay-side posts. The most striking and erroneous feature of the map is the distorted outline of Lake Winnipeg ... tilted on its axis, and stretching more than twice its actual length. Other errors include the longitudes generally ... The map bears all the marks of compilation from Indian descriptions and the vague reports of Company servants sent inland from York and Severn for some years previous."

7. Heckrotte, "Aaron Arrowsmith's Map," 20, observes: "There is no record, as far as I know, when Fidler prepared a map or sent maps or papers which would have served as the basis for the first 1802 issue (or for the previous issue [dated 1796, probably published in 1799] for that matter." The Hudson's Bay Company record of correspondence, remarkably complete, has no such record. But it is faintly possible that Fidler corresponded directly with Arrowsmith after publication of the 1795 issue in which the journey of 1792–93 was recorded. Arrowsmith's records were destroyed in the bombing of London during the Second World War—see Coolie Verner, "The Arrowsmith Firm and the Cartography of Canada," *Canadian Cartographer* 8, 1 (1971): 1 and passim 1–7.

8. As a commercial and explorational practice, wintering inland did not originate with Henday's journey in 1754–55. Henry Kelsey and William Stewart, HBC employees half a century before, had preceded Henday, and were themselves imitating a seventeenth-century French practice—see Conrad Heidenreich, "Mapping the Great Lakes: The Period of Exploration, 1603–1700," *Cartographica* 17, 3 (1980): 57: "With Champlain came the *coureurs de bois.* In a sense they were his idea. These were young men whom Champlain planted among various native groups to learn their language, persuade the natives to trap for furs and convey them to the St. Lawrence, learn native customs and learn something of the geography of the interior which the native groups were so reluctant to show the French."

9. Hudson's Bay Company Archives, Provincial Archives of Manitoba: Map endorsed *Moses Nortons Drt. of the Northern Parts of Hudsons Bay laid dwn on Indn. Information & bro^t Home by him anno 1760* (HBCA G.2/8); and a second map copied by Moses Norton, endorsed *Captain Mea.to.na.bee & I.dot.ly.a.zees, Draught. CR [Churchill], 1767–68* (HBCA G.2/27). The first map is reproduced in Ruggles, *A Country So Interesting,* pl. 8. Meatonabbee, one of the cartographers of the later map, led Samuel Hearne along the Coppermine River.

10. June Helm, "Matonabbee's Map," *Arctic Anthropology* 28 (1991): 54–87. In this article, Helm discusses both of the maps cited in note 9 above.

11. Even now, several respected historians of cartography are content to consider non-European maps in terms virtually unchanged since Leo Bagrow's *History of Cartography,* rev. R. A. Skelton (Cambridge, Mass.: Harvard University Press, 1964). Bagrow, 25–26, opines that "races given to stylisation of animal or human figures ... draw either no maps or very bad ones.... A primitive savage's drawing is often like a child's...Primitive peoples... know nothing of abstract maps, conventional generalisation, or data of a general kind.... They cannot portray the world, or even visualise it in their minds." In his *History of Topographical Maps* (London: Thames and Hudson, 1980), P. D. A. Harvey remarks on the ephemeral nature of "primitive" maps but with no mention of the very different power, requirements, and characteristics of memorial knowledge. Instead Harvey (p. 34) is convinced that "the more spectacular feats of cartography by primitive

people seem to have followed contacts which could have introduced them to more advanced forms of mapping." G. Malcolm Lewis has made close studies of the convention of North American native maps—see "Indicators of Unacknowledged Assimilations from Amerindian Maps on Euro-American Maps of North America," *Imago Mundi* 38 (1986): 9–34; "Misinterpretation of Amerindian Information as a Source of Error on Euro-American Maps," *Annals of the Association of American Geographers* 77, 4 (1987): 542–63; "La Grande Rivière et Fleuve de l'Ouest: The Realities and Reasons behind a Major Mistake in the Eighteenth-Century Geography of North America," *Cartographica,* 28, 1 (1991): 54–87; "Metrics, Geometries, Signs, and Language: Sources of Cartographic Miscommunication between Native and Euro-American Cultures in North America," *Cartographica* 30, 1 (1993): 98–106; as well as the article in Luebke, Kay, and Moulton, cited in note 4 above. Lewis nevertheless continues to describe native maps in terms of their failure to satisfy the standards of a topographical survey: native North Americans lacked the "levels or types of abstractions" necessary for writing and geometry, and native "cultures reveal little to suggest that they had a concept of grid." Christian Jacob, *L'Empire des cartes* (Paris: Albin Michel, 1992), follows the order of Bagrow's and Harvey's histories, briefly discussing "primitive" maps, then retreating to consider "classical" European cartography. Denis Wood, *The Power of Maps* (New York: Guilford Press, 1992), reiterates Bagrow's and Harvey's progressive-evolutionary view of cartographic history, from "nascent forms of protocartography" to "relatively simple mapmaking" and finally to "cartography per se" as developed by "nineteenth-century British and contemporary Americans." Wood's earlier table of children's hill drawings, showing a progression from elevation to plan, relies on the comparison between childhood and "primitive" cultures apparent in Bagrow; Harvey praises this "analysis" in *The History of Topographical Maps,* 26; and Wood in turn praises Harvey for his adoption of this scheme in "P. D. A. Harvey and Medieval Mapmaking," *Cartographica* 31, 3 (1994): 54. This congratulatory filiation simply continues the focus and bias which characterized the history of cartography earlier in the century. In fact, the assumptions of these historians reach back to Enlightenment dichotomies of primitive and civilized, simple and complex, concrete and abstract, which are then developed according to nineteenth-century notions of progressive evolution. The imperial energy which used a European grid of spatial coordinates to survey the world has not yet dissipated intellectually.

12. Jacob, *L'Empire des cartes,* 53–54.

13. Pierre Clastres, *La Société contre l'état* (Paris: Minuit, 1974), 15–16: "On aura depuis longtemps reconnu l'adversaire toujours vivace, l'obstacle sans cesse présent à la recherche anthropologique, l'*ethnocentrisme* qui médiatise tout regard sur les différences pour les *identifier* et finalement les abolir. . . . L'évolutionnisme, vieux compère de l'éthnocentrisme, n'est pas loin. La démarche à ce niveau est double: d'abord recenser les sociétés selon la plus ou moins grande proximité que leur type de pouvoir entretient avec le nôtre; affirmer ensuite . . . une *continuité* entre toutes ces diverses formes du pouvoir" (original emphasis).

14. David Woodward and G. Malcolm Lewis, eds., *Cartography in the Traditional African, American, Arctic, Australian, and Pacific Societies*, vol. 2, bk. 3 of *The History of Cartography* (Chicago: University of Chicago Press, forthcoming 1998).

15. Cf. Barbara Belyea, "Amerindian Maps: The Explorer as Translator," *Journal of Historical Geography* 18, 3 (1992): 267–77. This article is limited to explorers' treatment of Amerindian maps—see my discussion below of Fidler's "translation" of the Blackfoot maps. But historians of cartography should be wary of falling into the same habit of simplistic translation that served explorers' practical needs.

16. Amerindian ways of seeing, which may or may not be closely related to native experience during the contact period, are discussed by Barre Toelken and N. Scott Momaday in Walter Holden Capps, ed., *Seeing with a Native Eye* (New York: Harper and Row, 1976), 9–24, 79–85. William Sturtevant, "Tribe and State in the Six-

teenth and Twentieth Centuries," in Elizabeth Tooker, ed., *The Development of Political Organization in Native North America* (Washington D.C.: American Ethnological Society, 1983), 5, explains "the ethnohistorical technique of upstreaming from later and therefore better ethnographic and historical evidence ... Upstreaming presupposes that one can disentangle cultural persistences from the changes introduced in the intervening period." All the intercultural differences of the present are squared when they are applied to conditions of the past. When a culture has been profoundly disrupted, meaningful "upstreaming" may well be impossible.

17. Robert A. Rundstrom, "A Cultural Interpretation of Inuit Map Accuracy," *Geographical Review* 80, 2 (1990): 165. For problems of recording spoken arts in documentary form, i.e., translating from one medium to another, from one cultural tradition to another, see Dennis Tedlock, *The Spoken Word and the Work of Interpretation* (Philadelphia: University of Pennsylvania Press, 1983); and more recently, "From Voice and Ear to Hand and Eye," *Journal of American Folklore* 103 (1990): 133–56. Mary Carruthers, *The Book of Memory* (Cambridge: Cambridge University Press, 1990), describes the monastic/scholastic tradition of medieval Europe, a memorial culture in which orality and literacy were profoundly integrated. Given that mapping is in part a graphic process, Carruthers's book is a fine corrective to studies that simplistically oppose oral and written cultures. Carruthers also describes how memorial culture constantly renews itself: the memories of individuals are repositories of inherited textual knowledge which is enriched and modified by their personal experience.

18. Lewis makes this point in "Indicators of Unacknowledged Assimilations," 15. But in the same article (p. 17) he deplores "the absence on [Amerindian maps] of equivalents to graticules, grids or neat lines, their failure to conserve distance, or direction, or shape, and the absence—or at best, paucity—of unambiguous toponymic and environmental content." Cf. Matthew H. Edney, "Cartography without 'Progress': Reinterpreting the Nature and Historical Development of Mapmaking," *Cartographica* 30, 2 & 3 (1993): 54–68.

19. See Lewis, "Indian Maps," 64–65: a famous exception is the Pawnee star chart, dated as "pre-1906 and supposedly much older." The stars are marked as crosses on a symmetrical oval ground, bordered by a heavy line.

20. James Isham, *Observations on Hudson's Bay,* edited by E. E. Rich (Toronto: Champlain Society for the Hudson's Bay Record Society, 1949), 65, 102.

21. Lewis, "Indicators of Unacknowledged Assimilations," 22–31; "Misinterpretation of Amerindian Information"; "La Grande Rivière et Fleuve de l'Ouest"; D. W. Moodie and Barry Kaye, "The Ac Ko Mok Ki Map," *Beaver* 307, 4 (Spring 1977): 4–15, and Judith Hudson Beattie, "Indian Maps in the Hudson's Bay Company Archives: A Comparison of Five Area Maps Recorded by Peter Fidler, 1801–1802," *Archivaria* 21 (Winter 1985–86): 166–75.

22. Jacob, *L'Empire des cartes,* 147.

23. In 1869, for example, Kohklux, a Tlingit chief, drew two maps of the route he had taken to the Yukon seventeen years before. George Davidson, "Explanation of an Indian Map," *Mazama* (April 1901): 75–82, reprinted in [no author or editor], *The Kohklux Map* (Whitehorse: Yukon Historical and Museums Association, 1995), 14–24, prefaces his description of the larger of the two maps, drawn on the back of a topographical sheet, by stating briefly, and a bit cryptically: "At his own suggestion Kohklux proposed to draw upon paper his route ... The second attempt was upon a large sheet, 43 x 27 inches." A note on the back of the first map explains why it was left unfinished: "Kohklux started from his place at Klukwan and drew all around the paper for want of room. He asked for a big sheet of paper and I gave him the back of an old map on which he and his wife drew their routes etc. in 1852." Both maps are in the Bancroft Library, University of California at Berkeley (G4370, 1852, K61A; and G4370, 1852, K6D).

24. Philippe Buache, *Considérations géographiques et physiques sur les nouvelles découvertes...* (Paris: Académie Royale des Sciences, 1754), app. map 5. See Lucie Lagarde, "Philippe Buache, ou le premier géographe

français, 1700–1773," *Mappemonde* 87, 2 (1987): 26–30; and Lucie Lagarde, "Le Passage du Nord-Ouest et la Mer de l'Ouest dans la cartographie française du XVIIIe siècle: Contribution à l'étude de l'oeuvre des Deslisle et Buache," *Imago Mundi* 41 (1989): 19–43.

25. Hudson's Bay Company Archives, Provincial Archives of Manitoba: Ackomokki, "An Indian map of the Different Tribes that inhabit on the East & west side of the Rocky Mountains with all the rivers & other remarkable places, also the Number of tents," 1801 (HBCA G.1/25); Ackomokki, untitled map, 1802 (HBCA B.39/a/2, E.3/2); Akkoweak, untitled map, 1802 (HBCA B.39/a/2, E.3/2); Kioocus, untitled map, 1802 (HBCA B.39/a/2, E.3/2); ["Fall Indian"], untitled map, 1802 (HBCA E.3/2). All of these maps were copied by Peter Fidler, into two journals (B.39/a/2 and E.3/2) and onto a separate sheet (G.1/25).

26. The third of the three states dated 1796 is watermarked 1798 and indicates as Arrowsmith's address "N° 24 Rathbone Place," to which he moved in 1799.

27. Cf. Jacob, "Il faut qu'une carte soit ouverte ou fermée: Le tracé conjectural," *Revue de la Bibliothèque nationale* 45 (1992): 39: "Il est certains cas où le tracé se désigne lui-même comme conjectural . . . Le tracé est beaucoup plus simple que celui des régions déjà explorées." This may be true of some transcriptions of Amerindian maps—Clark's are a good example—but Arrowsmith's conjectural Missouri appears as complex as surveyed rivers.

28. Theodore Binnema, "Indian Maps as Ethnohistorical Sources," paper presented at the Northern Great Plains History conference, Brandon (Manitoba), September 27–30, 1995, and to the history department, University of Calgary, March 28, 1996.

29. See, for example, the instructions issued to Henday and Hearne: James Isham, "Instructions to Anth^y Hendy, Dated att York Fort, Feby ye 19th, 1754" (HBCA A11/114); and Samuel Hearne, *A Journey from Prince of Wales's Fort in Hudson's Bay, to the Northern Ocean* (London: Printed for A. Strahan and T. Cadell, 1795), xxxiii–xliv, 64–65. Similar instructions were given to Meriwether Lewis: see Donald Jackson, ed., *Letters of the Lewis and Clark Expedition* (Urbana: University of Illinois Press, 1962), 61–62.

30. Jean-Jacques Rousseau, *Essai sur l'origine des langues,* edited by Charles Porset (Bordeaux: Guy Ducros, 1968), 29. Cf. the paraphrase of Rousseau's argument in Jacques Derrida, *De la Grammatologie* (Paris: Minuit, 1967), 331: "La langue du geste et la langue de la voix . . . sont 'également naturelles.' Toutefois l'une est plus naturelle que l'autre, et à ce titre elle est première et meilleure. C'est la langue du geste qui est 'plus facile et dépend moins des conventions'. Il peut certes y avoir des conventions de la langues des gestes. Rousseau fait allusion plus loin à un code gestuel. Mais ce code s'éloigne moins de la nature que langue parlée." Gesture was privileged because of its immediacy: the author of the gesture must be bodily present, whereas words can be repeated, written down, divorced from their original speaker, and thus denatured. Derrida buys none of this, of course.

31. Garrick Mallery, quoted by David Murray, *Forked Tongues: Speech, Writing, and Representation in North American Indian Texts* (Bloomington: Indiana University Press, 1991), 21.

32. Rundstrom, "A Cultural Interpretation of Inuit Map Accuracy," 163–68. Rundstrom also mentions the artifactual extension of this bodily mimicry to string games, carving of small objects (*pinguak,* or imitation-things), man-shaped stone markers (*inuksuit*) and the drive lanes constructed for a traditional caribou hunt. The figures for the drive lanes, called "dead men" in Fidler's 1792–93 journal (HBCA E.3/2), were also part of traditional hunts for buffalo among Plains cultures. Rundstrom (personal communication) has warned against generalizing between Amerindian and Inuit cultures; he can only be right in urging cautious discernment. It is tempting nevertheless to draw certain parallels among the memorial cultures of North America. The parallels I suggest here are made very tentatively, as comparisons meriting further study.

33. Jack Nisbet, *Purple Flat Top* (Seattle: Sasquatch, 1996), 110.

34. Robert A. Rundstrom, "GIS, Indigenous Peoples, and Epistemological Diversity," *Cartography and Geographical Information Systems* 22, 1 (1995): 51–52.

35. Michel Foucault, *Surveiller et punir* (Paris: Gallimard, 1975), 9–21; Derrida, *De la grammatologie,* 173–202.

36. Denis Wood, "The Fine Line between Mapping and Mapmaking," *Cartographica* 30, 4 (1993): 55–56 and passim.

37. Roland Barthes, *L'Empire des signes* (Geneva: Albert Skira, 1970), 42; cf. Wood, "The Fine Line between Mapping and Mapmaking," 50–51.

38. Hugh Brody, *Maps and Dreams* (Vancouver: Douglas and McIntyre, 1988), 6–10.

39. Barry Came, "Fighting for Their Land," *Maclean's Magazine* (February 27, 1995): 16.

40. Cf. Wood, "P. D. A. Harvey and Medieval Mapmaking," 53–54: "What is it that makes a map a map? It is nothing other than the way [certain] relationships are embodied *in relatively durable graphic expressions*" (original emphasis). Both Harvey and Jacob remark that dialogue is a defining characteristic of "ephemeral" maps. Harvey, *History of Topographical Maps,* 31, 33: "We can learn practically nothing of maps of this sort that may have existed in distant ages. [He cites four examples of "ephemeral" maps.] ... not one of the four could possibly have been recognized [by Europeans] as a map without some explanation from the people who made it or used it." Jacob, *L'Empire des cartes,* 57–60: "le tracé rudimentaire . . . est indissociable du commentaire de qui le réalise . . . [il] n'a d'autre raison d'être que cet usage immédiat et éphémère." At the same time, Jacob is convinced "que le support, les formes et les tracés, la lettre même de la carte permett[ent] de retrouver, comme en creux, la trace, l'empreinte de ces gestes, de ces regards et de ces opérations intellectuelles" (348).

41. Peter Fidler to the HBC London Committee, July 10, 1802 (HBCA A.11/52), Hudson's Bay Company Archives, Provincial Archives of Manitoba.

42. Any constructive intercultural dialogue would have to turn these negatives into positive though foreign attributes—an important and still badly needed turnaround urged by Robert A. Rundstrom, "The Role of Ethics, Mapping, and the Meaning of Place in Relations between Indians and Whites in the United States," *Cartographica* 30, 1 (1993): 26: "A cross-cultural map comparison can be ethical and insightful only if the comparison is non-evaluative [i.e., recognizes the relativity of values, does not judge maps of one convention in terms of another] and facilitates the bridging of discourses, not the replacement of one discourse by another."

43. Cf. Ronda, "'A Chart in His Way,'" 89: "The story of Lewis and Clark Indian cartography has a curious and revealing epilogue. . . . the explorers drew a map . . . as Lewis put it, 'in their way.' The mapping ways of Hidatsas and Nez Perces had become at least partially an expedition way. Maps once formidable in structure and design could now be made and understood by the explorers themselves. Effort and understanding had made map encounters into common ground." Ronda is too sanguine.

44. Peter Fidler, journal entry for January 1, 1793 (HBCA E.3/2), Hudson's Bay Company Archives, Provincial Archives of Manitoba.

45. Cf. William Sturtevant, "Tribe and State in the Sixteenth and Twentieth Centuries," in Elizabeth Tooker, ed., *The Development of Political Organization in Native North America* (Washington D.C.: American Ethnological Society, 1983), 5: "The ethnohistorical technique of 'upstreaming' from later *and therefore better* ethnographical and historical evidence . . . presupposes that one can disentangle cultural persistences from the changes introduced in the intervening period. . . . But one can hardly subtract the later changes in the very cultural features in which one is interested, in order to discover a pre-existing sociopolitical system of the same type whose existence is precisely the question. One must, instead, examine the primary written sources . . . without reading into them features we know to have been present among the tribes of the region in [later] centuries" (my emphasis).

46. Gilbert Vaudet, "Du silence au dialogue: La fin des tribus," in Miguel Abensour, ed., *L'Esprit des lois sauvages: Pierre Clastres ou une nouvelle anthropologie politique* (Paris: Seuil, 1987), 145.

7

Native Mapping in Southern New England Indian Deeds

Margaret Wickens Pearce

Until the mid-eighteenth century, property mapping in southern New England was recorded primarily through the words of negotiated land transfers or deeds. Through written deeds, space was defined, interpreted, negotiated, bounded, translated, and defended. Occasionally, this process included graphic elements. More often, mapping occurred through the combination of written and spoken words and actions which comprised deeding processes.

In the early stages of colonial settlement, most land deeds were of a unique type known as native land transfers, or "Indian deeds," so called because they recorded land negotiations between colonists and Indians. Indian deeds as a body of legal documents included all types of land transactions, including conveyances, mortgages, leases, and testimonies. Hundreds of these deeds are still extant, preserved because of their value to colonial bureaucracies. Though the nature of the documents varied widely, the deeds provided a forum in which both Indians and colonists could potentially represent territory according to their own interests. Because the process sometimes documented both native and non-native conceptions of land, Indian deeds contain valuable evidence of native involvement in southern New England mapping processes during the seventeenth and eighteenth centuries.

Although native land transfers have been studied for answers to questions about land tenure, indigenous political and cultural boundaries, and European-native relations, little attention has been devoted to the deeds' capacity for territorial representation, and the implications of such representation for understanding cross-cultural mapping processes. This chapter explores the specific question of the place of Indian deeds in southern New England map history. In the first part, the forms and functions of native land transfers are introduced by way of the map traditions that informed them. In the second part, several deeds are analyzed as a way of illustrating the various forms of native mapping processes that may be produced by deed negotiations.

Looking for Native People in New England's Map History

When we think of New England map history, the conventional perception is that seventeenth- and eighteenth-century cartographies consisted primarily of a few types: coastal and maritime charts, colony maps developed for European investors, and a small body of property maps and surveys.[1] The maps that comprise this history have been included not so much because of their function, to represent colonial property, but primarily because of their form, that is, the degree to which they appear to resemble conventional Anglo-European notions of what a map, or map fragment, should look like. Thus, New England map history has traditionally emphasized printed maps produced by Englishmen, mostly royal explorers, hydrographers, and governors, and to a more limited extent, a few colonial surveyors and artists whose works document local or vernacular mapping activities. Whether produced by surveyor or governor, however, these maps mostly shared the same purposes: to portray to a European audience a New England in which colonization was successfully and inevitably overpowering the forces of wilderness, to advance the rhetoric of *vacuum domicilium*, and to downplay the presence of native people. To serve this end, maps portrayed a landscape in which colonial settlement advanced and became visible, and Indians and wilderness receded and were erased.[2]

Part of that which was erased was the notion that Indians could possibly influence the representation of territory themselves. Although native people may have had some type of cartographic tradition precontact, there has been no evidence of the existence of native cartography from the seventeenth century onward. The little that is known about native mapping in southern New England has been deduced from analyses of unacknowledged native contributions to the European printed maps mentioned above.[3] Thus, we have historically constructed an absence of native mapping in New England by examining the artifacts which, by definition, were designed to erase native landscapes and render Indians invisible. Because conventional map history depends heavily on the works that are products of colonial rhetoric, the assumption is made that colonial cartography inevitably obliterated or overlaid any existing native cartography in the same way that colonial settlement inevitably obliterated or overlaid existing native settlement: without native response or resistance.

To locate native mapping, then, we must look beyond the printed maps of colonial narratives and into the realm of daily mapping activities, where mapping was developed to meet other needs, both native and non-native, and where the forms of the maps may not resemble conventional European cartography. One of these mapping activities was the negotiation and registration of Indian deeds.

In order to explore this realm of mapping where deed negotiations occurred, a flexible definition of maps and mapping is required. In volume 1 of the *History of Cartography* series, J. B. Harley and David Woodward redefined maps in a way that would encompass different cartographies over time: "Maps are graphic representations that facilitate a spatial understanding of things, concepts, conditions, processes or events in the human world."[4] From this definition, it is necessary to enlarge the discussion to representations that may not be graphic. As will become evident in this history, the condition of graphicacy was optional for both native and non-native mapping *in this particular time and place*. For this study, then, maps are representations that facilitate a spatial understanding, and mapping is the process of creating and interpreting these representations. Such mapping may produce a wide variety of forms depending on the particular cultural context, and the historical period in which the mapping takes place.[5]

From the broad framework of mapping as spatial representation, this study is specifically concerned with spatial representation having a particular *function*: the mapping through which Indians and colonists represented property and territory to each other during the colonization of southern New England. A brief examination of the variety of early native and non-native mapping forms will facilitate an understanding of how these forms combined to create the mapping process in Indian deeds.

Forms of Native and Non-native Mapping

Though seventeenth-century English writers commented on such aspects of native culture as agriculture, clothing, and homes, on the subject of spatial representation they were silent. Roger Williams recorded some concepts of travel and direction in his study of Narragansett language, *A Key into the Language of America*. Williams noted several examples of Narragansett uses of direction words, from *Nopatin* for "East wind," to *Wompanand,* the name of the spirit dwelling in the east, to *Anan sowanakitauwaw,* "It goes to the South West." But when Williams recorded the words for the question "How far from hence?" *(Tounuckquaque yo wuche),* he failed to provide a conventional native response, and the reader is left unable to imagine how a seventeenth-century Narragansett person might have orally described distance over land.[6]

Some clues to native mapping practices have been found in research on the native use of toponyms, from place-naming to the use of places to record history. In the traditional indigenous relationship to mapping and landscape, toponyms are not arbitrarily assigned. Rather than make reference to a distant place or person, the place-name refers to *what is there*. Every toponym describes a site either in terms of its physical appearance, the way in which land is being used, the people who use the land, or a story or historical event that occurred at that place. Toponyms based on use may refer to the kinds of plants or animals that can be harvested, gathered, or hunted there. The stories that are associated with a place-name provide a moral, as well as historical and spiritual, character to the site.[7]

Some historians and linguists have concluded that the site-specific nature of native place-names recreates the landscape as an ecological and spiritual map, which is "read" through both the recitation of place-names, and experience. Thus, the recitation of place-names is a mnemonic device for recalling the landscape. Evidence of toponyms as a mnemonic device is found in the terse grammatical style with which they are recorded. In research on oral cultures, a terse grammar suggests a mnemonic because it is assumed that meaning is completed through experience, and that the recitation of the toponyms will act as a guide through that experience. For example, in Cruikshank's study of Athapascan toponymy, she found that "place names . . . are not decorative embellishments but structural markers, dividing the corpus into cognitive units and spatially anchoring stories so that they can be recalled by remembering the land."[8] Thus, in indigenous, nonarchival cultures, the web of place-names on the land comprises a map that orders physical, economic, and cultural information in a spatial framework, and which may be accessed through the combination of oral recitation and direct experience.

This pattern of mapping through naming has been documented to a limited extent among Eastern Algonquian-speaking peoples.[9] Eastern Algonquian place-names have been shown to have a nonarbitrary, place-specific nature. This place specificity is tied to both the cultural and ecological functions of the toponym. Bragdon has demonstrated the cultural function through the connection of toponyms to stories:

> These stories, often concerned with the doings of the Massachusett culture hero Moshup, a giant of enormous power, describe the origins of certain landscape features of southern New England and explain the appearance of some animals and fish. . . . The stories and their messages, tied as they are to prominent, existing landscape features, are called to mind by these features, which serve as daily reminders of good behavior.[10]

History was further memorialized, as Crosby writes, through experience of a site, by means of "brush heaps, stone piles, or holes dug into the ground, all of which served to remind those passing by of some significant event or occurrence."[11]

Cronon has shown further that Eastern Algonquian place-names carry an ecological function, denoting ownership but always in a collective, usufruct sense of ownership as *use*. "The purpose of such names," he writes, "was to turn the landscape into a map which, if studied carefully, literally gave a village's inhabitants the information they needed to sustain themselves."[12]

This native mode of mapping through place-names stands in contrast to the place-naming practices of English colonists. Upon receipt of a town patent from the General Court, English settlers renamed native places with names from home, such as Essex or Springfield. Also, English toponyms were based on the ownership or residence of an individual, or in homage to an individual person (for example, Winthrop, Mass., and Martha's Vineyard).[13] Only rarely did colonial English habitation-names denote the physical characteristics of a place, though this was somewhat more frequent for nature-names (landforms and bodies of water). Thus, English toponyms usually were arbitrarily placed, and not site-specific.

Still, property mapping in New England also centered on words, but in a very different style from native place-naming. Following a different approach from that which was practiced either in England or the Southern English colonies at that time, property mapping in colonial New England was a unique mode of mapping, created to meet the needs of specific circumstances, and heavily influenced by medieval practices.[14]

In the world of the medieval English open-field system, property was surveyed by written, rather than graphic, description. Surveyors perambulated the manor, making notes on the metes and bounds of the fields and holdings of tenants, and then the tenants were summoned to court for testimony concerning these holdings. All information collected was compiled, in Latin, in the lengthy documents known as court rolls, or terriers, which were "virtually written maps."[15] A map might be included in certain instances, but was not a normal component of the record. Thus when the first textbooks of surveying began to appear in the early sixteenth century, students of surveying were instructed to aim for a written product.

During the sixteenth and early seventeenth centuries, as enclosure progressed, graphics began to accompany words in representations of territory in property surveys. With the emphasis on graphic surveys came the rise of surveying as a profession, as well as the rise of survey-instrument and survey-textbook industries. By the early seventeenth century, improvements had been made to the cross-staff and geometrical tables; the plane table (particularly aiding graphic surveys), the circumferentor, and chain were also invented at this time. Scholars and mathematicians developed such practices as triangulation and back-sighting, and designed the theodolite for measuring angles more accurately than with the plane table. Distance measurement, however, continued to be in units of rods as it had in feudal survey.

In the midst of these developments, there was a time lag between theoretical developments and their application to new practices. Surveyors were not scholars, and were therefore unlikely to incorporate such innovations. Also, many surveyors and landowners were unwilling to pay the additional cost required by the new technologies. Early New England property mapping reflected this gap between theory and practice. The familiar practice of written surveys based on perambulation with cross-staff and rod was more likely the mode of property mapping that colonists expected to implement in the New World. The new triangulated and drawn surveys remained the realm of scholars and mathematicians in England or wealthy plantation owners in the Southern colonies who could afford the equipment and surveyors' fees necessary for such undertakings. Thus, in the New England colonies there was a general reluctance toward graphic property mapping, despite requests for such maps in England.[16]

In summary, these two mapping practices, Indian and English, each encoded maps primarily in words, although in very different ways. Native word maps most likely were encoded through a combination of the oral recitation and experience of site-specific names. English word maps were encoded primarily through written description of related points, and assigned names evocative of the old English landscape. Neither tradition would be easily recognized as mapping by conventional cartographic standards today. Both practices informed daily mapping activities, however, creating the framework from which eighteenth-century graphic mapping would arise. Indian deeds, as records of territorial negotiations between both colonists and Indians, document the interaction of these two different mapping practices.

An Overview of the Indian Deed-making Process

The origin of the New England Indian deed tradition has never been determined.[17] Most commonly, the earliest documents of native land transfers are attributed to Dutch practices in New Netherland. Historians have assumed that English colonists initiated a similar process in New England as a result of their colonial competition with the Dutch. However, there is evidence that native land tenure prior to colonization included specific borders between territories, specific ideas concerning rights within and across borders, and a protocol concerning ways by which such borders could change. Edward Winslow wrote, "Every sachim knoweth how far the bounds and limits of his own country extendeth," and Roger Williams noted that "the *Natives* are very exact and punctuall in the bounds of their Lands, belonging to this or that Prince or People."[18] Such notions of land tenure did not include the use of land transfers per se; however, it would be misleading to conceive of the deed-making process as created by and taught to the Indians by colonists alone. Rather, there is evidence that native people incorporated their own traditions concerning territorial rights, inheritance, and the definition of boundaries into deed-making.

Regardless of origins, by the mid-1600s Indian deeds were integral to English methods of land acquisition. With the increasing competition for land both between colonists and Indians and between colonists themselves, as well as the weakening power of claims based on the Crown, purchase from the Indians gained strength over time as a political and economic tool. The nature of the native land transfer process depended on the context of the transaction. Deed-making was influenced by political climate, the physical geography of the land, and the particular colony or native community involved in the transaction. It is possible, however, to make some generalizations about deed negotiation.

Indian deeds were negotiated between Indians and colonists, or between Indians themselves. The colonists involved in native land transactions were usually the town proprietors, who acquired title to land as a group, then allotted it to individuals for settlement. The proprietors sought colonial permission first, in the form of a grant. Only after this grant was received, and settlement begun, was native permission sought (sometimes years later). Because native purchase typically followed the receipt of the grant, colonists entered into native deed negotiations with a preconceived idea of the bounds of the land they envisioned as theirs, sometimes to the extent that the town bounds had already been surveyed and staked without native permission. Whether the land that was actually conveyed in the native deeds resembled this preconceived town map was another matter.

There were other patterns of native deed-making. Deeds from the early period of each town's history are often grants of permission to plant or otherwise make use of the natural resources in that region. Indian deeds could also be negotiated with individual land speculators, a practice that was particularly prevalent in Connecticut and western Massachusetts.

The native negotiators of a deed usually included a primary sachem or leader who acted as the overseer of the territory, as well as the consenting Indians living on or otherwise using the land in question. Some regions had no tradition of regional sachems; in these cases deeds were signed by all those with rights to the territory conveyed. Indians also negotiated deeds among themselves, a pattern that has been particularly documented among praying Indians. These negotiations could occur either parallel to or separate from dealings with colonists.[19]

The conveyance of land required the presence of not only sachems, speculators, and proprietors, but also witnesses and interpreters. Often there were a large number of native witnesses, perhaps influenced by an existing tradition of sachems requiring community consent for decision-making. The bargain was first agreed upon orally at the site of the territory being exchanged, a fee (if any) was exchanged in the form of goods, services, or money, and then the transaction was written down on paper, supposedly verbatim. Sometimes, a graphic map would be included. The deed was then signed, sealed, and witnessed. English colonists signed their names, or an "x" if illiterate. Native signatures varied from a pictographic mark depicting the symbol for the writer's particular clan, to an "x," regardless of the literacy of the writer.[20]

The deed was then taken to the colony or county clerk for registration. The clerk copied the deed, again supposedly verbatim, and returned the original. For the most part, originals have not survived. Most research, therefore, is limited to the examination of clerks' copies. Sometimes deeds were registered and copied repeatedly, signifying that the agreement was either based on tribute or being violated and in need of confirmation.[21]

From this framework, the particular form of the Indian deeds varied depending on the interests being represented. Sometimes an Indian deed was highly English in form, as in this description from Wethersfield, 1671:

> that whole Tract of Land. and all that Part of the Country & is within the Limmits of the Township of Weathersfield, that is to say six miles in Length By the Greate River tide; on the west side of the said River, which is Called Connecticot river; from the Tree marked: NF: the Boundary Tree Betwene Hartford and Weathersfield North: to the tree: WM So marked the boundary tree Between Weathersfield and Midltown South, the great River gap: and the whole Length to run Six Large miles into the wilderness west, in Breadth where Weathersfield and Midltown South. the Said Great River West the whole length to run three Large miles into the wilderness East, the which Lands as aforesaid hath been quietly Possessed by the English now for seuerall years passt. . . .[22]

Sometimes the document was heavily influenced by native tradition, as in this 1636 deed from Western Massachusetts:

> It is agreed between Commucke & Matanchan ancient Indians of Agaam for & in the name of al other Indians, & in particular for & in ye name of Cuttonus the right owner of Agaam & Quana, & in the Name of his mother Kewenusk the Tamasham or wife of Wenawis, & Niarum the wife of Coa. to & with William Pynchon Henry Smith & Jehu Burr their heirs & associates for ever to trucke & sel al that ground & muckeosquittaj or medows, accomsick, viz: on the other side of Quana; & al the ground & muckeosquittaj on the side of Agaam, except Cottinackeesh or ground that is now planted. . . .[23]

Sometimes, too, the deed was only marginally concerned with territorial representation, instead intended to establish relations of power among people.

This is not to say that the deeding process was fair. Indian deeds served colonial bureaucracies as tools for dispossessing people from their land. Settlement of land transactions through native deed were sometimes forged, or the result of coercive force, of violence, and the desperation of a people marginalized into poverty, landlessness, and powerlessness. But within the condition of being colonized, native people continued to speak, to articulate territory, to contest and resist English encroachment. This study focuses on those deeds in which the representation of territory not only plays a dominant role in the deeds' function, but also suggests a particularly native influence or character. It is these kinds of articulations of territory that provide a framework from which we can address questions about the nature of native mapping among Algonquian-speaking peoples. What follows is an exploration of both verbal and graphic mapping types, and the circumstances that created them. In each case, because of the high influence of native mapping, the native signatories did not provide what the colonists sought, and the deed was either ignored or renegotiated with other Indians.

Mapping through Words

As mentioned above, deeds usually represented territory through words only, perhaps because this was the mode of mapping most familiar to English, most familiar to Indians, or both. In these two examples, the native use of toponymy as a mnemonic map is evident. In the first example from Mattatuck, there is a strong use of native mapping through words, interspersed with English mapping of seemingly different boundaries. In the second example from Ridgefield, Indian and English mapping have become more closely integrated, and native mapping distinctly compromised in the translation.

Mattatuck 1685

Located in the Naugatuck River valley, Mattatuck lay at the eastern edge of the territory of the Paugussett-affiliated peoples in the late seventeenth century (see fig. 7.1).[24] Until their colonization by Connecticut settlers, Paugussetts had lived throughout the Housatonic watershed, practicing a mixed economy of agriculture, hunting, and fishing. Those living at Mattatuck comprised the northern edge of the Paugussetts of the Naugatuck and lower Housatonic drainages, estimated to have been between 1,200 and 4,800 people in 1600.[25]

English colonists from Farmington first began seeking title to land in the Naugatuck Valley in 1657, at "A Tract of Land Called matetacoke."[26] In April 1674, a Farmington settlement

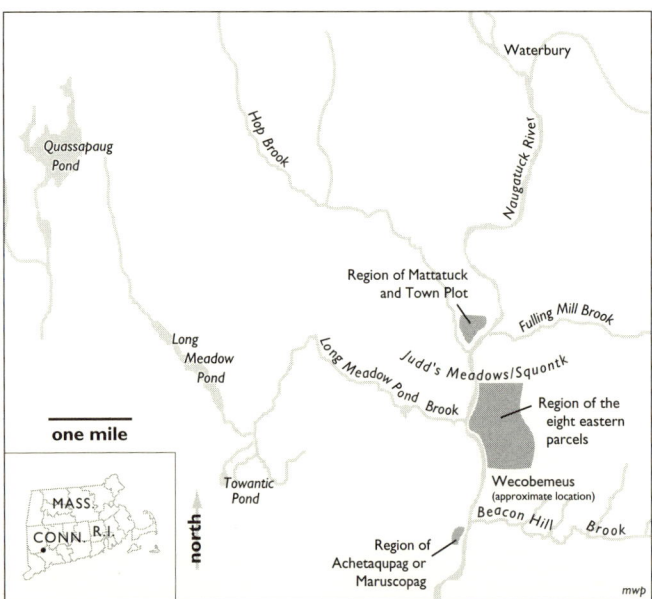

7.1 The Mattatuck deed related to a nineteenth-century topographic map. By the author.

committee reported to the General Court, describing the economic potential of the land, and noting in particular the existence of many meadows well suited for grazing cattle, on both sides of the river.[27] When the court granted permission to colonize Mattatuck, the committee wrote up a list of "Articles of Agreement" for their settlement, began to prepare the layout of the town center, and finally, sought native title to the land.

When the colonists finally filed a purchase from the Indians in August, however, it was not with the Paugussetts. Rather, they chose the signatures of the people with whom they had been negotiating deeds for over forty years in Farmington, the Tunxis people. The deed described the land as the colonists wanted it described, and probably as the Tunxis people were asked to describe it:

> one persell of land att matatuck set[ ] on eich side of mattatuck River being ten miles in length north and south and six miles in bredth bredth [sic] buting upon the bounds of farmington one the east upon pagisut and pagatauck west and pomparag west and one pagisutt south and west and upon the open willderness north.[28]

The town was laid out, but apparently not settled. King Philip's War interrupted in 1675, requiring colonial resources and halting land transactions between colonists and Indians. By 1677, however, the settlement of Mattatuck resumed. By 1681 the plantation at Mattatuck was bound, divided into thirty-nine lots, and home to 145 Farmington colonists settled without permission on Paugussett lands.

In February 1684, however, a deed was finally negotiated with the Paugussetts. The chief negotiators were Cockapatana and Awawas, representing the Paugussett people. It was Cockapatana's first deed negotiated independently from his sachem predecessors. By this time a majority of the Paugussett territory had already been ceded to whites, through twenty-six deeds, converted into the townships of Milford, Derby, and Stratford.[29] Despite the eroding Paugussett land base, there was a strong native mapping voice in the deed of February 1684, differentiating it from previous agreements about Mattatuck with the Tunxis. The deed described "Twenty parcels of land, by their names distinguished as follows:

> Wecobemeus, that land upon the brook, or small river that comes through the straight northward of Lebanon, and runs into Naugatuck river at the south end of Mattatuck bounds, called by the English Beacon Hill Brook, and Packawackuck, or Agawacomuck, and Watapeck, Pacaquarock, Mequnhattacke, Musquauke, Mamusqunke, Squapmasutte, Wachu, which nine parcels of land lie on the east side of Naugatuck River southward from Mattatuck town, which comprises all the land below, betwixt the forementioned river, Beacon Hill Brook and the hither end of Judd's meadows, called by the name Squontk, and from Naugatuck River eastward to Wallingford and New Haven bounds, with all the lowlands upon the two brooks forementioned.
>
> And eleven parcels on the west side; the first parcel called, Suracasko; the rest as follows: Petowtucki, Wequarunsh, Capage, Cocumpasuck, Megenhuttack, Panooctan, Mattuckhott, Cocacoko, Gawuskesucko, Towantuck, and half the cedar swamp, with the land adjacent from it eastward; which land lies southward of Quasapaug pond; we say to run an east line from there to Naugatuck river; all of which parcels of land forementioned lying southward from said line, and extend or are comprised within the butments following; from the forementioned swamp, a straight line to be run to the middle of Towantuck Pond or the cedar swamp, a south line which is the west bounds

towards Woodbury, and an east line from Towantuck pond, to be the butment south, and Naugatuck river the east butment, till we come to Achetaqupag, or Maruscopag, and then to butt upon the east side of the river upon the forementioned lands,—these parcels of land lying and being within the township of Mattatuck, bounded as aforesaid, situate on each side of Naugatuck and Mattatuck rivers.[30]

The first interesting aspect of this description is that the land was conveyed not as a single tract but as twenty "parcels," perhaps coinciding with the river meadows in such demand by colonists at that time, and for which Mattatuck was renowned. The separate parcels also suggest that the land being conveyed was not contiguous, and that Paugussetts still claimed rights to land in between the parcels.

As the narrative begins, Mattatuck is defined through native toponym as nine parcels on the east side of the Naugatuck river, and eleven on the west side. Wecobemeus, the first parcel, is situated in the context of two features with English names, Lebanon and Beacon Hill Brook, perhaps to establish a beginning bound familiar to all negotiators of the deed. The next eight of the eastern parcels or places are given no explanation, conveying their location through Paugussett name. They seem to be located in the shaded area on the eastern banks of the river, where the meadows would have been, but the description continues as though these places stretched all the way to Wallingford.

The nine parcels on the western shore also convey their meaning through native toponym, linking to English mapping conventions at the end of naming. It is more difficult to locate these place-names, apart from knowing that they lie on the western side of the Naugatuck, and south of Quassapaug pond. Towantuck, however, must be in the vicinity of Toantic Pond, and Mattuckhott is likely the same parcel from which the plantation name of Mattatuck was originally derived. Megenhuttack is probably contiguous to the eastern meadow called Mequnhattacke.[31]

The deed then continues with English mapping conventions with the linking phrase, "all of which parcels of land forementioned lying southward from said line, and extend or are comprised within the butments following." The territory is then remapped through an English description of the bounds, "till we come to Achetaqupag, or Maruscopag."

In the first, native mapping, the township is described in terms of named land areas, the meaning of which probably emphasizes a spatial description of the place, rather than the place's linear boundary. This is followed by a remapping of the bounds using English emphasis on place as defined by the line that circumscribes it. It is not quite finished by the line of English metes and bounds, however, for it is the last corner, in the southeast section of the tract, which becomes Achetaqupag/Maruscopag, and the eastern parcels. Thus, the 1684 deed seems to have been an attempt to preserve colonists' interests in the outer town boundary in the face of Paugussett interests in conveying the land by place-name. The two mappings do not seem to fit, entirely: though the naming of purchased native places is detailed, whether or not the meadow places extend all the way to the prescribed town boundaries seems dubious.

Why did Cockapatana and the Mattatuck people map the land in this fashion, after several decades of deeding, and nine years after the war?

There are several possible answers. Perhaps Cockapatana felt that a new political strategy was required in his dealings with the English. Paugussetts had filed complaints in the General Court of fraud and intentional abuse of land agreements since at least 1649, so native people were well aware of the consequences of deed-making. Alternatively, the territorial representation may have been dictated by the people who lived there and for whom Cockapatana and Awawas spoke, seven of whom (both men and women) are named in the deed. Perhaps in this instance the Paugussetts believed that they could convey only land to which they had rights, perceiving their rights only to those meadows or other places that they used, and that they could not convey rights to those areas that they did not use. Another possibility is that Cockapatana and the other Paugussetts may have found themselves in an unusually strong position of power, knowing that the colonists were being pressured to file legal documentation of their land transfers in order to secure a patent from the new administration of James II. In any case, this different method of mapping was never used again by the Paugussetts, even though both Cockapatana and Awawas went on to preside over later deeds.[32]

Most problematically, the Paugussett deed did not clearly convey rights to the bounds already ratified by the court. The purchase was only for the parcels in the southern half of the prescribed township, leaving a large area as yet unobtained. Two more Indian deeds were obtained in that year, one in April and one in December; both were signed by Tunxis, and not Paugussett, peoples.[33] The first added a section of land to the north side of Cockapatana's deed, and the second summarized the whole as comprised by deeding since 1674. Both deeds described the land by English conventions rather than by named parcels. The deeds in order, the paperwork was submitted to the General Court. In May 1686 the proprietors of Mattatuck received their patent, and the name of the town was changed to Waterbury.[34]

The names of the mapped parcels never appeared in the colonial records again. In a sense, the mapping of Cockapatana and Awawas was ignored by the town, with the exception perhaps of Toantic, which survived because it marked a border between towns, as opposed to describing an aspect of the town's interior. The 1684 deed from the Paugussetts had not provided what the colonists wanted. It provided something else: a detailed mapping of native use and rights in the river valley.

Ridgefield 1708

Some Indian deeds show evidence of the accommodation of native mapping by place-name to fit the emphasis on lines connected by points required by English mapping. This is illustrated by the first native deed for Ridgefield, negotiated twenty-three years after Mattatuck on land west of the Paugussett territory, and east of the colony of New York.

Norwalk colonists first came to view this land in 1706. In May 1708, they petitioned the General Assembly for permission to purchase from Indians the parcel of land that they identified as north of Norwalk and west of Danbury (see fig. 7.2).[35] Four months later, the Ramapo sachem Catoonah met with John Copp, Joseph Seeley, and John Holmes, Jr., to convey a tract of land for settle-

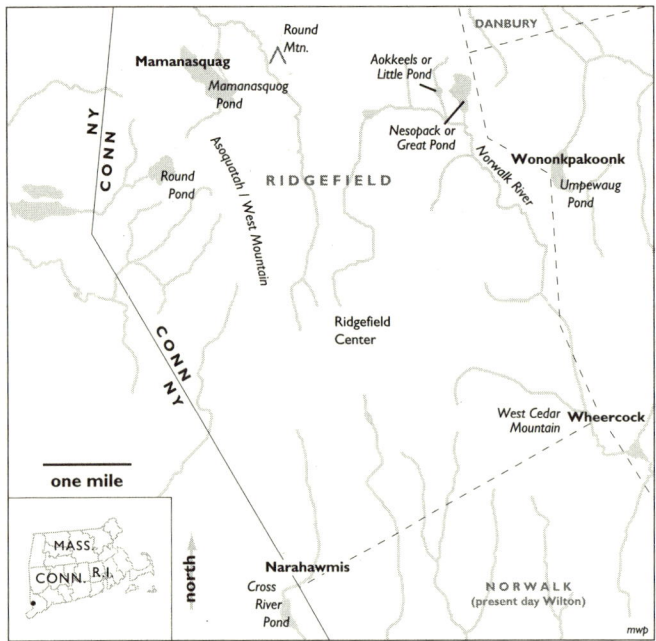

7.2 The Ridgefield deed related to a nineteenth-century topographic map. By the author.

ment by Norwalk and Milford colonists. The territory was mapped as follows:

> at a Rock with stones Lay'd thereon that lyeth upon ye west side of Norwalk River about twenty rod northward of the Crossing or where Danbury old Cart path Crosseth the River which said Rock is the South East Corner and from said Corner a line Runneth upwards unto Umpewange pond to a White Oak tree, Standing by the Northwest Corner of said Pond, the said tree being marked and Stones Lay'd about it and is the North East Corner, and from the said Corner Tree, another line Running Near Two points to the North of West into a pond called Nesopack and continues ye Same Coarse untill it meets with a second pond Called Aokkeels, Crossing by ye south End of both ponds, and from thence Running Near West untill it Extends to a place called Mamanasquag, where is a Oak Tree Marked on ye North Side of the outlet of water that Comes out from a sort of grassy pond, which is known and Called by Said Name, which tree is the North West Corner, and from said Tree another line Running South bearing to ye East About one mile and a half. Runing by ye East side of another Mountain Called Asoquatah untill it meets Stanford Bound line, about a quarter of a mile to ye Eastward of Cross River pond, where stands a Marked White Oak Tree with Stones about it, and is ye South West Corner, and from Said Marked Tree a long by Stanford line untill it comes to Norwalk purchase and so by Said Purchase Bounds to the Said Rock at the South East Corner. Containing by Estimation Twenty Thousand Acres, be it more or less. The Four Corners of Said Tract of Land being Called by the following Indian Names South East Corner "Wheer cock" North East Corner "Wononkpakoonk" North West Corner "Mamanasquag" South West Corner "Narahawmis."[36]

The named features of this deed are shown on the map in figure 7.2. The deed begins with English mapping conventions. The bounds are described in terms of boundary lines and the points or corners that anchor them. These lines and points combine natural (rocks, ponds, brooks) as well as artificial (Stanford Bound line) features in the map, as was the English tradition at the time. When native words are used for reference, they refer to features that the English perhaps perceived as having unambiguous boundaries or identities (a mountain "Called Asoquatah," and several ponds), and thus acceptable as natural points for corners. The exception to this is one reference at the northwest corner to "a place called Mamanasquag," though even this is immediately redefined by two points lying within Mamanasquag (a tree and a pond) as the true corner markers.

At the end of the deed, the territory is remapped through native toponymy. As in Mattatuck, the toponyms refer to tracts of land, the size and limits of which cannot be interpreted without knowledge of the language or the map from which they were drawn. Though unbounded, the places are still named as "corners." Thus, a transition is being made in the native mapping mode, from mapping the territory according to its interior spaces to mapping according to intersections of outer boundaries. No name or collection of names is given to the place purchased, other than "A Certain Tract of Land," reinforcing the lack of native identity of the interior lands (and thus the lack of native presence or threat), and the comparative importance of the colonized lands of the exterior.

The following year, the General Assembly appointed three men to survey and lay out the newly purchased land. Their instructions were to "to make a survey of the tract of land granted by the General Court in May, 1708, . . . and to lay it out for a town plat."[37] This order ignored the land as defined by Catoonah in favor of the earlier definition conceived by the viewing committee. The surveyors' report was as follows:

> all the aforementioned tract or parcel of land, butted and bounded as followeth, that is to say, on the south or southerly with the said town of Norwalk, on the west or westerly with the line or boundary between this Colony and the Province of New York, on the east or easterly partly with a line to be continued and run like unto the line between the said town of Norwalk and the town of Fairfield, from the north end thereof unto a certain black oak tree marked with letters and having stones layed about the same, standing upon the mountain commonly called the West Cedar Mountain, and partly with a direct and straight line to be run from the said black oak tree to a certain large white oak tree marked and having Stones layed about it, standing at or near the north west corner of Umpewange pond; and on the north or northerly with a direct straight line to be run from the said white oak tree to the southwest corner of the town of Danbury, and continued unto the said line or boundary between this Colony and the Province of New York, be the said tract of land more or less. . . .
>
> And this Assembly do hereby enact and grant that the said tract of land shal be an entire township of itself, and shall be called and known by the name of Ridgfield.[38]

The grant significantly remapped the land, by removing all native toponymy except for "Umpewange pond," an established reference point on the boundary between Ridgefield and Danbury. As in Mattatuck, the Indian deed did not clearly convey the bounds for which colonists had already assumed rights. Instead, it conveyed territory as the Indians chose to define it. In response, the Ridgefield proprietors ignored the content of Catoonah's deed, buried the place-names, and reasserted their vision of the map of Ridgefield.

Six more purchases from the Indians would map the expanding limits of Ridgefield's borders, the last deed recorded in December 1739. As with the Waterbury deeds, these later purchases would always be English in style, with native place-names relegated to names of hills, ponds, and brooks.

Mapping through Graphics

As shown above, property mapping in colonial southern New England could combine the traditions of English and native mapping styles, recorded in words through registered Indian deeds. Occasionally, however, circumstances dictated that a graphic map should be included in the transaction. These circumstances are not only rare for Indian deeds; they are rare to colonial property deeds as a whole until the mid-eighteenth century.[39] The first two graphics in this section were explicitly identified as drawn by native people. The third, from Wecabaug, may not be native; it is the product of native testimony and survey, however, and thus highly influenced by native people in its content.

With such a small number of examples of graphic maps, it is difficult to generalize about the nature of such maps on Indian deeds. It is notable, however, that two of the maps identified as native were drawn in order to clarify which land was for English purchase, and which land retained some degree of native rights or possession. In the Wecabaug testimony, the map seems to be present in order to clarify the geography of a place small in extent but large in its political significance for tribal and colonial boundaries. Thus, in each case,

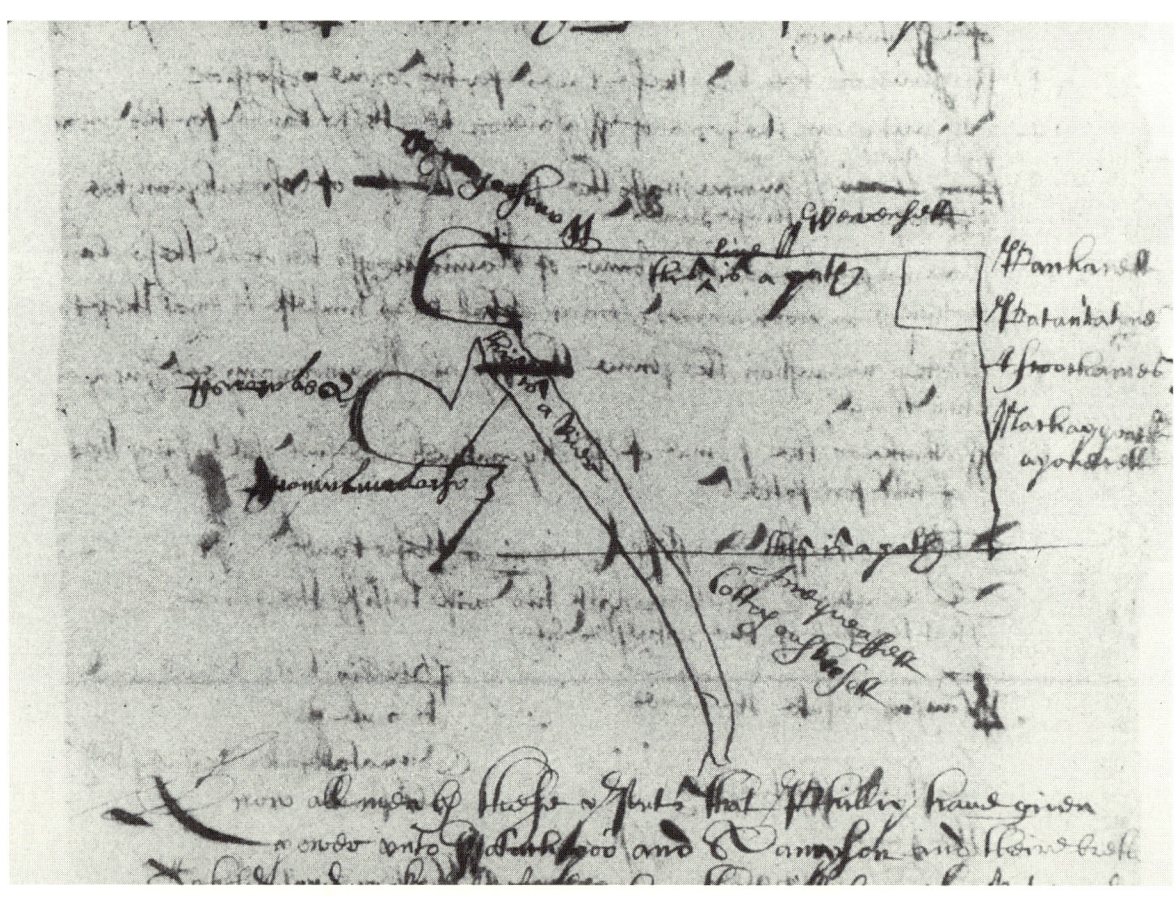

7.3A The Sepecan deed. Courtesy of Plymouth County Commissioners.

a graphic map was drawn in order to clarify a point that was somehow not sufficiently clear or powerful if encoded in words.

A further note of caution concerns the forms of the deeds as objects. The maps of Wecabaug and Sepecan are *clerk's copies;* it must be assumed, then, that the maps were altered during the copying process. This circumstance should be taken into account particularly in interpretations of the design of the maps. In the case of the Weantinock map, the extant document is at least two generations removed from the original, a historian's copy of the clerk's copy.

Sepecan 1666

The first example is from Plymouth Colony (see fig. 7.3A–C).[40] Though the location of the original deed is unknown, the clerk's copy was registered in the Plymouth Colony Records, dated 1668 (see copy of the map portion in figure 7.3A; the graphic is partially obscured by writing on the reverse side of the document). In 1839 J. W. Barber included a later, adapted version of the map in his *Historical Collections . . . of Every Town in Massachusetts* (fig. 7.3B), and in 1861 the sketch was again interpreted for the printed version of the *Records of the Colony of New Plymouth in New England: Book of Indian Records for Their Land* (fig. 7.3C). The sachem conveying land in this deed was Metacom, or "King Philip," sachem of the Pokanoket people (later known as the Wampanoag). In the deed that accompanied the sketch, Metacom's words were recorded as follows:

> This may informe the honord Court that I Phillip ame willing to sell the Land within this draught; but the Indians that are vpon it may liue vpon it

still but the land that is [waste] may be sold and Wattachpoo is of the same mind; I haue set downe all the principall names of the land wee are not willing should be sold.[41]

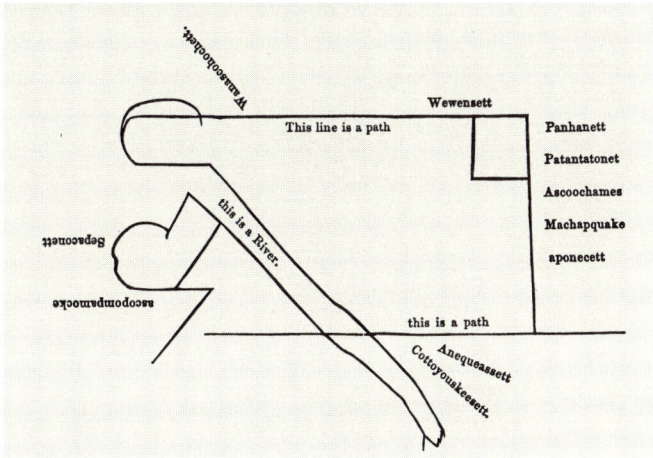

7.3B The Sepecan deed as redrawn in J. W. Barber, *Historical Collections . . . Relating to the History and Antiquities of Every Town in Massachusetts* (Worcester: Dorr, Howland & Co., 1841), p. 526. Courtesy of the Edward E. Ayer Collection, the Newberry Library.

7.3C The Sepecan deed as redrawn in *Records of the Colony of New Plymouth in New England* (Boston: William White, 1861), 12:237. Courtesy of the Newberry Library.

The map was originally drawn oriented with south approximately toward the top left corner. In figure 7.4 the graphic map elements have been inverted to align with north at the top, and related to a topographical map.[42] The lands portrayed are part of the larger area of land known as Sepecan, which in the seventeenth century covered the region of the coast bounded by Middleborough on the north side, Dartmouth on the west, and Agawam Plantation (Wareham) on the east. Today this is Mattapoisett, Rochester, and Marion, Massachusetts.[43] Two of the native place-names set down by Metacom can be located on the topographical map in figure 7.4. Sepaconett refers to Sippican Neck (#3), and Aponecett can be related to Aponequet, now Long Pond, in Lakeville (#6).[44] The words "this is a River" refer to Sippican Harbor (#2).

The meanings or locations of the other toponyms are unknown. These places may have been the names of old planting grounds, meadows, settlements, or hunting or fishing places that Watachpoo or Metacom sought to protect. Similarly, the meanings of the straight lines and the corner square cannot be known for certain. They may represent paths or roads through the area, or they may represent the boundaries between sequestered and unsequestered lands.

Metacom's map of Sepecan shows characteristically native mapping style in both its written and graphic elements. The graphic elements include the representation of irregular paths, bounds, some of the coastline, and river as straight lines with angular intersections. This portrayal of irregular line features as straight has been demonstrated by Lewis as characteristic of native graphic style.[45] Also, the lines are not differentiated by symbol, implying that the reader is able to differentiate between lines signifying water and lines signifying political boundaries. This suggests a mnemonic approach to symbolism similar to the mnemonic approach to place-naming described in the first section of this chapter.

In addition, the text indicates that Metacom "set downe all the principall names" of the land not to be sold (though more likely this was done by Metacom's secretary John Sassamon, as discussed below). This statement emphasizes the importance of mapping by words, the recording of the toponymic map in the native tradition. That the names are "principall" is particularly interesting, implying that there are many more named places in the area, though the area represented is only a few miles square. Thus, the outer circle of place-names represents the land that

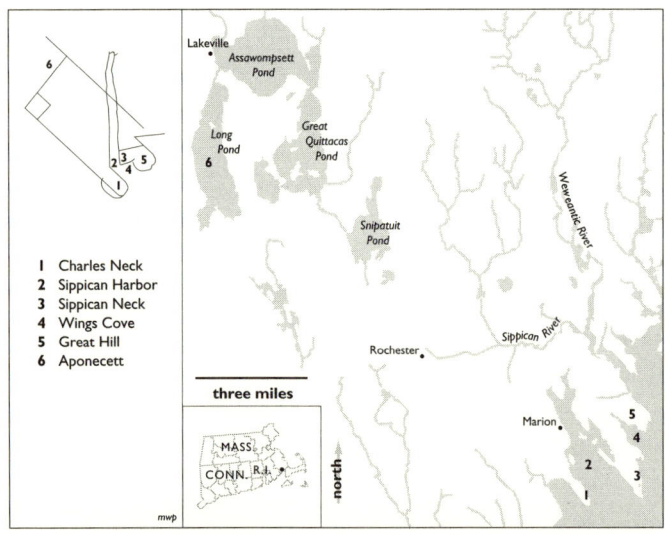

7.4 The Sepecan deed related to a nineteenth-century topographic map. By the author.

Philip does *not* wish to sell, with the inner, unnamed space representing land available for sale to Plymouth colonists.[46]

Sepecan was one of the last land holdings that remained in Pokanoket or Wampanoag possession during the years preceding the war in 1675. At the time of the deed, the competition for land had relegated Wampanoag claims to fringe settlements along the coast, where the soil was not as conducive to agriculture, while colonists controlled the prime bottomlands and transportation routes of the major interior waterways.[47]

The deed's principle signer, Philip or Metacom, was not a sachem of Sepecan, but served as a primary sachem for Pokanoket peoples. He was the son of Ousamequin (Massasoit), sachem of Pokanoket at the time of the colonists' arrival in 1620. Though Ousamequin had signed a treaty of peace with Plymouth Colony, Metacom found himself sachem at a time of increasing tension between Indians and colonists, as English settlement steadily infringed on and eliminated Pokanoket territories.

Metacom's words that "Watachpoo is of the same mind" refers to the sachem of Sepecan, Watachpoo (also known as Totosin). Native testimony in the Plymouth Court Records indicates that Watachpoo had inherited the land as a sixth-generation sachem:

> Wassauwon hee had these lands for his owne possession.
> Vspauhquan the sonne of Wassauwon held these lands for his owne.
> Naunaumasso the sonne of Vspauhquan, hee likewise held these lands.
> Maumoowampees the sonne of Naunaumosso hee had these lands which Maumoowampees Amawekkett knew himselfe to enjoy these lands.
> Pohquantauson the sonne of Maumoowampees Injoyed this land;
> Wattachpoo the sonne of Pohquantauson desires still to Injoy the land of his forefathers.[48]

Despite continued native control of Sepecan, Plymouth colonists coveted the lands, and began vying for rights to them as early as the 1630s. Repeatedly, during the seventeenth century, colonists attempted to purchase Sepecan lands, and repeatedly their requests were denied (though settlement of the area continued, illegally, anyway).

In 1665, however, the beginnings of the successful purchase from Watachpoo can be seen in the records:

> Wheras formerly Richard Bourne and Willam Bassett were appointed by the Court to purchase a pcell of land desired by Thomas Butler, and that it doth appeer vpon tryall that the Indians will not pte with it, a further libertie and order is graunted to the said Richard Bourne and Willam Bassett, in the behalfe of the said Thomas Butler, to purchase other land desired by him, and that they make reporte thereof to the Court, that they may doe therein as they shall meet.[49]

It was Bourne who was eventually successful in negotiating the purchase of part of Sepecan for the English. In the deed's transaction, Philip, as the primary sachem for Wampanoag territories, gave permission to Watachpoo to sell off some of his land. Philip's secretary, John Sassamon, was recorded as the witness; he was most likely also the writer of the original deed, and the person who wrote in the place-names representing sequestered lands (educated at Harvard Indian

College, Sassamon could both read and write English).

Perhaps most interesting is that the document as a whole portrays not only a sale of land but also the symbols of the forces playing against Metacom which would eventually develop into the events leading to the war of 1675. Known as "King Philip's War," it would be the last major organized resistance to English colonization in southern New England.[50]

The first of these symbols is the presence of John Sassamon, the secretary and witness. Sassamon came increasingly into suspicion as a spy for the English, and years later was dismissed from Metacom's service. At some point, he moved back to Assawompsett (now Lakeville). In January 1675, he overheard the plans of the Wampanoags and other Indians to conspire against the English in war, beginning with an attack on Swansea. This information he reported directly to Josiah Winslow, a report apparently not taken seriously. When on January 29 his body was found in Assawompsett Pond (in northwest corner of map, fig. 7.4), the victim not of an accident but murder, his report to Winslow was considered more seriously. In June, on the basis of a single Indian's testimony, three Indians were hanged for Sassamon's murder. The hangings caused a collective anger among the Wampanoags, bringing Indians into war against Plymouth prematurely.

Another piece of Metacom's fate can be read in the sketch map's inclusion of Sepecan Neck as part of the lands protected by Philip and Watachpoo. A year after Sassamon's murder, Sepecan Neck served as the site for a further betrayal of Metacom. Awashonks, sachem of Sakonet and Metacom's cousin, met the colonist Benjamin Church on the Great Hill (#5 in fig. 4) in July 1676, to seal an alliance between the Sakonet Indians and the English against the Wampanoag sachem. When Church left, he took with him some Sakonet men to serve as soldiers.[51] The following month, Metacom was dead. Thus, though Sepecan Neck was a place protected by Metacom for native use, ten years later that same land would be used by others to plot against him.

Indians living at Sepecan maintained control over most of the lands up until the war, continuing to document their claims in the Plymouth Records.[52] After the war, numerous complaints of trespass were filed by both Indians and colonists claiming rights to Sepecan. Multiple entries in the Plymouth Records indicate English settlement on sequestered lands, and resistance to such settlement particularly by the Sepecan Indians Charles and Connet. These disputes continue to surface in the colonial records until the 1680s. Sepecan was finally incorporated as an English town in 1686. With the patent came the inevitable replacement of the native toponym with the English transplant, and Sepecan was remapped as Rochester.

Weantinock 1703

Another graphic map was drawn on an eighteenth-century deed, at a place called Weantinock. Translated as "where the water swirls around," this was (and is) the homeland of the Weantinock and Schaghticoke peoples, part of the Paugussett confederacy. Weantinock was also one of the three larger settlements in the region, and the site of the primary Council Fire. It is estimated that in the seventeenth century, between 500 and 1,000 native people lived here. The primary settlement was on the west bank of the Great River (to the English, the Housatonic), at the eastern base of a hill overlooking the river. At the top of the hill a stockade fort was maintained, a place now known as Fort Hill. Stretching down the floodplain were about 200 acres of ancient planting fields, as well as scattered, smaller settlements, "at least five centuries old" in the 1700s.[53] As Richmond has written,

> Weantinock people planted corn, beans, and squash in small gardens and community-worked fields. We hunted deer in forests and along swamp edges; gathered seasonal nuts, edible plants, and curing herbs; and fished for shad and eels. Paths linked settlements to one another and to the places where important resources could be found. Our ancestors also traveled down the Housatonic River, sometimes as far as the coast, to visit and trade with their kin.[54]

About two miles downriver was Metichawon, a site that served as both an economic and spiritual center. Seventeenth-century Metichawon was an area of land overlooking the Great Falls at the confluence of the Great River and what is now the Still River. In the pools below the Great Falls was the primary spot for fishing for shad and eels; on the cliffs above, the traditional burial ground. Metichawon was also an early center for trade. From 1643 to 1658, Stephen Goodyear maintained a trading post just below the falls, where the goods he obtained from native people were sold to English, Dutch, and West Indian markets.[55]

Though both English and Dutch colonists had settled in the area since the 1630s, pressure to sell land to English colonists began in the 1670s and quickly intensified. In 1700, a committee of Milford colonists reported on 100 acres of land that they had viewed at Weantinock and deemed suitable for a plantation. Permission to plant was granted in 1702.[56] The colonists also received permission to negotiate a grant from the Weantinock Indians. So, with the help of John Minor, "interpreter allowed both by the English and the Indians," an Indian deed was drawn up at Metichawon on February 8, 1703.[57] The result was a document compiling both written and graphic mapping of the territory being conveyed. The location of the original deed, which contained the only original copy of the map, is unknown (fig. 7.5).[58]

The text of the deed makes reference to this graphic on several occasions during the verbal description of the territory, noting specifically that the "draught" was provided by the Indians:

> A Certain Tract of Land called Weeantenock with all and singular its Rights members and appurte-

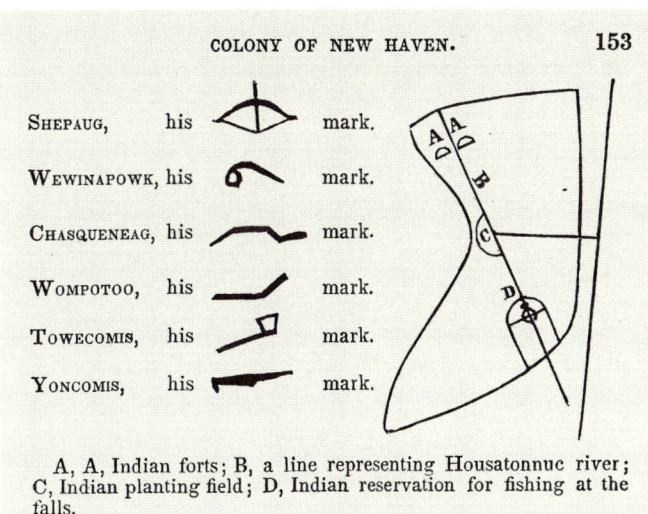

7.5 The Weantinock deed as redrawn in Edward R. Lambert, *History of the Colony of New Haven* (New Haven: Hitchcock and Stafford, 1838), p. 153. Courtesy of the Newberry Library.

nances together, ... according to the draught given of said Land by the above named Indians; the abovesaid Land being bounded easterly with woodbury bound, and a parallel line running northward and southward according to the draught above mentioned and is bounded westerly with the mountain as appears by said draught, and from the abovesaid Brook at the northwest corner — the northernmost bounds run to the eastward, to the parallel line abovementioned according to said draught and with Danbury Bounds southwesterly according to said draught, and Southeast with Woodbury bounds.[59]

The "above named Indians" who provided the draught are listed in the deed as Papetoppe, Rapiscotoo, Wampotoo, Hawwasues, Yoncomis, Shoopack, Parameshe, Nokopurrs, Paconaus, and Wewinapouch. This is the closest that the deed comes to identifying the map's artist or artists.

The last line notes that not all rights to Weantinock are being sold: "It is to be noted that the above named Indians do reserve for their own use their present planting field and a privilege of fishing at the Falls."[60] Below the transcription of the deed, the clerk noted, "Here follows upon the Record a diagram of Wiantinogue as drawn by the indians marking their reservations."[61]

The native map of Weantinock was thus drawn to indicate the critical sites of the transaction, which sites are related to a topographical map in figure 7.6. The "present planting field" was located on the west bank of the Housatonic, beside Fort Hill (#1 in fig. 7.6). The "Falls" refers to the Great Falls at Metichawon (#5). These were the two sites of reserved native rights. Some of the symbols on the map are more ambiguous. The circles, for example, may represent hills or the locations of Indian forts.[62] The + symbol seems to rep-

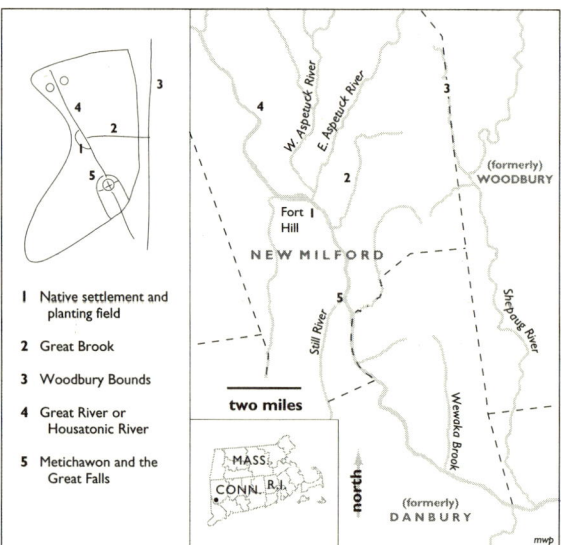

7.6 The Weantinock deed related to a nineteenth-century topographic map. By the author.

resent Metichawon, and may have been used to indicate the site at which the negotiation of the deed (and the drawing of the map) was occurring or occurred.

Like the map of Sepecan discussed above, lines are used without symbolic differentiation to describe rivers, forts, or mountains, town boundaries, and the bounds reserved for native use. All lines are solid, and the reader of the map must have the knowledge to interpret the significance of each line to features on the land.

Partial native rights to the planting field were sold to a Woodbury man in 1705 by "Shamenunckqus, alias Bapistoo," again through Minor's interpreting.[63] The bounds of the field were left ambiguous, and Shamenunckqus also noted that he was not selling the apple trees in that area. As a result, rights to use and ownership of the planting field continued to be controversial throughout the first half of the eighteenth century.[64]

The reservation of rights to fishing at the Great Falls were never sold. They continued to be claimed by Paugussett peoples until the 1870s, although this claim was continuously contested by non-Indians. In 1877, however, the Housatonic Water Company dammed the Housatonic at Derby, ending the fishing at the falls. Although Housatonic Water was forced to pay damages to the Cove Fishing Company, a group formed of non-native shareholders who fished at the falls, native people were never compensated. Thus the native rights to fish at the falls were not relinquished, but rather, extinguished.[65]

By 1733, all other native claim to Weantinock lands had been sold. As more colonists came to purchase and settle land, many Weantinock people (along with Pootatuck people from the south, who found themselves in similar circumstances) moved north to Pishgachtigok, becoming part of the Schaghticoke people. Today the Schaghticoke Reservation is located in this same area on the Housatonic River in Kent, Connecticut. Weantinock and Metichawon, however, remained (as they remain today) important spiritual sites for Paugussett peoples, both for those who stayed at Weantinock, and those who continued to travel there each year.[66]

Wecabaug 1662

The third example, though chronologically the earliest, indicates a type of native influence over the graphic different from the first two examples. In a 1662 testimony from Uncas to the court in New London, a graphic map was used as part of the testimony over questions of ownership and rights to lands at Wecabaug.[67] The testimony was part of a land battle between the southern New England colonies, the Narragansetts, and the Mohegans. Originally the territory of the Eastern Niantics, Wecabaug became the site for both native and colonial pursuits of land and power in the years following the Pequot War of 1637. The mapping of Wecabaug became crucial to the establishment of both native and non-native land claims, illustrating a specific instance in which the definition of colonial borders was dependent on exact delineations of native territorial borders.

Although the Mohegans and Narragansetts both aligned with the colonies against the Pequots in 1637, after the war they turned against each other in the struggle for power. Wecabaug became one site of these disputes, claimed by both Uncas and Miantonomi based on their assistance in the Pequot War. These disputes were mediated in part by the Commissioners of the United Colonies. In 1643, tension between the Mohegans and Narragansetts erupted into war and the murder of Miantonomi.[68] Hostilities between the two nations continued with Pessacus as the primary Narragansett sachem.

Meanwhile, English colonists had begun to settle the region, claiming rights to Wecabaug which hinged on each colony's native allegiances. Rhode Island made a claim for the land based on Narragansett deeds negotiated from 1644 to 1660 which documented the colony's rights to land stretching as far as the Pawcatuck River. To strengthen their claim, Rhode Islanders moved to the east bank of the Pawcatuck to establish the settlement of Misquamicut. Massachusetts settlers claimed the same area, but after the Mohegan-Narragansett war Charles II ruled against it on the basis that Massachusetts lacked native allegiances in the area. Massachusetts continued to claim the land by right of conquest over the Pequots, and settled the town of Southington, or Sotherton, on the Pawcatuck's eastern bank. Connecticut also claimed the land on the basis that it was Pequot territory, refusing to recognize the Massachusetts claims. These conflicting native and colonial claims were further complicated by the additional claims of private land speculation companies, the Atherton Company and the Misquamicut Company, both of whose membership came from all three southern New England colonies.[69]

In a September 1658 entry in the *Acts of the Commissioners of the United Colonies*, a report was filed concerning the activities of two Pequots, Cashawasset and Casasinamon. Cashawasset (also known as Harmon Garrett) was a Pequot then known as a Niantic sachem; like Uncas, he had also fought against the Pequots in 1637, and, as United Colonies' appointed governor of the Pequots at Paquatucke and Wequapauge, actively sought to expand rights to former Pequot territories in the years following. Casasinamon (also known as Robin) was the Pequot sachem at Noank, on the west bank of the Mystic River, west of Pawcatuck. He was also a former servant of John Winthrop, Jr., another Atherton Company member.[70] In the records, he and Cashawasset "appeared to giue an account of their Gou^rment the last yeare; and to Request some further portions of land for theire settleing and planting." The commissioners agreed

> that Cashwashett and his Companie shall haue a meet proportion of land att Squamscutt necke on the East side of parketuck Riuer and Cashasinnimon and his Companie shall haue a fitt proportion of land alowed them att Wawarramoreke near the path that leads from misticke River; and it is Comended to the Generall Court of Conecticott to appoint as soon as may bee some meet psons to lay out and bound the said lands for them; and Capt: Gorge Dennison and Tho: Stanton and sarjeant Minor or any two of them are desired appointed to bee aiding and assisting to the said Gou^rning Indians according to the orders and Instructions giuen them and by theire Councell and Countenance to Conteine the Rest in obeidience to them.[71]

George Denison (along with then governor of Connecticut Colony, John Winthrop, Jr.) had been a resident in the Pawcatuck area since 1653, and was as well a member of the Atherton Company. Stanton and Minor were probably "aiding and assisting" as interpreters.

According to the Mohegan land claims following the Mohegan-Narragansett war, Pequot country extended as far east as a place called Wecabaug.[72] Thus the legitimation of both native and colonial claims to Wecabaug depended on the establishment of its exact location. In February, 1659, the survey of the lands at Wecabaug took place, carried out by both Indians and colonists interested in rights to the territory:

> We, whose names are underwritten being appointed by the inhabitants of Southern Towne, to set the line at Weakapauge. For to help us to understand where Weakapauge is, we desired some Poquatuck Indians to go with us, whose information as followeth, that Hermon Garrett did charge them that they should not goe any farther than the east side of a little swampe, near the east end of the first greate pond where they did pitch down a stake, and told us that Hermon Garrett said that, that verie place was Weakapauge, that he said it and not them and if they should say that Weakapauge did goe any further Hermon Garrett would be angry, for he was the governor. Further they told us that Her-

mon Garrett said, that the land next adjoining, namely, called Muqutak liing betwixt the two ponds was Hermon Garrett's owne land, but upon further debate about Weakapaug. How far Eastward it did extend, Cassasinamon with the other Indians affirme that the lands betwixt the two ponds to their owne knowledge was ever accounted Pequit lands, called Weakapauge. Cassasinamon further says for confirmation of the same, that here-ae to fore there was a whale cast ashore upon this neck of land and the Pequit Sachem came with a company of men and fetched it away. Further; Thomas Stanton afirmes that he heard Hermon Garrett say, that the neck of land called Quinicuntauge, was his land. The names of the Indians that was with us and gave us this information, was as follows:—Cassasinamon, Wissqunch, Johnequamapatah.

GEORGE DENISON,
THOMAS MINER,
SAM. CHEESEBROUGH
THOMAS PARKE
THOMAS STANTON
NEHEMIAM PALMER

The whole from Weakapauge to Misticke River is 10 miles and 28 poles.

From Weakapauge to Mr. Stanton's, is 3 miles and 300 rods. From Mr. Stanton's to Goodman Cheesebrough's, is 2 miles and 123 rods. From Goodman Cheesebrough's to Misticke River by Capt. Denison's house, is 4 miles.[73]

In April 1662, the status of the land at Wecabaug became even more vague with the issue of the Connecticut charter by Charles II. The charter, which transferred rights to the Pequot country from Massachusetts to Connecticut, defined the colony's eastern border as the "Narragansett river." With no river so named, Connecticut chose to interpret this as meaning Narragansett Bay. Rhode Island protested, interpreting the Narragansett river as the river that Rhode Island believed to define the boundary between the Mohegans and Narragansetts, the Pawcatuck. It was thus now in Connecticut's interest to document more clearly its eastern borders.

Four months later, Denison requested a group of Indians to gather at the General Court in New London to document their testimony concerning Wecabaug and Squamicut (see fig. 7.7).

> In New London this 4th of August 1662 Woncass Sagamore of Mohegan by Request of Capt George Denison Appeared before me, and this aboue draght being drawne he declared to my vnderstanding and affirmed that at what time the English did Conquer the Pequids, theire Country did Reach to a brooke called weex-co-da-wa which brooke falls into the end of that water or pond called nekeequoweese & that the land falling betweene that & the pond called teapanocke, Called by them muxquota is & was then Pequit Land, the same it afirmed by Casasinamon & that he being then a boy vsed there to driue theire (to say for the Pequids) deere into that neck of Land, allso weesawegun, affirmeth the same, and that eastward of that brook weexcodawa, is & was Naraganset Land belonging to Ninagrads and his heires by marriage of Harmon Garets sister.[74]

The testimony was signed by Casasinamon, Wesawegun, and Uncas. Wesawegun is referred to in a 1651 document as an Indian from Nameag, or New London, and Ninigrads, or Ninigret, was the Niantic sachem of Misquamicut.[75]

Above the testimony, as shown in figure 7.7, a graphic map was drawn of the lands at issue. The map is related to a topographical map in figure 7.8; the numbered place-names refer to the toponyms as listed on the original map. As it was copied onto the testimony, the map is oriented with north at the top. In the text, Uncas declared that Pequot

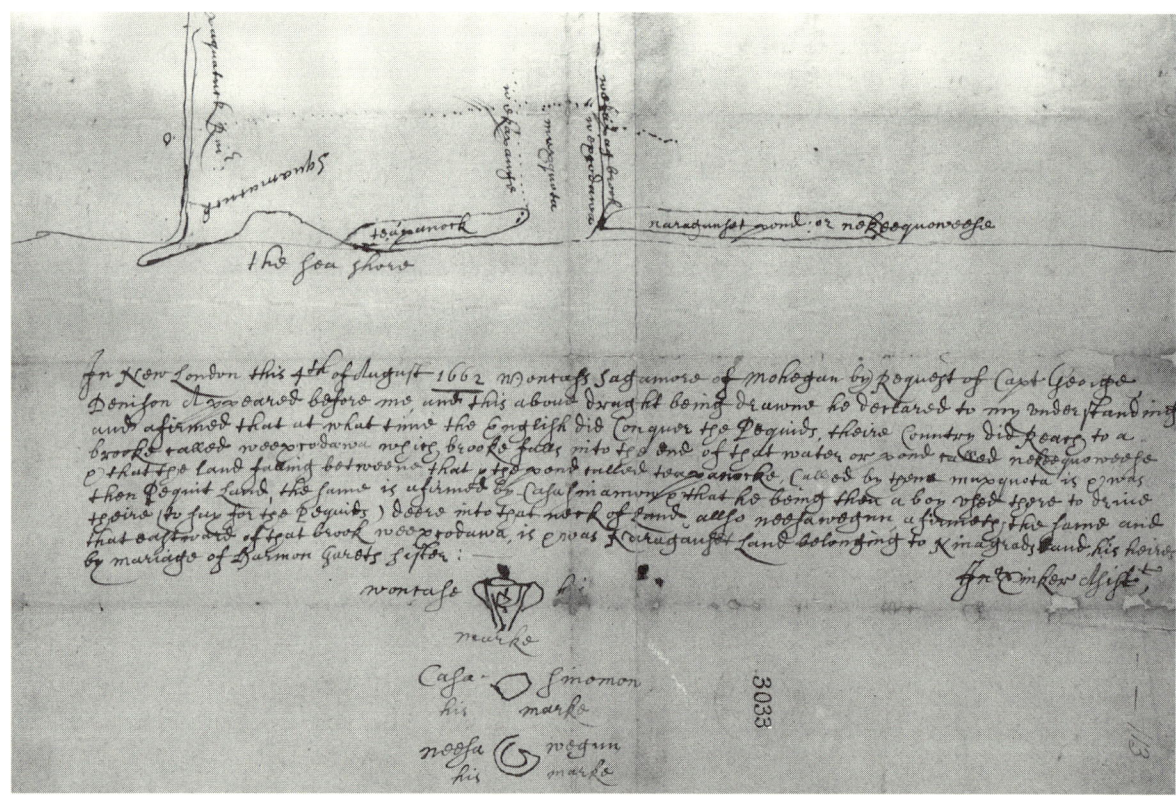

7.7 The Wecabaug testimony original (1662). Courtesy of the Massachusetts State Archives, Massachusetts Archives Series, ser. SC1 #45X, vol. 30, p. 113.

lands extended to the east as far as the brook at #6 and #7. He further pointed out that this eastern limit includes the land of Muxquota, lying between the two ponds at #3 and #8. The circle on the western edge of the map (across the river from #1) probably represents a Pequot settlement on the Pawcatuck River. Interestingly, there are two line symbols used in this map: a straight line indicating water (both shoreline and brooks), and a dotted line that seems to represent paths or roads (this differentiates the Wecabaug map from the maps of Sepecan and Weantinock, below).

Establishing Wexcodawa Brook at the eastern limits of Pequot territory confirmed the earlier findings of the 1659 survey report, and addressed the confusion over the land at Muxquota. In the survey, Casasinamon and two Indians of Pawcatuck had stated "that the lands betwixt the two ponds to their owne knowledge was ever accounted Pequit lands, called Weakapauge." They further noted that Cashawasset defined Wecabaug's eastern bounds at the eastern shore of the pond (#3, Te,ápanock), and Muqutak (or Muxquota, #5) as his own land, a boundary definition with which

Casasinamon disagreed. Thomas Stanton then further discredited Cashawasset's boundary with the statement that Cashawasset's land was actually the neck of land called Quinicuntauge, not Muqutak; this neck was probably the strip of land south of the pond at #8, today called Quonochontaug Pond. Thus, in the 1662 testimony and graphic map, Uncas testified that Muxquota was Pequot territory in 1637, and thus should be considered part of the "Pequot country" to be inherited from the war.

The testimony and map comprised a persuasive document against Rhode Island and Narragansett claims, particularly in its absence of Narragansett testimony or mapping of Wecabaug. Because of the animosities between the Mohegans and Narragansetts, it was certainly in Uncas's interest to recognize Pequot rights over Narragansett/Niantic rights to land. It was also in the interest of the United Colonies, since the period was marked by increasing fear of Narragansett anticolonial conspiracy.[76]

The following year, Rhode Island received its charter, specifying a western border of the Pawcatuck River. Although the Connecticut–Rhode Island boundary was not finalized until 1728, the disputed lands at Wecabaug were eventually awarded to Rhode Island.

The Pequots' land rights and subsequent histories were delineated according to their seventeenth-century division under Casasinamon at Noank and Cashawasset at Quonochontaug. Casasinamon never successfully obtained title to the Noank lands. He died in 1692, after which the Noank Pequots were moved to a reservation in what is now Ledyard, to become the Mashan-

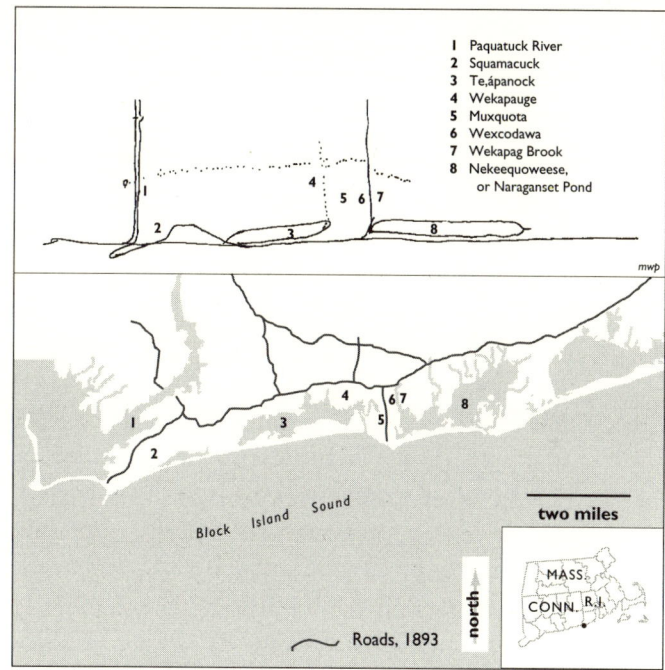

7.8 The Wecabaug testimony related to a nineteenth-century topographic map. By the author.

tucket Pequots. Indians living under the jurisdiction of Cashawasset were moved to a reservation on Long Lake at Lantern Hill, to become the Eastern or Pawcatuck Pequots, though they no longer have rights to Pawcatuck.[77]

In summary, there is considerable evidence of native mapping in the documents comprising the native land transfers or Indian deeds of southern New England. This evidence is found in both written and graphic form, as was typical of mapping among Indians and colonists at that time. Native mapping surfaces in deeds through the use of toponym in a mnemonic style, in the shapes of parcels conveyed, and in the use of characteristic native graphic devices such as straightened line features, symmetrical networks, and a mnemonic approach to symbolism. By differentiating between written and graphic styles, I do not intend to polarize them as separate types or traditions of mapping. Rather, I differentiate between the two styles as a way of showing that mapping may emphasize words to the exclusion of graphics, and still be part of a culture's map tradition. To overlook written maps in favor of graphic (and thus "more map-like") forms is to ignore a major part of property mapping in colonial southern New England.

The above examples only scratch the surface of the native mapping information contained within these documents. Indian deeds were negotiated throughout the English and Dutch colonies of New England and New Netherland. A comprehensive inquiry is needed into mapping in the deed process as a whole, by both native and non-native people. Such an inquiry would not only illustrate a distinct mode of colonial map communication, it would also greatly benefit our understanding of how the mapping of New England has taken place.

Notes

This study is based on research in part supported by fellowships at the Newberry Library, Chicago, and John Carter Brown Library, Providence. In addition, I would like to acknowledge the guidance and support of Malcolm Lewis, Harry Steward, Nanepashemet, Sarah Deutsch, Mark Warhus, and the staffs of Goddard Library, at Clark University, the Connecticut State Library, and the American Antiquarian Society.

1. For overviews of New England map history, see in particular William P. Cumming, "The Colonial Charting of the Massachusetts Coast," in *Seafaring in Colonial Massachusetts* (Boston: The Society [Charlottesville]: distributed by the University Press of Virginia, 1980); and Jeanette D. Black, "Mapping the English Colonies in North America: The Beginnings," in Norman J. W. Thrower, ed., *The Compleat Plattmaker* (Berkeley: University of California Press, 1987), 101–25. For a collection and discussion of vernacular mapping, see Peter Benes, *New England Prospect* (Boston: Boston University for the Dublin Seminar for New England Folklife, 1981).

2. This has been well documented by William Boelhower, "Inventing America: A Model of Cartographic Semiosis," *Word & Image* 4, 2 (April-June 1988), 475–97; Graham Huggan, "Decolonizing the Map: Post-Colonialism, Post-Structuralism, and the Cartographic Connection," *Ariel* 20, 4 (1989), 115–31; and J. Brian Harley, "New England Cartography and the Native Americans," in Emerson W. Baker et al., eds., *American Beginnings: Exploration, Culture, and Cartography in the Land of Norumbega* (Lincoln: University of Nebraska Press, 1994), 287–313.

3. The phenomenon of unacknowledged contributions is analyzed in detail in G. Malcolm Lewis, "Indicators of Unacknowledged Assimilations from Amerindian Maps on Euro-American Maps of North America," *Imago Mundi* 38 (1986), 9–34; see also Harley, "New England Cartography," 288–96.

4. J. Brian Harley and David Woodward, eds., *Cartography in Prehistoric, Ancient, and Medieval Europe and the Mediterranean*, vol. 1 of *The History of Cartography* (Chicago: University of Chicago Press, 1987), xvi.

5. Briefly, the significance of both process and historical period in the definition of mapping is an intentional reference to two premises in theoretical cartography. The first is that the interpretation of indigenous cartography must necessarily emphasize *processual* activity because overemphasis on maps as texts and as end-products is a specifically Western approach to cartography, as theorized by Robert A. Rundstrom, "Mapping, Postmodernism, Indigenous People and the Changing Direction of North American Cartography," *Cartographica* 28, 2

(1991): 1–12. Second, map history may be conceptualized not as a single, evolutionary progression from primitive to civilized mapping, but rather as a series of mapping modes that receive different emphases in different historical periods based on the function of the map; see Matthew H. Edney, "Cartography without 'Progress': Reinterpreting the Nature and Historical Development of Mapmaking," *Cartographica* 30, 2/3 (1993): 54–68.

6. Roger Williams, *A Key into the Language of America* (London: Gregory Dexter, 1643). Although there is no evidence that native people conceived of land measurement in an abstract sense similar to English colonists' conceptions of rods, perches, and miles to denote distance, this is not to say that measurement was unknown. As in any culture, measurement permeated daily life in the practices of architecture, agriculture, village layout, and art, to name only a few. The deeds also indicate that, when circumstances dictated, Indians had the ability to adapt to English concepts of measure early in the deed transactions.

7. This is explored by Julie Cruikshank, "Getting the Words Right: Perspectives on Naming and Places in Athapascan Oral History," *Arctic Anthropology* 27, 1 (1990): 52–65; by Keith Basso, *Wisdom Sits in Places: Landscape and Language among the Western Apache* (Albuquerque: University of New Mexico Press, 1996); and by Kathleen J. Bragdon, *Native People of Southern New England, 1500–1650* (Norman: University of Oklahoma Press, 1996), 125–29.

8. Cruikshank, "Getting the Words Right," 55.

9. I am using the term "Algonquian" as a convenient label for all native people of the Northeast. It is not the only label in use: Kathleen Bragdon uses the term "Ninnimissinuok." See Bragdon, *Native People*, xi.

10. Kathleen J. Bragdon, "Vernacular Literacy and Massachusett World View, 1650–1750," in Peter Benes, ed., *Algonkians of New England: Past and Present* (Boston: Boston University for the Dublin Seminar for New England Folklife, 1993), 34.

11. Constance A. Crosby, "The Algonkian Spiritual Landscape," in Benes, ed., *Algonkians of New England*, 36.

12. William Cronon, *Changes in the Land: Indians, Colonists, and the Ecology of New England* (New York: Hill & Wang, 1983), 65–66.

13. Ibid.

14. For a comprehensive account of property mapping in England during the sixteenth and seventeenth centuries, see A. W. Richeson, *English Land Measuring to 1800: Instruments and Practices* (Cambridge: MIT Press, 1966). For the Southern English colonies, see Sarah S. Hughes, *Surveyors and Statesmen: Land Measuring in Colonial Virginia* (Richmond: Virginia Surveyors Foundation, Virginia Association of Surveyors, 1979).

15. Roger J. P. Kain and Elizabeth Baigent, *The Cadastral Map in the Service of the State: A History of Property Mapping* (Chicago: University of Chicago Press, 1992), 6.

16. P. D. A. Harvey, "Local and Regional Cartography in Medieval Europe," in Harley and Woodward, eds., *Cartography in Prehistoric, Ancient, and Medieval Europe and the Mediterranean*; John W. Reps, *The Making of Urban America* (Princeton, N.J.: Princeton University Press, 1965), 121. For a discussion of the reluctance to incorporate graphic maps in England, see also David H. Fletcher, *The Emergence of Estate Maps: Christ Church, Oxford, 1600 to 1840,* Christ Church Papers no. 4 (Oxford: Clarendon, 1995).

17. The problem of the origin of native land transfers is reviewed in Alvin H. Morrison, "Indian Land-Deeds in the Northeast: Some Ethnohistorical Basics," in W. Cowan, ed., *Papers of the Twenty-third Algonquin Conference* (Ottawa: Carleton University, 1992), 298–309.

18. Edward Winslow, *Good Newes from New England* (1624; reprinted in Alexander Young, ed., *Chronicles of the Pilgrim Fathers. . . . ,* 2d ed., Boston: Little, Brown, 1844), 361 (page citations are to the reprint edition); Williams, *A Key into the Language of America*, 167.

19. For a study of deeding without primary sachems, see Peter A. Thomas, "In the Maelstrom of Change: The Indian Trade and Cultural Process in the Middle Connecticut River Valley, 1635–1665" (Ph.D. diss., University of Massachusetts, 1979). In other regions, the absence of

primary sachems in a deed negotiation has been interpreted as evidence that a deed was forged or otherwise fraudulently obtained. See for example Frank Lorens Wojciechowski, *Ethnohistory of the Paugussett Tribes* (Amsterdam: De Kiva, 1992), 82. For the deed process among praying Indians, see Jean M. O'Brien, *Dispossession by Degrees: Indian Land and Identity in Natick, Massachusetts,* Cambridge Studies in North American Indian History (New York: Cambridge University Press, 1997).

20. Ives Goddard and Kathleen J. Bragdon, *Native Writings in Massachusett* (Philadelphia: American Philosophical Society), 1:16; American Indian Archaeological Institute, *As We Tell Our Stories*, n.d.

21. In native tradition, tribute is the process of giving a small offering repetitively over time as a means of conveying respect. This is described in relation to the deed process by Robert Steven Grumet, "'We Are Not So Great Fools'" (Ph.D. diss., Rutgers University, 1979).

22. Deed for the Town of Wethersfield, Connecticut, 1671, Ayer Collection, MS 407, Newberry Library.

23. "Part of Agawam, Chicopee, Longmeadow, Springfield and West Springfield," in Harry Andrew Wright, ed., *Indian Deeds of Hampden County* (Springfield: 1905), 11.

24. The maps in these figures are based on USGS topographical quadrangles from the late nineteenth century. These maps were chosen because they portray the southern New England physiography before twentieth-century engineering projects altered the courses of rivers, removed hills and ridges, and created artificial lakes. Thus they may not resemble the present geography of each town.

25. Wojciechowski, *Ethnohistory of the Paugussett Tribes*, 47, 54.

26. Henry Bronson, *The History of Waterbury, Connecticut*... (Waterbury: Bronson Brothers, 1858), 2.

27. Connecticut Archives, Towns and Lands, 1st ser., vol. 1, document 164.

28. Waterbury Land Records, 2:224–25.

29. Wojciechowski, *Ethnohistory of the Paugussett Tribes*, 104–29.

30. Ibid., 131; Waterbury Land Records, 2:226–27.

31. Joseph Anderson, *History of Waterbury* (New Haven: Price and Lee, 1896), 1:41.

32. Wojciechowski, *Ethnohistory of the Paugussett Tribes,* 131, 149–51; Anderson, *History of Waterbury,* 193.

33. Waterbury Land Records, 2:228–31.

34. *Public Records of the Colony of Connecticut* (Hartford: Case, Lockwood & Brainard Co., 1894–), vol. 3, ed. Charles Hoadley and J. Hammond Trumbull, 197.

35. Ibid., 4:197 and 5:42.

36. Ridgefield Land Records, 1:2.

37. *Public Records of the Colony of Connecticut*, vol. 5, ed. Hoadley and Trumbull, 101.

38. Ibid., 5:120–23.

39. Besides the three deeds containing graphic maps mentioned here, I know of only three others in all of southern New England: Original deed for Warwick, R.I., from Miantonomo, January 12, 1642, John Carter Brown Library; Petition written by an unknown person on behalf of Tunxis Indians, 1672, Wyllys Papers, Connecticut Historical Society; Woodbury Land Records, 2:136 (this latter deed is thought to be a forgery, however).

40. Untitled map by King Philip and John Sassamon, 1668, Records of Plymouth Colony [MPLms I(1):21]. Size of the original 20 1/2 x 13 in (52 x 33 cm).

41. *Records of the Colony of New Plymouth in New England* (Boston: William White, 1859–61), 12:237.

42. The relating of these two maps is primarily due to the expertise of Nanepashemet, former director of the Wampanoag Indian Program at Plimoth Plantation, through communication with the author, 1992.

43. Laurie Lee Weinstein, "Indian vs. Colonist: Competition for Land in Seventeenth Century Plymouth Colony" (Ph.D. diss., Southern Methodist University, 1983), 178.

44. Maurice Robbins, *An Archaic Ceremonial Complex at Assawompsett* (Attleboro: Massachusetts Archaeological Society, 1968).

45. Lewis, "Indicators of Unacknowledged Assimilations," 22.

46. In the history of the interpretation of this map,

there has been some confusion about the transcription of this testimony. Barber transcribes the land which Philip is *"now* willing to be sold," an interpretation which considerably changes the meaning of the document; see John Warner Barber, *Massachusetts Historical Collections of New England*.... (Worcester: Door, Howland and Co., 1839). However, the copy of the registered deed, and the transcription published in 1861, confirm that the land is that which Philip is *"not* willing to be sold"; see *Records of the Colony of New Plymouth*, 12:237.

47. Weinstein, "Indian vs. Colonist," 186.
48. *Records of the Colony of New Plymouth,* 12:236.
49. Ibid., 4:109.
50. For various interpretations of this war, see Douglas Edward Leach, *Flintlock and Tomahawk* (N.Y.: Norton, 1958); and Russell Bourne, *The Red King's Rebellion* (N.Y.: Atheneum, 1990). Benjamin Church's account of his experiences is reprinted in *Diary of King Philip's War, 1675–1676*, edited by Alan and Mary Simpson (Tiverton, R.I.: Lockwood Publications, 1975).
51. Benjamin Church, *Diary of King Philip's War*.
52. Weinstein, "Indian vs. Colonist," 178.
53. Trudie Lamb Richmond, "A Native Perspective of History: The Schaghticoke Nation, Resistance and Survival," in Laurie Weinstein, ed., *Enduring Traditions: The Native Peoples of New England* (Westport: Bergin and Garvey, 1994); Wojciechowski, *Ethnohistory of the Paugussett Tribes*, 85; Claire C. Carlson, "Hidden Histories and Engendered Archaeologies: Excavations at the Weantinock Indian Planting Fields" (M.A. thesis, University of Massachusetts–Amherst, 1994), 2.
54. American Indian Archaeological Institute, *As We Tell Our Stories*, n.d.
55. Carlson, "Hidden Histories," 28–29.
56. *Public Records of the Colony of Connecticut,* Connecticut Archives, Towns and Lands, 1st ser., 2:85; PRCC v. 4:389.
57. New Milford Land Records, 9:269–70; Peck, *Howard Peck's New England,* 4–5.
58. Deposited for a period in the Milford Town Records, it is no longer a part of this collection. Two of the three volumes of early Milford records were destroyed by the colonial historian Edward R. Lambert in the early nineteenth century, the original and first copy of the deed possibly among them. The present copies of the Milford and New Milford town records omit this graphic element. Lambert, however, included a redrawn copy of the graphic in his *History of the Colony of New Haven*, reproduced in figure 7.5. Rutheva B. Brockett, City Historian of Milford, Connecticut, communication with the author, May 2, 1995; Edward R. Lambert, *History of the Colony of New Haven* (New Haven: Hitchcock and Stafford, 1838).
59. Reprinted in Wojciechowski, *Ethnohistory of the Paugussett Tribes*, 237.
60. Samuel Orcutt, *History of the Towns of New Milford and Bridgewater, Connecticut, 1703–1882* (Hartford: Case, Lockwood and Brainard, 1882), 7.
61. *History of Litchfield County* (Philadelphia: J. W. Lewis, 1881), 426.
62. Orcutt, *History of the Towns,* 7; *History of Litchfield County,* 426.
63. New Milford Land Records, 2:3.
64. Orcutt, *History of the Towns,* 11.
65. Russell G. Handsman, "What Happened to the Heritage of the Weantinock People," *Artifacts* 19, 1 (Spring/Summer 1991): 89.
66. Richmond, "A Native Perspective of History," 107; Handsman, "What Happened to the Heritage," 6.
67. *Plan of the Pequot Country*, Connecticut or Rhode Island, 1662, Massachusetts Archives. Size of the original 9 x 13 3/4 in.
68. For a discussion of the Pequot War and its aftermath, see Alfred A. Cave, *The Pequot War* (Amherst: University of Massachusetts Press, 1996).
69. Francis P. Jennings, *The Invasion of America: Indians, Colonialism, and the Cant of Conquest* (Chapel Hill: University of North Carolina Press, 1975), 257–59; 278.
70. Kevin A. McBride, "Historical Archaeology of the Mashantucket Pequot, 1637–1900," in Lawrence Hauptman and James Wherry, eds., *The Pequots of Southern New England* (Norman: University of Oklahoma Press, 1990), 104, 105; *Records of the Colony of New Ply-*

moth, 7:142.

71. *Records of the Colony of New Plymoth,* 7:199.

72. Francis Manwaring Caulkins, *History of New London, Connecticut* (New London: F. M. Caulkins, 1852), 19.

73. Elisha Potter, *Early History of Narragansett,* 2d ed. (Providence: Marshall, Brown, 1886), 56–57.

74. *Plan of the Pequot Country,* in *Records of the Colony of New Plymoth,* 7:450.

75. The reference to Wesawegun is in Samuel G. Drake, *Biography and History of the Indians of North America,* 7th ed. (Boston: Antiquarian Institute, 1837), 2:110; a biographical sketch of Ninigret is in the *National Cyclopedia* 9 (1907): 218–19.

76. Neal Salisbury, "Indians and Colonists in Southern New England after the Pequot War," in Hauptman and Wherry, eds., *The Pequots of Southern New England,* 87.

77. Jack Campisi, "The Emergence of the Mashantucket Pequot Tribe, 1637–1975," in Hauptman and Wherry, eds., *The Pequots of Southern New England,* 118.

8

Eighteenth-Century Arkansas Illustrated: A Map within an Indian Painting?

Morris S. Arnold

In the Musée de l'Homme in Paris, there is a splendid painted buffalo hide (see fig. 8.1) that may well do much to illuminate the colonial history of Arkansas.[1] Measuring about 7 1/2 feet by 5 1/2 feet, it features two feathered calumets, four Indian villages, a French village or fort (see fig. 8.3), and representations of the sun and moon. Above the three villages that are grouped together are written the words Ackansas, Ouzovtovovi, Tovarimon, and Ovqappa. Ackansas (Arkansas), of course, is the generic name that the Illinois Indians (and thus the French) applied to the Quapaw Indians;[2] the other words are the names of the three individual Quapaw villages of the eighteenth century—more usually written these days as Osotouy, Tourima, and Kappa.[3] (The Quapaws were frequently referred to as the Quapaws of the three villages.)[4] Though there are a number of objects on the skin, the main events portrayed are a battle between the Quapaws and another Indian tribe (from which the Quapaws emerged victorious) and a scalp dance performed in the Quapaws' villages in celebration of their victory (see fig. 8.2).

A line, which runs through the French village or fort, connects the Quapaw villages with the battle site.

I believe, for reasons that will appear at the end of this chapter, that it is altogether possible that the events depicted in this Indian painting are placed on it in a way that roughly preserves their original spatial relationship, and that therefore the painting may well have a deliberate cartographic content. If this is true, we have in this painting the earliest surviving example of an original North American Indian map. In order to understand why this is so, it is first desirable to attempt to establish, within the limits of the evidence that survives, what event the painting probably portrays, and its approximate date. It will be argued that it is likely that the battle depicted on the skin took place between the Quapaw and Chickasaw tribes in 1740–50. Before that argument is developed, however, it seems relevant to inquire into the provenance of the painting itself. If, as I believe and indeed hope to show, it is highly likely to be of Quapaw origin, there would be more

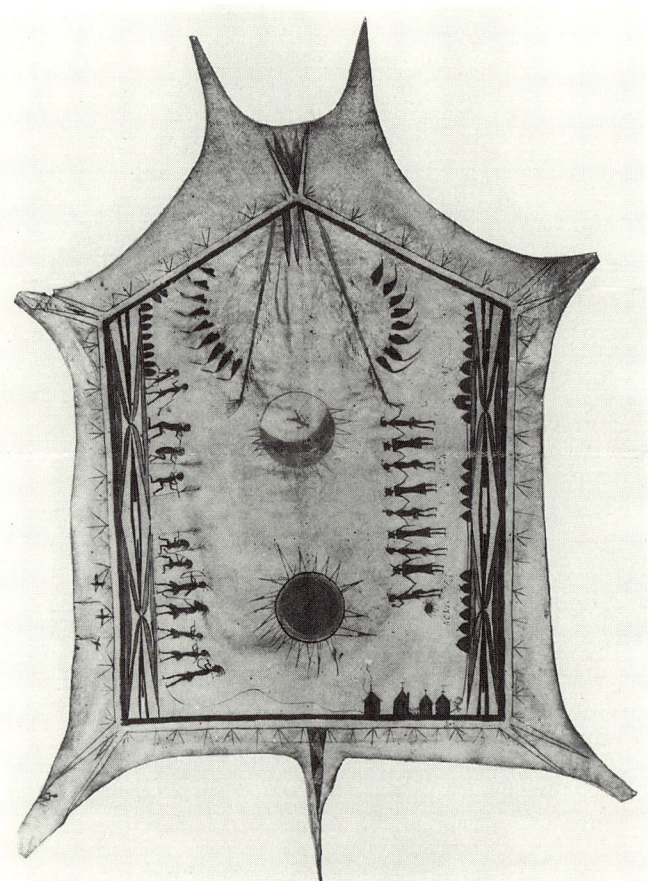

8.1 The Quapaw/Arkansas painted buffalo hide. Courtesy of Musée de l'Homme, Paris.

Establishing the Provenance

A recent publication suggests, albeit tentatively, that the painting on this skin might well be the work of Illinois Indians.[5] In support of this supposition the author adduces, first of all, the tradition that the skin was part of a larger collection of Indian painted hides assembled by French travelers, notably Father Marquette, in the Illinois country. It is not easy to put much stock in an uncorroborated oral tradition; and in fact, all that we know certainly about the early custody of the skin is that it was in France before 1789.[6] But the surest indications of an Illinois origin, the argument continues, are four of the elements that the painting features, all of which were very much at home in eighteenth-century Illinois: conical Indian dwellings covered with cane mats, European houses like the ones occupied by French colonial missionaries and merchants, dancers whose faces are painted red and black, and calumets.[7] It is certainly true that these elements are consistent with an Illinois origin. They are, however, equally consistent with a Quapaw origin, and, indeed, it would seem evident from other considerations that will be identified and discussed that such a provenance is much more likely.

First of all, some of the Indian houses that are pictured (those of the Quapaws' enemies) may indeed be conical like those of the Illinois. There is nothing, however, that justifies a conclusion that they were covered with mats; from all appearances, they might just as well have been bark-covered, as were those of the Quapaws.[8] The other houses, the ones in the Quapaws' villages, moreover, conform

reason to think that the painting has a cartographic component: because the Quapaws were more familiar with the geography of the relevant area and with the events in which they were involved than any other Indian tribe, they would be more likely to be interested in these events and their spatial relationship and in getting that relationship right when executing the painting.

We turn first, then, to a consideration of the painting's provenance.

8.2 Detail of dancers and villages, from the Quapaw/Arkansas painted buffalo hide. Courtesy of Musée de l'Homme, Paris.

8.3 Detail of French village, from the Quapaw/Arkansas painted buffalo hide. Courtesy of Musée de l'Homme, Paris.

exactly to the descriptions left by some late-seventeenth- and early-eighteenth-century travelers who described Quapaw houses as round;[9] and even if, as some scholars have posited, Quapaws lived in long huts with curved roofs,[10] what the skin may be showing is the gable end of a long hut. The profile section of such a building will be virtually indistinguishable, if at all, from the section of a circular dome, and, moreover, the French buildings on the skin seem to be shown with their gable ends forward. (By resorting to this strategy, incidentally, the artist can make room for more buildings.) Additional support for the proposition that the skin is showing long huts in section is that only one of the Indian buildings shows a doorway; and if the huts were dome-shaped it would seem to have been natural to depict the parts of them that contained entrances rather than their featureless elements. The large house with the door, shown with deliberate prominence in the village of Kappa (see fig. 8.2), could well be "the large council cabin built . . . in the middle of the village" of which Bossu wrote in 1750,[11] the same structure, presumably, that he spoke of twenty years later as "the large cabin of the Council of the Nation."[12] It would seem, therefore, that there is no particular reason to think that the houses in the villages marked with Quapaw names are anything but Quapaw houses.

Houses of vertical logs were common throughout colonial Louisiana from the Gulf Coast to Indiana and from the earliest times. There is therefore no reason to assume that this feature necessarily connects the skin to the Illinois country. Since the French houses are shown hard by the Quapaw villages, they would in any case almost certainly have to be those of Arkansas Post, not some Illinois settlement—whatever the origin of the painting. So much must surely be obvious.[13] But, as luck would have it, we may be able to say a great deal more about these buildings than that they were in Arkansas sometime during its colonial period. In 1732, Pierre Petit de Coulange reestablished the garrison at Arkansas Post in order to counter attempts by the Chickasaws to frustrate and undermine the French colonial effort.[14] He built a military establishment there that, according to a description written in 1734, consisted of four buildings: A barracks, a prison, a dwelling house with a fireplace (no doubt for Coulange and his wife), and a powder magazine, all built in typical French colonial fashion using vertical logs.[15] Our skin shows a French establishment with four buildings, only one of which has a chimney, and clearly illustrates the vertical-log *bousillage* construction of one of the buildings (the one on the far left of fig. 8.3). This almost eerie correspondence would seem to exceed the bounds of mere coincidence, especially since no other eighteenth-century Arkansas military establishment of which we have a detailed account so exactly matches this description. This position was abandoned and the post was moved after 1749 when the Chickasaws attacked it, killed some of the *habitants,* and carried off a number of women and children into slavery.[16] So it would seem that what we may have here, incredibly, is a relatively realistic picture of the Post of Arkansas as it existed in the 1730s and 1740s. There is, therefore, no apparent reason to believe that the French buildings shown on the painting have any connection with the Illinois country. Indeed, there is every reason to think that what is depicted is Arkansas Post.

The Illinois Indians, it is true, did paint their faces red and black—not to mention other colors.[17] But so did the Quapaws, in common with most of the other Indians of Louisiana.[18] In 1687, Henri Joutel encountered a Quapaw chief who, "to the end he may appear the finer, . . . never fails to besmear himself with Clay, or some red or black Colouring they make use of"; he also observed some Quapaw dancers "all besmear'd with Clay, of Black or Red, so that they really look'd like a Company of Devils or Monsters."[19] In 1699 the Jesuit Father Gravier gave a chief "a box of vermilion to daub his youth."[20] Virtually a full century after that, John Pope happened on an aging Quapaw who "was in Mourning, having his Face blacken'd over with a Commixture of Bear's Oil, Charcoal and Turpentine: just under his Jowls was two Streaks of red and white, which ran parallel to each other."[21] Red, the color of blood, is the color that warriors chose to paint their faces, as Bossu is careful to tell us twice. "The war dance is very interesting to see," he reports. "All the young men are painted red."[22] The lead dancer in the painting, moreover, is shown holding a rattle (see fig. 8.2), a favorite rhythm instrument of the Quapaws that European visitors frequently mentioned. Father St. Cosme, whose baroque ears found the Quapaws' music "not the most agreeable," said that these rattles were "gourds with pebbles in them,"[23] though others asserted that seeds and glass or enamel beads were used to produce the sounds.[24] (There are accounts, too, of Quapaws employing drums, bells, and even reed flutes in their dancing,[25] though none of these instruments is in evidence on the skin.) The women dancers, moreover, are wearing their hair in cylindrical rolls around the ears, a style that the naturalist Thomas Nuttall says was preferred by unmarried Quapaw women.[26] There therefore appears to be no reason to assume that the dancers are anything other than Quapaws.

Finally, the Illinois Indians, it needs saying, were hardly the only tribe to whom the calumet was of importance: it figured centrally in Quapaw rituals of welcome, alliance, adoption, prayer, war, and peace from the very beginning of European contact with them.[27] The black and white feathers attached to the calumets on the skin may well be those of an eagle, which is what the Quapaws used to adorn their calumets. It may not be irrelevant that the calumets depicted have red tips: Father Gravier said in 1700 that red was the color of the war calumet among the Quapaws.[28]

There seems no very good reason, therefore, to posit that our painted skin has a connection with the Illinois Indians or the Illinois country. Its content conforms perfectly with what we know of the Quapaws and of Arkansas Post in the eighteenth century. The painting literally has Arkansas written all over it and it concerns itself with events that feature the Quapaws, some of which certainly occurred in Arkansas. These are all facts that tend to indicate a Quapaw origin for the painting. It is noteworthy too that Quapaw folklore included a story about the moon being inhabited by a man who held a trophy head in his hand,[29] and our skin features a man on the moon who seems to be clutching something.

Furthermore, the Quapaws were famous throughout Louisiana for their hide paintings. As early as 1687, Henri Joutel had been moved to remark on the "buffalo hides that the [Quapaws] have the industry to dress and paint with a kind of red coloring that is quite pretty,"[30] which the Quapaws hung in their dwellings to divide them into compartments. Thirty-five years later, Diron d'Artaguette reported that Quapaw men had by then achieved a wide renown for "the dressing of buffalo skins, upon which they paint designs with vermilion and other colors." These skins, he added, "are very highly prized among the other nations."[31] Father du Poisson, who owned at least one of these Quapaw painted hides, called them *matachés*, which he defined as skins "painted in different colors, and on which they represent calumets, birds and beasts." Quapaw *matachés* made of deerskins "can be used as tablecloths," he noted, while those of buffalo hides served as bedcovers.[32] He also attested to the Quapaws' interest in artistic matters: "They are in ecstasies when they see the picture of Saint Régis that I have in my room ... *Oukantoqué*, they exclaim, *it is the Great Spirit!* ... They place themselves in different parts of my room and say, each time smiling: *He is looking at me; he almost speaks, he needs only a voice.*"[33] The Quapaws' artistic ability furnishes further evidence that our painted skin is of Quapaw origin.

But the most probative clue to the painting's Quapaw provenance is its subject matter. One has to ask why the Illinois Indians would go to the trouble of painting a picture that shows the Quapaw Indians defeating an enemy and celebrating a victory, for this, as we have said, is the substance of the principal events that the skin depicts. This has been recognized as putting an obstacle in the way of an Illinois provenance for the painting, but to counter this difficulty the observation is offered that the "Illinois and Quapaws were constantly at war."[34] As it happens, that is not true: The Illinois and Quapaws were fast friends, as what follows will make clear.[35]

As early as 1680, a Frenchman familiar with the Illinois country recorded that the Illinois tribes and the Quapaws had at some previous time formed a kind of confederation for the purpose of making war on the Iroquois, and he recalled having visited an Illinois village of 400 huts and 1,800 warriors, some undisclosed number of whom were Quapaws.[36] In the next century, the Illinois and the Quapaws stood resolutely with the French during both of their disastrous wars against another common enemy, the Chickasaws.[37] After France abandoned Louisiana in the wake of the Seven Years' War, moreover, the two old Indian allies participated in the pan-Indian resistance usually, if somewhat misleadingly, called Pontiac's rebellion,[38] and they coordinated their opposition to the imposition of English rule in their respective territories.[39] After the Spanish took possession of the Arkansas country, various groups of the Illinois (including Kaskaskias and Peorias) took refuge there, and a large contingent of them was even incorporated for a time into one of the Quapaw villages.[40] Far from being "constantly at war," therefore, the tribes of the Illinois confederacy and the Quapaws were instead constant and unshakeable allies.

Even if it were true that the Illinois and Quapaws were enemies, that would tend rather to deepen a mystery than to solve one: why would the Illinois want to record a victory by their enemies? If the painting is of Indian origin, as it seems certainly to be, it would seem far more likely that it was the product of a Quapaw artist, since its evident aim is to record and laud a successful Quapaw military adventure. We know that Indian paintings frequently aimed to celebrate and memorialize the exploits of an individual or a group. They often served as mnemonic devices that "enabled a successful warrior to advertise his achievements to the community at large."[41] It is beyond unlikely that a tribe or some member of it would want to recall a disgrace or defeat, or even record the virtues and accomplishments of another tribe. While it is of course impossible to prove a negative, I know of no recorded instance in which that occurred.

For all of the reasons adumbrated, based on what we can deduce from this painted hide itself as well as what we know generally about Indian painting, a Quapaw provenance for the painting is highly likely. Indeed, it would seem to me to be all but certain.

Places and Events

What else can we say more or less reliably about this remarkable artifact? If we assume for the moment, for reasons already indicated, that it illustrates events involving the fort constructed in 1732 and attacked in 1749, one can make a reasonable guess concerning the identity of the Quapaws' defeated enemies. They would almost surely be the Chickasaws, with whom the Quapaws were already at war when La Salle treated with the Quapaws in 1682,[42] and against whom they struggled for most, if not all, of the French colonial period. The Chickasaws' complicity with the Natchez who destroyed the French settlement at Natchez in 1729 and their steadfast attachment to the English aim of subverting the French effort in Louisiana prompted the colonial government to strengthen its presence at the Arkansas. After the Natchez turned back an attempt by an expedition of more than seventy men to reach the Arkansas country in 1731, Pierre Petit de Coulange and twelve soldiers at last established themselves early the next year on the Arkansas River.[43] French troops had been absent from the Arkansas for about seven years, so it is a safe bet that the Quapaws would have engaged in their favorite calumet ceremony to resurrect the military alliance. Perhaps it is this event that the calumets on the skin commemorate.[44]

We know, at any rate, that Coulange succeeded early on in galvanizing the Quapaws, for by 1733 Governor Périer could write enthusiastically that he "had made himself so beloved that [the Quapaws] have been at war [against the Chickasaws] for two years, although I have had nothing to give them except promises." In fact, one of the Quapaw war parties to which Périer referred had killed two Chickasaws and burnt another to death in one of the Quapaw villages. There were other successful operations against the Chickasaws that yielded two Chickasaw scalps in 1744, five in 1746, and two more in the spring of 1747, for which the Quapaws were paid generously.[45] Any one of these excursions would have furnished an appropriate occasion for decorating a buffalo hide.[46]

A troublesome embarrassment in the way of being completely confident about the proposed reading of the skin is that the order in which the Quapaw villages are arranged (or named) on it does not correspond with the configuration that early-eighteenth-century travelers ascribed to them. The sources for the 1730s are in unequivocal agreement that the first village encountered on the Arkansas River on the way up to the Arkansas Post was Tourima (with Tonguinga), followed by Osotouy, and then Kappa, this last directly across from the post.[47] The skin, on the other hand, portrays Kappa first, followed by Tourima and Osotouy. The skin, in fact, depicts the order in which the villages appeared on the river in 1777, when the post was located close to the Mississippi in Desha County.[48] It is possible, however, since the names of the villages may have been added sometime after the skin was painted,[49] that a mistake was made when they were affixed. (We have no idea who did this, of course, or how much, if anything, he knew of the region.) It is also possible that the skin deals with a time after 1748, when, on account of flooding, the Quapaws moved their villages above the fort to *Écores Rouges*.[50] There is no record of exactly where the Quapaws were located at that time, nor do we know the order in which they arranged themselves on the river. We do know that they remained hostile to the Chickasaws after the 1749 attack on the post: Later that year, they pursued the attackers down the Mississippi and produced twenty-five Chickasaw scalps,[51] and the next year they sent the governor eighteen more.[52] In 1751 Governor Vaudreuil exulted that the "Arkansas still continue to be attached to us and are making frequent raids on the Chickasaw, from whom they have quite recently brought back some scalps."[53] Since the line on the painting shows the Quapaws moving from their villages, through the post, and on to the battle, the skin may well have to do with engagements that occurred after the attack of 1749, when the fort in fact lay between the Quapaws and the Chickasaws. Despite a great deal of searching and reconnoitering, only one archaeologist has ever ventured to claim that he had discovered the site of an eighteenth-century Quapaw village,[54] and that identification has not gone unchallenged.[55] The present state of our archaeological knowledge, therefore, can shed no light on the question of how the Quapaw villages were arranged relative to each other at any particular time.

But whatever the exact date of the occurrences recorded, the conclusion that the vanquished foe is the Chickasaws is more or less irresistible because there is simply no other tribe to which the events illustrated can be reasonably connected. There were, it is true, Quapaw raids against the Koroa in 1702 and 1730,[56] but at the earlier of those two dates there was no French settlement at the Arkansas, and most probably there was no settlement there in 1730 as elaborate as the one pictured.[57] Other Quapaw engagements of a minor sort with the Tunicas and the Natchez find mention in the records of the 1730s, but they were of a limited character.[58] There were also difficulties between the Osages and French hunters from Arkansas in about 1720, and a few Osages attacked some Quapaws in 1751;[59] there is, however, no record of Quapaw retaliation in either case. It is doubtful, then, that the skin has any connection to these events.

According to Bossu, as late as 1770 the Quapaws still abhorred "the *Chikachas* dogs who have become our implacable enemies since they killed and burned some Frenchmen, along with the *chief of prayer* (a missionary)."[60] (This was an allusion to the disastrous denouement of the First Chickasaw War in 1736, when a large number of Frenchmen were taken captive and burned to death in the Chickasaw villages.) We cannot therefore completely discount the possibility that the skin deals with a time after 1759, when the downriver fort that has come to be called Fort Desha was finally completed.[61] It, too, may have had four buildings, but the punctuation in Capt. Philip Pittman's 1765 description of it makes it difficult to know exactly how many buildings it contained; and he does not record the number of fireplaces.[62] But there is no record of Quapaw war parties being sent against the Chickasaws, or anyone else, from this location during the French colonial period, which makes this possibility less likely.[63] Of course, if the skin was painted after the French period, which is not impossible, the enemy could well have been Osages, against whom the Quapaws sent numerous war parties from Fort Desha.[64] But further corroboration that the enemy were Chickasaws is provided by the fact that the Chickasaws' houses were square or rectangular with thatched roofs,[65] and the profiles of the enemies' houses on the skin are consistent with that.

Ethnography and Possible Acculturation

I offer here some detailed observations on the matters illustrated on the skin.

Nicolas de La Salle claimed that in 1682 the Quapaws were "completely naked, like all the other nations."[66] But Henri de Tonty, reporting what he observed at virtually the same time, indicated that the Quapaws wore buffalo skins,[67] and all the later French sources confirm this.[68] The men also sometimes wore breechclouts of spun buffalo wool that the women made "to cover their husbands' nakedness."[69] The women and girls, according to Father St. Cosme, like the Illinois Indians, wore a skin that covered them from their waists to their knees, and some also sported a small deerskin draped around their shoulders in the manner of a scarf or shawl.[70] Jean-Bernard Bossu's *Nouveaux voyages dans l'Amérique septentrionale*, which records his adventures in Arkansas early in the Spanish colonial period, contains illustrations depicting bare-breasted Quapaw women.[71] This was perhaps not merely an artistic convention, for Bossu claims that he once had occasion to explain to a Quapaw envoy "that it would be the vilest impropriety for a French woman to expose her bosom, although it was very natural for an *Akanças* lady to do so."[72] Bossu thus at the least intimates, if he does not say directly, that eighteenth-century Quapaw women frequently wore nothing above the waist. In 1791, some Quapaw women piqued John Pope's interest because they "very innocently displayed their Navels, and the curious Eye might have explored other Parts which civilized Nations industriously conceal."[73] Even as late as the end of the colonial period, therefore, it seems that indigenous Arkansas women had not yielded completely to European sartorial convention.

Given all this, the breechclouts on the dancing men are not surprising,[74] but it is more than a little curious that the women are quite evidently dressed in chemises or shifts that cover them from neck to knee (see fig. 8.2). Perhaps these are indications that the Jesuit missionaries had succeeded, at least temporarily, in imposing their ideas of modesty on native Arkansans. (It is not impossible, however, that Quapaw women wore longer, more formal attire on important ceremonial occasions like the one pictured.)[75] An alternative possibility is that a priest (or some other European) exerted influence on the artist not to depict the reality of the situation. This latter hypothesis is rendered perhaps less likely by the fact that the fleeing Chickasaws are shown completely naked, though this may have been done to indicate (who knows?) their moral inferiority in the eighteenth-century Christian mind.[76] It is clear, in any case, that there was some European influence in the making of the skin, the most obvious indicium being the names affixed to the Indian villages in the Romanesque letters that the Jesuits regularly employed. It may be that Father Avond or Father de Vitry, both of whom labored at the post in the late 1730s, had a hand in this enterprise.[77] But it is just possible that a native American artist drew these letters with his own hand: the Jesuits were keen to learn the language of their Indian neighbors in order to have a chance at converting them, and they may also have taught them how to write their own names and the names of their villages.

The Cartographic Component

The representations of the sun and the moon in the painting might have served a dual purpose. First of all, they are likely to be of religious significance. Diron d'Artaguette reported in 1723 that the Quapaws "worship the moon, to which they are accustomed to pray every evening,"[78] and Bossu later asserted that "they fear the devil, called the Evil Spirit, and worship the sun and the moon."[79] Father Gravier remarked that the Quapaws regarded the calumet as the "pipe of the sun, and in fact they proffer it to him to smoke when they wish to obtain calm, rain, or fair weather."[80]

Most important for the purposes of this volume, however, it may be that the representations of the sun and moon were aligned in a way intended to provide the viewer with an east-west directional axis with which to read and interpret the skin.[81] If so, then the Quapaws are shown venturing up the Arkansas River (more or less east to west), turning north through Arkansas Post (on Lake Dumond in present-day Arkansas County), and then heading east to engage the Chickasaws (see fig. 8.4). The Quapaws regularly employed a land route to raid the Chickasaws that ran north

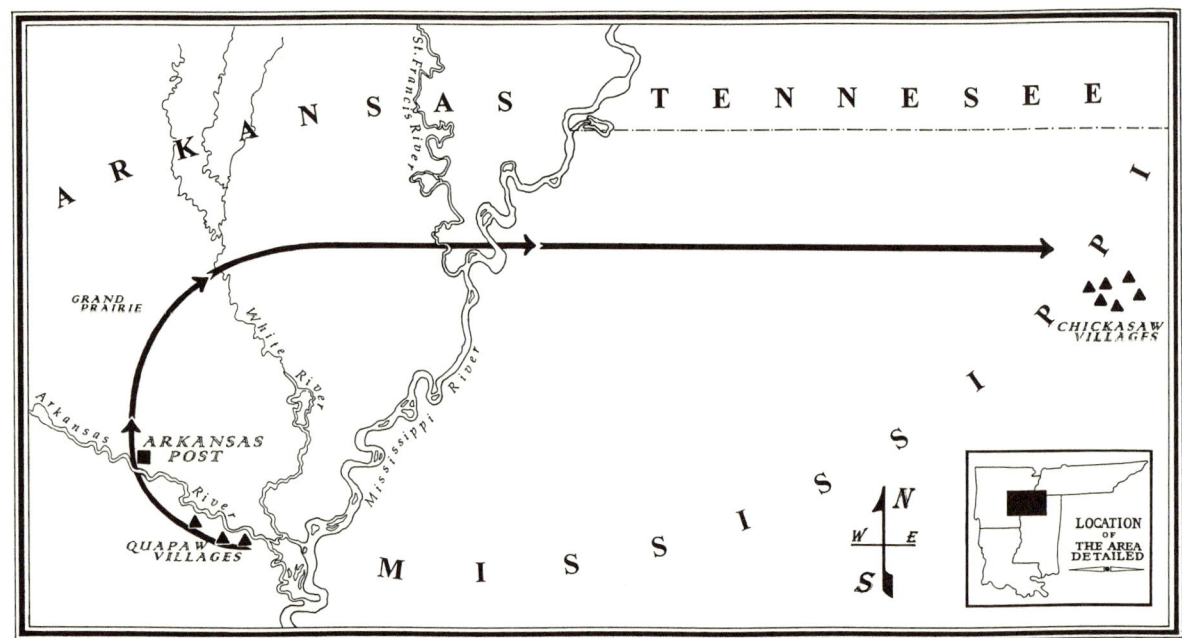

8.4 Cartographic interpretation of the Quapaw/Arkansas buffalo hide. Cartography Susan Corrotto Lieux.

and then east of the post and crossed the Mississippi River just above the mouth of the St. Francis River. In the 1730s, the French spent a great deal of time scouting this so-called *chemin des Chickachas* for use in their abortive schemes to conquer the Chickasaws.[82] If this interpretation is correct, this skin would be the oldest known original North American Indian map,[83] and it establishes a clearly European influence on the arrangement of the contents of the painting. If, on the other hand, the skin portrays a sortie against the Osages during the time that Arkansas Post was located downriver, the proposed cartographic content of the skin is entirely undone, since the Osages lived northwest of the Quapaws, and Arkansas Post was east of the Quapaw villages. In that case, the skin does not maintain cardinal direction at all, and the line that unites the Quapaw villages, the fort, and the battle site is merely the thread of a story in a diachronic picture,[84] without directional significance, indicating only that the Quapaws first went to Arkansas Post from their villages and then proceeded to engage the enemy.

It would not in any case be at all surprising if the Quapaws were producing maps in the eighteenth century. Indeed, it is well established that "[s]ome if not all of the Indians of North America could draw maps at the time of their first contact with white people";[85] and from "the earliest contacts, Amerindians have transmitted to Euro-Americans spatially arranged information about the lands, coasts, waters, places, routes and resources of North America."[86] It might nevertheless reasonably be objected that if the hide does portray events that took place, say, ca. 1750, it would be remarkable if sufficient cultural amalgamation could have occurred by that time to allow the Quapaws to have produced a graphic that would have roughly preserved cardinal direction in its configuration. That is because Native Americans do not seem to have produced maps of this sort during the early contact period. But, in the first place, even if the events depicted occurred ca. 1750, that date marks only a *terminus a quo,* the painting could well have been executed some years later. Second, mixed-blood *(métis)* denizens of Arkansas are reported as early as the seventeenth century, and it is certain that by 1750 Quapaws and Europeans had been coexisting on the Arkansas River more or less continuously from 1686 on, that is, for upward of three generations. Third, the Quapaws would have seen and comprehended the utility of the maps and surveying instruments that priests, commandants, military engineers, and exploratory expeditionaries frequently brought with them to the post beginning at the latest about 1720, and almost certainly before. Finally, and perhaps most important, one needs to recognize that a general cultural assimilation between Quapaws and Frenchmen is not a necessary condition for a Quapaw to have produced an item with what we might term a quasi-cartesian content. All that is required is for one person to have had an idea and to have acted on it. Thus what we have here is quite possibly "a graphic representation that facilitate[s] a spatial understanding of . . . events in the human world"[87]—that is, a map.

But all of this hypothetical rumination about the Quapaws' potential for mapmaking is rendered virtually unnecessary by events that occurred at Arkansas Post in 1737. On October 21 of that year, Bernard de Vergès, a French engineer, arrived at the post to secure the Quapaws' help in locating and surveying a road to the Chickasaws to facilitate a French attack on that tribe. This was an effort to which the Quapaws were ideally suited to make a contribution, for, as we have seen, they were making frequent raids on the Chickasaws and were therefore in a position to know the best and most direct routes to their villages.

During their talks with Vergès, various Quapaw chiefs described three different ways to reach the Chickasaw villages from Arkansas Post. A Quapaw village chief named Buagrès then told Vergès that "he knew an excellent road a little above the St. Francis River to which he would guide me if I wished, and that the other two roads were not good for rolling stock [*charrois*]." Buagrès, Vergès reveals, "at the same time traced on the drawing board a map of this country for me, and gave me a description of it, with a great deal of good sense and intelligence." It is not likely that Vergès could have had this much enthusiasm for the chief's map if he did not at least have some confidence that it preserved cardinal direction (a confidence that he would have developed from his discussion with Buagrès), for otherwise the sketch would have been useless to him.[88]

Because of the combination of Native American and European elements, some evident and others more conjectural, which we have identified in the content of this marvelous painted skin, we are left to wonder whether it was not, at least in design if not in execution, a collaborative effort between a Quapaw Indian and a Frenchman. Perhaps it even represents the work of what may, as a cultural matter, be the same thing, namely, one of those French and Indian *métis* for whom Arkansas Post was famous in the eighteenth century. Such an artifact would come down to us as an appealing epitome of the cultural symbiosis that Indians and Europeans achieved in the six generations that they lived and died together on the Arkansas River during Louisiana's colonial period. Most satisfying of all, it appears that a native American (or *métis*) artist has bequeathed us the only known depiction from the colonial era of the first European settlement in what became Jefferson's Louisiana. De Batz, Dumont de Montigny, and Le Bouteux, all Frenchmen, provided us with views (some sophisticated, some amateurish) of early-eighteenth-century French dwellings in lower Louisiana,[89] but none of these stooped to record the sights available in the remoteness of the Arkansas region. Now we can take advantage of the profoundly evocative power of pictures to help reconstruct lives that were passed in eighteenth-century Arkansas, and to comprehend the significance of the events that shaped them and gave them content.

Notes

Portions of this chapter were previously published in "The Significance of the Arkansas Colonial Experience," *Arkansas Historical Quarterly* 51, 1 (Spring 1992). The author gratefully acknowledges the help of Jay Edwards, Michael Hoffman, Malcolm Lewis, Dan Morse, George Sabo III, William C. Sturtevant, Carrie Wilson, and Phillip Zane in the preparation of this chapter. He is especially grateful to Anne Vitart for generously allowing him to view the Quapaw skins in her custody at the Musée de l'Homme during a trip to Paris in 1994. Responsibility for errors remains, of course, with the author.

1. See *Robes of Splendor: Native American Painted Buffalo Hides* (New York: New Press, 1993), 28, 56–57, 91, and 136–37, for illustrations and discussion of this skin. There is another skin in the Musée de l'Homme, depicting hunters, animals, and dancers wearing buffalo heads, that probably also has an Arkansas connection. See ibid., 65, 72, 73, 134–35. This skin may have captured the dance of the wild animal hunt described by Jean-Bernard Bossu or the buffalo dance described by Lahontan. For an account of these dances, see Samuel D. Dickinson, "Quapaw Indian Dances," *Pulaski County Historical Review* 23 (Fall 1984): 46.

In the print department of the Bibliothèque Nationale in Paris, there are two volumes that include an engraving of the painted hide that is the subject of this article. See Of.4b tome I/folio and Of.4, Pet. fol. Neither volume, however, contains any indication of when the engravings were executed or who executed them. They appear to belong to the last quarter of the eighteenth century.

2. W. David Baird, *The Quapaw Indians: A History of the Downstream People* (Norman: University of Oklahoma Press, 1980), 5.

3. See, e.g., ibid., 10–11. There was formerly a fourth village, Tonguingas, but it was consolidated with Tourima by 1727. See "Letter from Father du Poisson, Missionary to the Arkansas, to Father Patouillet," in Reuben Gold Thwaites, ed., *Travels and Explorations of the Jesuit Missionaries in New France* (Cleveland: Burrows Brothers, 1900), 67:319. Father Charlevoix had reported this consolidation as early as 1720; but Benard de La Harpe, writing of conditions in 1722, said the two villages still had a separate existence. Louise Phelps Kellogg, ed., *Journal of a Voyage to North America* (Chicago: The Caxton Club, 1923) 2:230; Ralph A. Smith, "Exploration of the Arkansas River by Benard de La Harpe, 1721–22," *Arkansas Historical Quarterly* 10 (Winter 1951): 350.

4. See, e.g., the 1779 map of the Post of Arkansas in Morris S. Arnold, *Colonial Arkansas, 1686–1804: A Social and Cultural History* (Fayetteville: University of Arkansas Press, 1991), 14–15, where the legend refers to *Arkansas des 3 Villages*.

5. George Horse Capture, "From Museums to Indians: Native American Art in Context," in *Robes of Splendor*, 91. In contrast, several earlier publications had identified the robe as Quapaw. See Anne Vitart, "Chronique d'une rencontre en terre de Canada," in Daniel Lévine, ed., *Amérique continent imprevu: La rencontre de deux mondes* (Paris: Bordas, 1992), 106. See also, for what seems to be the earliest published picture of our painting, Manuel Ballesteros Gaibrois and Paul Kirchhoff, *Arte antiguo norteamericano: Pieles de bisonte pintadas* (Madrid: Tipografía de Archivos, 1934), where the authors attribute the robe to the Quapaw. Gordon Brotherston, *Book of the Fourth World: Reading the Native Americas through Their Literature* (Cambridge: Cambridge University Press, 1992), 26, 181–82, identifies the skin as Quapaw. But Brotherston misreads one of the village names as "Cahokia," and my interpretation of the events portrayed on the skin differs considerably from his. See also Marius Barbeau, *Indian Days on the Western Prairies* (Ottawa: National Museum of Canada, 1960), 225–26, for a short notice of the skin that is the subject of this chapter.

6. Anne Vitart, "From Royal Cabinets to Museums: A Composite History," in *Robes of Splendor*, 54. The hide was in the collection of the public library in Versailles by 1869; and if it was part of the collection of the Marquis de Sérent, from which much of this library's collection came, then it was in France by "about 1786" when the Marquis formed his collection. See *Cabinet de curiosités et d'objets d'art de la Bibliothèque Publique de la Ville de Versailles, Catalogue* (Versailles, 1869), 3:21.

The suggestion made in Araceli Sánchez Garrido, "Plains Indian Collections of the Museo de América," *European Review of Native American Studies* 6 (1992): 21, that the Musée de l'Homme robe collection antedates 1713 is not based on any evidence. Sánchez speculates that painted hides in the Museo de América in Madrid might at one time have been part of the same collection as the painted hides in the Musée de l'Homme, that the Madrid robes might have come from Cardinal Luis de Borbón, who might have inherited them from Felipe V, who came to Spain in 1713 and might have brought the hides with him because he might have inherited them from some supposed Dauphin of France. The difficulty with all this is that not a single one of the links in this fanciful chain has any support for it. There is no evidence even that the Madrid skins came from Luis de Borbón. See Paz Cabello Carro, *Coleccionismo americano indígena en la España del siglo XVIII* (Madrid: Ediciones de Cultura Hispánica, 1989), 167–68. Sánchez relies for his views on a previous article (see Manuel Ballesteros Gaibrois, "Pieles de bisonte pintadas: Tres ejemplares de Museo Arqueologico Nacional," *Tierra Firme* 65 [1935]), but in that article Ballesteros simply made the reasonable suggestion that the Madrid hides and the Paris hides had been part of the same collection at one point and that the Madrid hides were sent from Paris because of the Bourbon family connection. He makes no suggestion about when that might have happened. The simple truth is that none of the Madrid robes "has any documented pre-1865 collection history." Garrido, "Plains Indian Collections," 21, 29 n. 21.

7. George Horse Capture, "Gallery of Hides," 136.

8. Henri de Tonty, "Relation de Henri Tonty," in Pierre Margry, ed., *Découvertes et établissements des français dans l'ouest et dans le sud de l'Amérique septentrionale,* 6 vols. (Paris: D. Jouaust, 1876–86), 1 (1876): 599; Louise Phelps Kellogg, ed., *Early Narratives of the Northwest, 1634–1699* (New York: Barnes and Noble, 1917), 298.

9. See, e.g., Minet, "Voyage Made from Canada Inland Going Southward during the Year 1682," translated by Ann Linda Bell and edited by Patricia Galloway, in Robert S. Weddle, ed., *La Salle, the Mississippi, and the Gulf* (College Station: Texas A&M University Press, 1987), 62; and Jean-François Dumont de Montigny, *Mémoires historiques sur la Louisiane* (Paris: J. B. Bauche, 1753), 1:142. On the general subject of the shape of Quapaw buildings, see Michael P. Hoffman, "Quapaw Structures, 1673–1834, and Their Comparative Significance," in Hester A. Davis, ed., *Arkansas before the Americans* (Fayetteville: Arkansas Archeological Survey, 1991), 40–54.

10. Hoffman, "Quapaw Structures"; George Sabo

III, *Paths of Our Children* (Fayetteville: Arkansas Archeological Survey, 1992), 33–34.

11. Seymour Feiler, ed. and trans., *Jean-Bernard Bossu's Travels in the Interior of North America, 1751–1762* (Norman: University of Oklahoma Press, 1962), 62. Bossu's books contain a lot of tall tales, so one needs to be cautious about relying on him.

12. Samuel Dorris Dickinson, ed. and trans., *New Travels in North America* (Natchitoches, La.: Northwestern State University Press, 1982), 37. For Quapaw council houses, see also Sabo, *Paths of Our Children*, 38. The sources also mention an important house of the Quapaws called the *cabanne de valeur* (house of valor) where religious ceremonies evidently took place. See Arnold, *Colonial Arkansas*, 132–33. Whether this building was different from the council house is not clear.

13. After I first came to this conclusion in Morris S. Arnold, "Eighteenth Century Arkansas Illustrated," *Arkansas Historical Quarterly* (Summer 1994): 119, 123, I discovered that I was hardly the first to do so. The files in the Musée de l'Homme themselves contain notes made by Paul Kirschoff in 1934 that identify the European settlement depicted in our painting as "without doubt the French fort that was situated . . . on the lower course of the Arkansas River." Similarly, in 1948 an eminent architectural historian wrote of Arkansas Post that a "picturesque view of this little settlement, showing four French houses, may be seen in an Indian painting on deerskin [sic] still preserved in Paris." Charles E. Peterson, "French Landmarks along the Mississippi," *Magazine Antiques* 53 (April 1948): 286.

14. Arnold, *Colonial Arkansas*, 31, 99–101.

15. Ibid., 31.

16. Ibid., 31, 105–6.

17. Newton D. Mereness, ed., *Travels in the American Colonies* (New York: Macmillan, 1916), 72.

18. Vermilion pigment (mercuric sulfide) was one of the commonest items traded or given as presents to the Indians in French Louisiana. See Gregory A. Waselkov, "French Colonial Trade in the Upper Creek Country," in John A. Walthall and Thomas E. Emerson, eds., *Calumet and Fleur-de-Lys* (Washington: Smithsonian Institution Press, 1992), 40. Carrie Wilson informs me that the Quapaws used black paint to signify death.

19. Henry Reed Stiles, ed., *Joutel's Journal of LaSalle's Last Voyage, 1684–7* (Albany, N.Y.: Joseph McDonough, 1906), 182, 183.

20. John Gilmary Shea, ed., *Early Voyages Up and Down the Mississippi* (Albany, N.Y.: Joel Mansell, 1861), 129.

21. John Pope, *A Tour through the Southern and Western Territories of the United States of North America; the Spanish Dominions on the River Mississippi, and the Floridas; the Countries of the Creek Nations; and Many Uninhabited Parts* (Richmond: John Dixon, 1792), 26.

22. Feiler, *Bossu's Travels*, 63. To the same effect, see ibid., 65.

23. Jean-Marie Shea, ed., *Relation de la Mission du Missisipi du Seminaire de Québec en 1700* (New York: Cramoissy Press 1861), 41.

24. Dickinson, *New Travels*, 38 and n. 3; Feiler, *Bossu's Travels*, 63.

25. See, e.g., Jean-Marie Shea, *Relation de la Mission du Missisipi*, 41 (drums, per St. Cosme in 1699); Weddle, *La Salle*, 47 (drums, per Minet in 1682); Feiler, *Bossu's Travels*, 63 and n. 6 (drums and bells, per Bossu in 1751); Dickinson, *New Travels*, 81 (drums and reed flutes, per Bossu in 1770). The drums were clay pots over which a skin was stretched.

26. Thomas Nuttall, *A Journal of Travels into the Arkansas Territory during the Year* 1819 (Philadelphia: Thomas H. Palmer, 1821), 87. Father Marquette had made a similar observation almost 150 years earlier. See Reuben Gold Thwaites, ed., *The Jesuit Relations and Allied Documents* (New York: Pageant Book Co., 1959), 59:157.

27. Early accounts of the Quapaws' use of the calumet are extremely numerous. There is a detailed description of it in 1699 by Father St. Cosme in Jean-Marie Shea, *Relation de la Mission du Missisipi*, 40–42; John Gilmary Shea, *Early Voyages*, 71–72; and Kellogg, *Early Narratives*, 358–59. For an illuminating discussion of the

meaning of the calumet and the various Quapaw ceremonies associated with it, see George Sabo III, "Inconsistent Kin: French-Quapaw Relations at Arkansas Post," in Davis, ed., *Arkansas before the Europeans,* 105–30. There is an excellent reconstruction of a Quapaw calumet ceremony in Dickinson, "Quapaw Indian Dances," 42–44.

28. John Gilmary Shea, *Early Voyages,* 180.

29. James Owen Dorsey, "Kwapa Folk-Lore," *Journal of American Folk-Lore* 8 (January-March 1895): 130. Smithsonian Institution, Ms. 4800, Dorsey Papers: Quapaw (3.2.4)(2), 274.

30. Henri Joutel, "Remarques de Joustel sur l'ouvrage de Tonty relatif à la Louisiane," Service Hydrographique Archives (Paris), 115–19, no. 12, f, 16.

31. Mereness, ed., *Travels in the American Colonies,* 58. It appears from this passage that Quapaw painted skins were the object of trade. It is possible therefore that the hide with which this article deals was intended for export (that is, for sale) at the time that it was painted. Whether the skin was intended to be worn is not altogether clear. Although it is perhaps not technically a robe because the hair has been removed, Indians used such skins as wrappers during warm weather. See John Canfield Ewers, *Plains Indian Painting* (Stanford: Stanford University Press, 1939), 1. Among the Plains Indians, there is evidence that women produced the skins that exhibited geometric designs, while men produced those containing representative figures. Ibid., 7.

32. "Letter from Father du Poisson," 319. *Mataché* may derive from an Algonquian word that passed into Canadian French, whence it traveled to Louisiana in the eighteenth century. See William A. Read, *Louisiana-French*(Louisiana State University Press: Baton Rouge, 1931), 95–96. It survives in Louisiana French in such expressions as "un chien mataché" (a spotted dog) and in modern English in the word "matchcoat."

33. Ibid., 325.

34. Horse Capture, "From Museums to Indians," 91.

35. The Quapaws and the Mitchegamas (an Illinois tribe) may have been at odds briefly in the late seventeenth century, see Samuel Dorris Dickinson, "Lake Mitchegamas and the St. Francis," *Arkansas Historical Quarterly* (Autumn 1984): 197, 203; but in general the Quapaws and the Illinois tribes were firmly allied in the French interest throughout the French colonial period, as the text that follows demonstrates.

36. Pierre Margry, *Découvertes et établissements des francais dans l'ouest et dans le sud de l'Amérique septentrionale* (Paris:1887), 2:96; Stanley Faye,"Indian Guests at the Spanish Arkansas Post," *Arkansas Historical Quarterly* 4 (Summer 1945): 93.

37. For the parts played by the Illinois and the Quapaws in the First Chickasaw War, see Stanley Faye, "The Arkansas Post of Louisiana, French Domination," *Louisiana Historical Quarterly* 26 (July 1943): 633, 675; Clarence Walworth Alvord, *The Illinois Country, 1673–1818* (Urbana: University of Illinois Press, 1987), 176–80. For the parts played by the Illinois and the Quapaws in the Second Chickasaw War, see Jean Delanglez, "Journal of Father Vitry of the Society of Jesus, Army Chaplain during the War against the Chickasaws," *Mid-America* (1941), 30; *Journal de la guerre du Micissippi contre les chicachas* (New York: Cramoissy Press, 1859), Clarence Walworth Alvord, *The Illinois Country*, 182–83; Morris S. Arnold, *Colonial Arkansas, 1686–1804: A Social and Cultural History* (Fayetteville: University of Arkansas Press, 1991), 101–4.

38. For the Quapaws and Pontiac, see Clarence Walworth Alvord and Clarence Edwin Carter, eds., *The Critical Period, 1763–1765* (Springfield: Illinois Historical Library, 1915), 352, 456.

39. Ibid., 175.

40. Stanley Faye, "The Arkansas Post of Louisiana, Spanish Domination," *Louisiana Historical Quarterly* 27 (July 1944): 629–43; Stanley Faye, "Indian Guests at the Spanish Arkansas Post," *Arkansas Historical Quarterly* 4 (Summer 1945): 93; Carmen González López-Briones, "Spain in the Mississippi Valley: Spanish Arkansas, 1762–1804" (Ph.D. thesis, Purdue University, 1983), 95, 123–24.

41. Arni Brownstone, *War Paint: Blackfoot and Sarcee Painted Buffalo Robes in the Royal Ontario Museum*

(Toronto: Royal Ontario Museum, 1993), 11. See also John Canfield Ewers, *Plains Indian Painting: A Description of an Aboriginal American Art* (Stanford: Stanford University Press, 1939), 17.

42. Henri de Tonty, "Relation," 1:599.

43. Arnold, *Colonial Arkansas*, 31, 99–101.

44. While it is probable that the hide was painted to commemorate and record a past raid against the Chickasaws, it is just possible that it portrays a plan of action to be undertaken in the future. In fact, though this seems even less likely, it might have been executed to memorialize the making of a treaty and the obligations of the Quapaws under it. In that case, this buffalo skin would be both a work of art and a legal document.

45. Arnold, *Colonial Arkansas*, 101, 105.

46. If this skin is meant to be celebratory, as seems likely, what is plain is that it cannot be related to the two all-out campaigns that the French undertook in the 1730s against the Chickasaws, and in which the Quapaws participated, because both of these ended in disaster for the French and their allies. For a synopsis of these campaigns, see ibid., 101–4.

47. W. H. Falconer, ed., "Arkansas and the Jesuits in 1727: A Translation," in *Publications of the Arkansas Historical Association* (Little Rock: Arkansas Historical Association, 1917), 4:368–69; Jean Delanglez, ed., "Journal of Father Vitry," 34.

48. See Archivo General de Indias, Seville, Papeles Procedentes de Cuba, legajo 190: 112–13, a census of the Quapaw nation that indicates the positions of the Indian villages relative to the Spanish fort.

49. This suggestion is made in Horse Capture, "Gallery of Hides," 136. The letters are of a color somewhat different from the black paint used elsewhere on the skin, and may well have been drawn in ink.

50. Arnold, *Colonial Arkansas*, 105.

51. Bill Barron, ed., *The Vaudreuil Papers* (New Orleans: Polyanthos, 1975), 60.

52. Ibid., 250.

53. Patricia Kay Galloway, ed., *Mississippi Provincial Archives: French Dominion, 1749–1763,* vol. 5 (Baton Rouge: Louisiana State University Press, 1984), 76.

54. James A. Ford, "Menard Site: The Quapaw Village of Osotouy on the Arkansas River," *Anthropological Papers of the American Museum of Natural History* 48 (1961): 131–91.

55. See Hoffman, "Quapaw Structures."

56. Baird, *The Quapaw Indians*, 28, 31.

57. Arnold, *Colonial Arkansas*, 27, 30–31, 180 n. 6.

58. Dunbar Rowland and A. G. Sanders, eds., *Mississippi Provincial Archives: French Dominion*, 5 vols. (Jackson: Mississippi Department of Archives and History, 1927–84), 1 (1927): 222, 367.

59. Willard H. Rollings, *The Osage: An Ethnohistorical Study of Hegemony on the Prairie-Plains* (Columbia: University of Missouri Press, 1992), 118, 128.

60. Dickinson, *New Travels*, 40.

61. Rochemore to the minister, June 23, 1760, Archives Nationales, Paris, Archives Coloniales, sous-série C^{13A} 42:121.

62. Pittman reports "a barrack . . . , commanding officer's house, a powder magazine, and a magazine for provisions, and an apartment for the commissary." Philip Pittman, *The Present State of European Settlements on the Mississippi River* (London: J. Nourse, 1770), 40. This might be read to mean that there were five buildings in the fort. But in a letter written on December 17, 1765, the magazine for provisions and the apartment for the commissary are not separated by a comma, indicating, perhaps, that the apartment was in the magazine. This would mean that the fort contained only four buildings. See Philip Pittman, *Captain Philip Pittman's The Present State of the European Settlements on the Mississippi,* edited by John Francis McDermott (Memphis: Memphis State University Press: 1977), LIV.

63. Perhaps some specialist may notice something in the design of the flintlocks, or the style of the painting, or the manner in which the enemy's hair is arranged, that will confirm my conclusions as to the date and content of this skin, or provide further clues with respect to these matters. Perhaps, too, a chemical and physical analysis might help fix the date that this work was executed and

identify the kinds of paint that were employed. I would also welcome help on the identity of the extraordinary cross-like finials on the eaves of the French buildings!

64. For a synopsis of some of the Quapaw activities against the Osages, see Arnold, *Colonial Arkansas*, 112–24.

65. David I. Bushnell, "Native Villages and Village Sites East of the Mississippi," *BAE Bulletin* 69 (1919).

66. *Récit de Nicolas de La Salle*, in Margry, *Découvertes et établissements*, 1:554.

67. Henri de Tonty, *Relation*, 1:599. Henri Joutel said that in 1687 the Quapaws had "some otters of which they made robes or blankets to cover themselves when they are cold." Henri Joutel, *Remarques*, no. 12, f. 16.

68. For St. Cosme's comments on Quapaw dress in 1699, see Jean-Marie Shea, *Relation de la Mission du Missisipi*, 43. There is a translation of the relevant passage in John Gilmary Shea, *Early Voyages*, 73. (The translation in Kellogg, *Early Narratives*, 360, is somewhat truncated.) See also Mereness, *Travels in the American Colonies*, 57 (Diron d'Artaguette writing of conditions in 1723); Feiler, *Bossu's Travels*, 42 (Bossu writing of conditions in 1770).

69. Feiler, *Bossu's Travels*, 42.

70. Jean-Marie Shea, *Relation de la Mission du Missisipi*, 43; Jean Gilmary Shea, *Early Voyages*, 73; Kellogg, *Early Narratives*, 360.

71. These pictures are most readily available to the modern reader in Dickinson, *New Travels*, xvi, 43, and 60; and Arnold, *Colonial Arkansas*, 63, 81, and 84.

72. Dickinson, *New Travels*, 106.

73. Pope, *A Tour through the Southern and Western Territories of the United States*, 26.

74. Perhaps these are not breechclouts at all. In 1687, Joutel observed some Quapaw dancers who "had three or four Calabashes or Gourds, hanging at a leather Girdle about their wastes, in which there were several Pebbles, . . . so that when they ran, the Gourds made a ratling Noise." Stiles, *Joutel's Journal*, 181. Note, however, that the Quapaws engaged in battle are pictured with the same apparatus attached to their waists; and one can hardly think that they would attire themselves with gourds to wage war.

75. I am indebted to Michael Hoffman for this insight.

76. Some Indians, however, apparently went to war naked even in the eighteenth century, and one of the famous Segesser hide paintings features nude Indian attackers. See Gottfried Hotz, *The Segesser Hide Paintings: Masterpieces Depicting Spanish Colonial New Mexico* (Santa Fe: Museum of New Mexico Press, 1991), 157, 162. Thus the skin may be portraying the reality of the situation.

77. The Segesser hide paintings were probably the work of multiple artists who had the assistance of witnesses to the events that they depict. See Charles Bennett, "The Segesser Hide Paintings: Revelations about the Southwest's Colonial Past," *Terra* 30 (Summer 1992): 27–28. The Plains Indians seem also to have produced hide paintings for which more than one artist was responsible. Ewers, *Plains Indian Painting*, 6. We need to consider the possibility that the Quapaw skin was the product of a collaborative effort as well.

78. Mereness, *Travels in North America*, 57.

79. Feiler, *Bossu's Travels*, 65.

80. John Gilmary Shea, *Early Voyages*, 130. See also Samuel D. Dickinson, "Shamans, Priests, Preachers, and Pilgrims at Arkansas Post," in Davis, ed., *Arkansas before the Americans*, 96.

81. Interestingly, in the cosmology of the Osages (Siouan cousins of the Quapaws), the sun represented the east and the moon the west. Garrick A. Bailey, ed., *The Osage and the Invisible World from the Works of Francis La Flesche* (Norman: University of Oklahoma Press, 1995), 33. The same was true of the Cherokees. See Charles Hudson, *The Southeastern Indians* (Knoxville: University of Tennessee Press, 1976), 132.

82. For these activities, see Arnold, *Colonial Arkansas*, 101–2.

83. I am indebted to Malcolm Lewis for pointing this out to me.

84. For the diachronic character of much Indian art, see Arni Brownstone, *War Paint: Blackfoot and Sarcee Painted Buffalo Robes in the Royal Ontario Museum*

(Toronto: Royal Ontario Museum, 1993), 26–30.

85. G. Malcolm Lewis, "The Indigenous Maps and Mapping of North American Indians," *Map Collector* 9 (1975): 25.

86. G. Malcolm Lewis, "Indicators of Unacknowledged Assimilations from Amerindian Maps on Euro-American Maps of North America: Some General Principles Arising from a Study of La Vérendrye's Composite Map, 1728–29," *Imago Mundi* 38 (1986): 9. On the matter of North American Indian maps, see also Gregory A. Waselkov, "Powhatan's Mantle," in Peter H. Wood, Gregory A. Waselkov, and M. Thomas Hatley, eds., *Powhatan's Mantle: Indians in the Colonial Southeast* (Lincoln: University Nebraska Press, 1989).

87. This is the definition of a map suggested in J. B. Harley and David Woodward, eds., *Cartography in Prehistoric, Ancient, and Medieval Europe and the Mediterranean*, vol. 1 of *The History of Cartography* (Chicago: University of Chicago Press, 1987), xvi.

88. Archives Nationales, Paris, Archives Coloniales, sous-série C^{13A} 25:279, 288v.

89. For these, see Samuel Wilson, Jr., "The Drawings of François Benjamin Dumont de Montigny," and Samuel Wilson, Jr., "Louisiana Drawings by Alexandre de Batz," in Jean M. Farnsworth and Ann M. Masson, eds., *The Architecture of Colonial Louisiana* (Lafayette: Center for Louisiana Studies, University of Southwestern Louisiana, 1987), 108, 261. The Le Bouteux drawing of New Biloxi is reproduced in Arnold, *Colonial Arkansas,* at 10–11.

9

Indian Maps of the Colonial Southeast: Archaeological Implications and Prospects

Gregory A. Waselkov

North American archaeologists have, as a rule, tended to overlook the informative potential of indigenous maps, preferring instead to rely on more familiar forms of data—the comfortable objectivity of artifacts and the, by now, well-accustomed narrative biases of visiting Europeans' accounts—in their efforts to understand the nature of prehistoric and early historic Indian societies. In fact, there are some compelling reasons for this apparent oversight. Native maps are both rare and difficult to interpret. The unique perspectives that these maps offer of long-vanished social landscapes, however, should tempt archaeologists to explore these little-known portrayals of early historic America. Also, data from native cartography and archaeology are largely complementary; our knowledge of the past is enhanced by considering both sources together, identifying and attempting to reconcile different interpretations derived from each. This chapter is intended to redress the archaeological limitations of my earlier work on the cartographic record of Southeastern North America (the area I know best) produced by native inhabitants during the colonial period, by considering both the archaeological implications of these maps and some of the interpretive difficulties involved with their use by archaeologists.

Traditional Southeastern Indian maps of the seventeenth and eighteenth centuries were often ephemeral things, sketched in the ground with sticks or, somewhat more permanently, drawn on bark or painted on skins as visual aids to complement a recitation of tribal history, or describe the locations of allies and enemies, or simply to plot a route. (More durable images were carved or painted on rock, but no surviving petroglyphs or pictographs in the region convey geographical information, so far as we know.)[1] Few of these inherently impermanent records have survived, except as European copies on paper filed in ministry archives. Of the half dozen extant, colonial-period Indian maps of Southeastern North America, all but one are European copies.[2] About an equal number of maps drawn by colonists acknowledge considerable input from Indian sources, although many times that number undoubtedly incorporated substantial, unacknowledged Indian contributions.[3] So only a few rare documents record, perhaps imperfectly due to virtually inevitable copyists' errors or linguistic misunderstandings, a fragment of the sum of Southeastern Native American cartographic knowledge.

From this small sample we can discern two types of maps that served very different functions. One sort of Indian map related village locations to river courses, paths, and other landscape features.

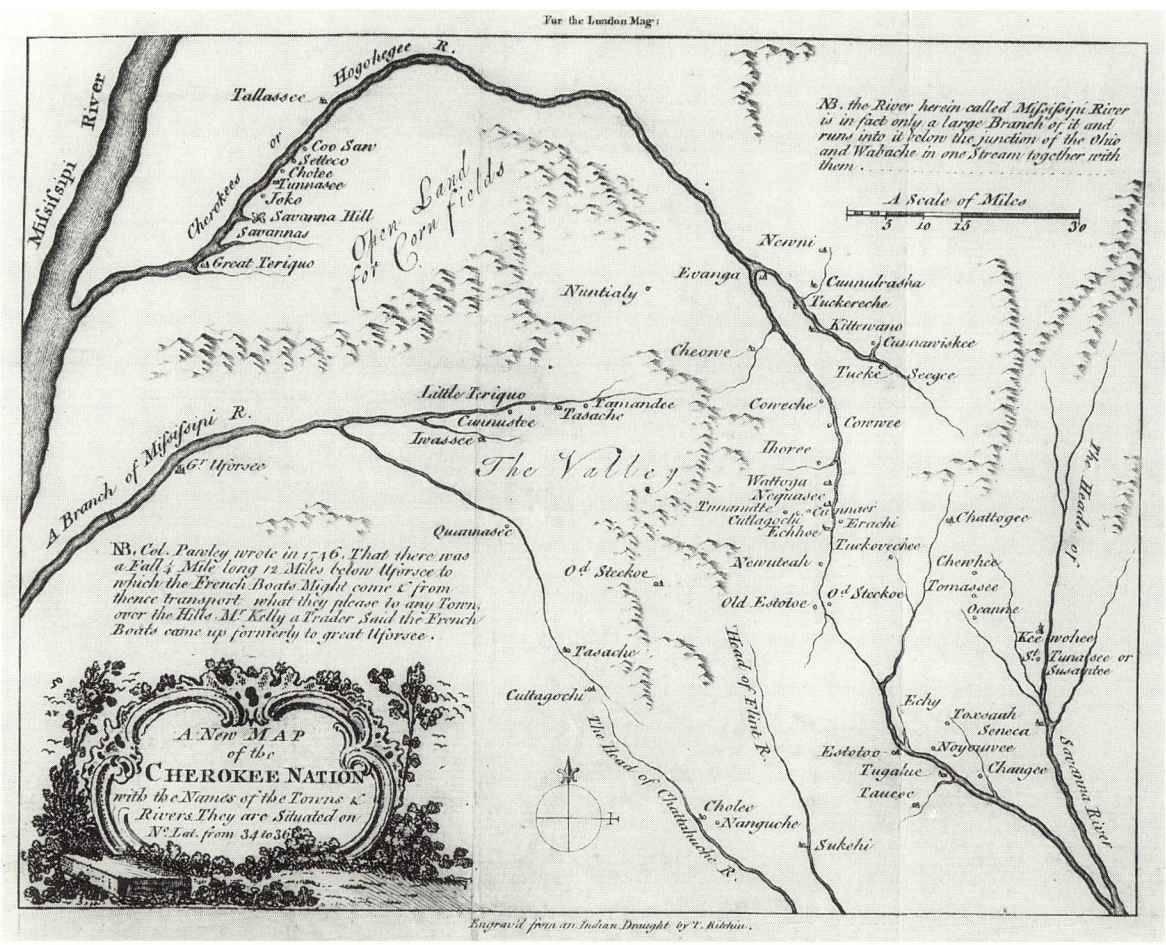

9.1 Thomas Kitchin, "A New Map of the Cherokee Nation with the Names of the Towns and Rivers: They are Situated on No. Lat. from 34 to 36," from *London Magazine* 29 (February 1760): opp. p. 64; 8 3/4 x 6 5/8 in. Kitchin's map was "Engrav'd from an Indian Draught" (after an unknown original, probably Cherokee) that evidently portrayed the topographic relationship between rivers, mountains, and Cherokee towns. Courtesy of the Newberry Library.

These were the maps that Europeans most readily understood; they eagerly solicited such drafts to discover new trade routes, plan military campaigns, and locate already-cleared lands for potential colonial settlement. The cartographic information obtained in this form from Indians was also easily appropriated and incorporated into European maps of the Southeast (fig. 9.1).[4] Native American leaders frequently recognized that such information in European hands could pose a threat to their own societies' well-being. By the late eighteenth century, few native Southeasterners were any longer offering to draw maps of their country for their obtrusive colonial neighbors. On at least one occasion, Indian headmen accompanied a colonial surveying party to insure that the resultant treaty map accurately delineated the extent of lands retained and the limits of lands ceded.[5]

A second kind of Indian map conveyed primarily social and political relationships; by varying the relative size of circles—each representing a village or some larger social group—and by manipulating distances and directions between the circles, native cartographers could graphically evoke degrees of ethnic relatedness, limits of political control, and networks of cooperating or competing groups (fig. 9.2). In such maps, Indian mapmakers spoke a symbolic language unfamiliar to Europeans. Their reliance on a flexible, topological, non-Euclidean view of landscape space also effectively excluded Europeans, whose mapping tradition prescribed that each map have an unvarying distance scale and compass orientation. Colonists found the content of these social maps difficult to grasp, let alone to coopt, so only a few copies were preserved as ethnographic curiosities.

Both sorts of Southeastern Indian maps, as well as Indian cartographic information imbedded in contemporary European maps of the Southeast, will undoubtedly prove to have archaeological utility. But I know of only two instances where this potential has been demonstrated. The first example is the interpretation of seventeenth-century archaeological sites depicted on Capt. John Smith's map of Virginia, first published in 1612, which contains substantial Indian content, both acknowledged and unacknowledged (fig. 9.3).

Immediately upon arrival in Virginia in April 1607, John Smith and his fellow invaders from England, the original colonists at Jamestown, expended much energy exploring and mapping the environs of Chesapeake Bay and its tributary rivers. That region's inhabitants, the Algonquian-speaking Powhatans and their eponymous leader, apparently at first doubted that this small band of sickly, improvident strangers posed a serious threat to their way of life. So natives of all ranks repeatedly shared with the Englishmen a wealth of geographical information about their own and more distant lands. On the initial English voyage up the James River, Capt. Gabriel Archer found an Indian man who "offred with his foote to describe the river to us: So I gave him a pen and paper (shewing first yᵉ use) and he layd out the whole River from Chesseian [Chesapeake] bay to the end of it so farr as passadg was for boates."[6] Upon arriving at this man's village, Archer sought chiefly confirmation of his guide's cartographic veracity. "I caused now our kynde Consort that described the River to us, to draw it againe before kyng Arahatec, who in every thing consented to this draught, and it agreed with his first relaty-

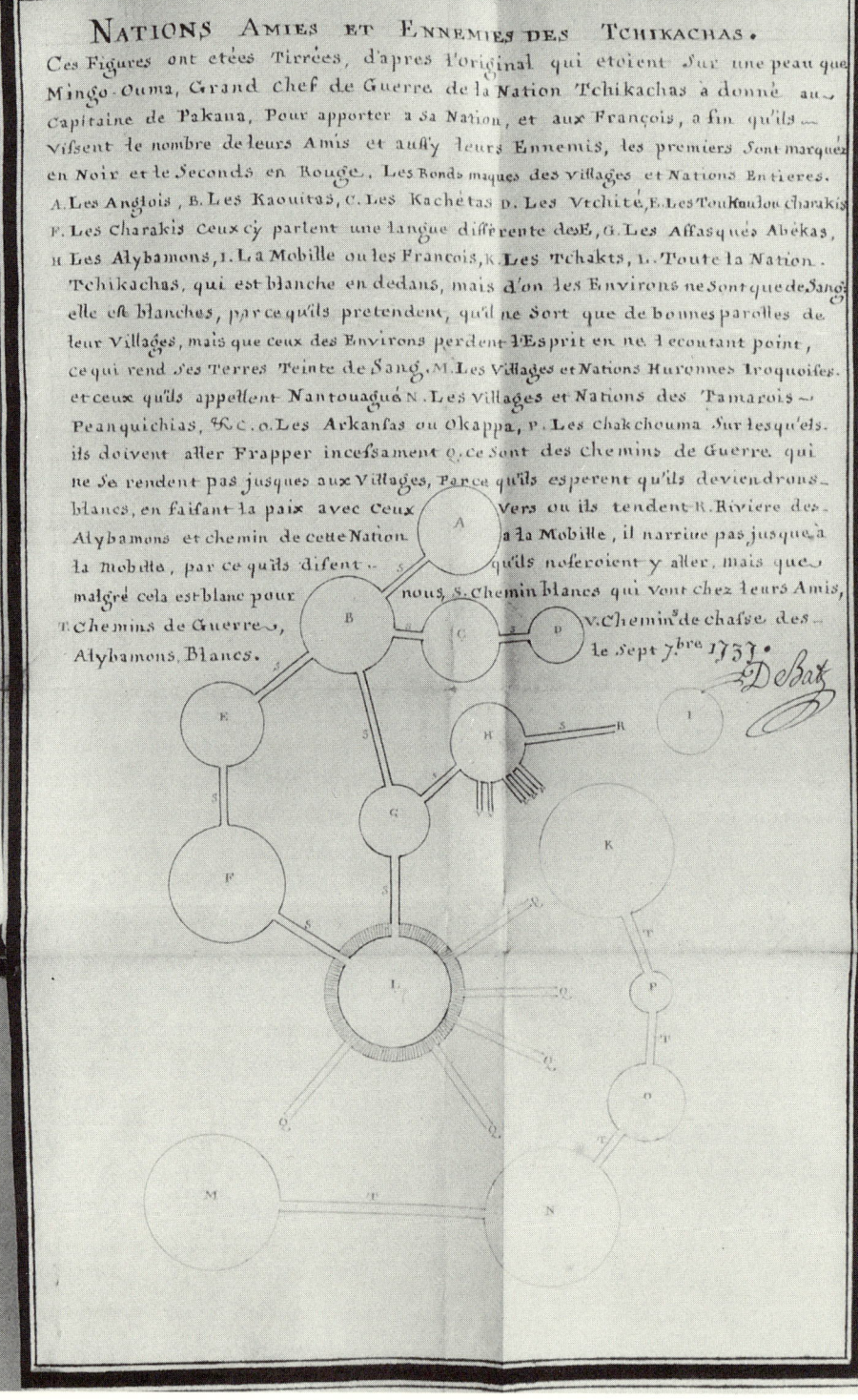

NATIONS AMIES ET ENNEMIES DES TCHIKACHAS.

Ces Figures ont etées Tirrées, d'apres l'Original qui etoient sur une peau que Mingo-Ouma, Grand chef de Guerre de la Nation Tchikachas à donné au Capitaine de Pakana, Pour apporter a sa Nation, et aux François, a fin qu'ils vissent le nombre de leurs Amis et aussy leurs Ennemis, les premiers sont marquéz en Noir et le seconds en Rouge. Les Ronds marques des villages et Nations Entieres.

A. Les Anglois, B. Les Kaouitas, C. Les Kachétas D. Les Vtchité, E. Les Toukoulou Charakis F. Les Charakis Ceux cy parlent une langue différente des E, G. Les Assasqués Abékas, H Les Alybamons, I. La Mobille ou les Francois, K. Les Tchakts, L. Toute la Nation. Tchikachas, qui est blanche en dedans, mais d'on les Environs ne sont que de sang, elle est blanches, par ce qu'ils pretendent, qu'il ne sort que de bonnes parolles de leur Villages, mais que ceux des Environs perdent l'Esprit en ne l'ecoutant point, ce qui rend ses Terres Teinte de Sang. M. Les Villages et Nations Huronnes Iroquoises. et ceux qu'ils appellent Nantouaguo N. Les Villages et Nations des Tamarois-Peanquichias, &c. o. Les Arkansas ou Okappa, p. Les chakchouma sur lesquels ils doivent aller Frapper incessament Q. ce sont des chemins de Guerre, qui ne se rendent pas jusques aux Villages, parce qu'ils esperent qu'ils deviendront blancs, en faisant la paix avec Ceux Vers ou ils tendent R. Riviere des Alybamons et chemin de cette Nation a la Mobille, il n'arrive pas jusque a la Mobille, par ce qu'ils disent qu'ils n'oseroient y aller, mais que malgré cela est blanc pour nous S. Chemin blancs qui vont chez leurs Amis, T. Chemins de Guerres, V. Chemin de chasse des Alybamons, Blancs.

le sept 7bre 1737. DeBat

on."⁷ These maps of river courses certainly belonged to the first category of indigenous maps, the sort from which Europeans appropriated so much information when compiling their initial charts of unexplored Indian lands.

In mid-December 1607, John Smith was captured during an exploration of the Chickahominy River and brought before the paramount chief, Powhatan, who "began to draw plots upon the ground" to illustrate the regional political scene, indicating the various people living beyond the limits of his chiefdom.⁸ Powhatan's drafts probably corresponded to the second type of native Southeastern maps, with their characteristic social circles (representing different tribes and chiefdoms) arranged roughly in geographical order. One of the great surviving examples of Southeastern Indian art, a large leather artifact decorated with shell beads arranged in disks and known as "Powhatan's Mantle," may resemble in design the ephemeral sketches drawn for Smith.⁹

During his captivity, John Smith also observed a lengthy ceremony performed by seven priests who hoped "(as they reported) to know if any more of his countrymen would arive there, and what he there intended." According to Smith, the priests first

> made a faire fire in a house ... and about the fire, they made a circle of [corn] meale. That done the chiefe Priest ... began to shake his rattle, and the rest followed him in his song. At the end of the song, he laid downe 5 or 3 graines of wheat [maize seeds or kernels] and so continued counting his songs by the graines, til 3 times they incirculed the fire, then they divided the graines by certaine numbers with little stickes, laying downe at the ende of every song a little sticke.¹⁰

Three published versions of Smith's account of his captivity differ considerably in the details. Only the last, edited by Samuel Purchas, includes an interpretation of these "conjurations." We can legitimately question how well Smith understood the ceremony performed before him, since he

---

9.2 *Opposite*, *Nations amies et ennemies des Tchikachas* [Nations Friendly and Hostile to the Chickasaws], the Chickasaw/Alabama map of the Southeast, 1737, as redrawn and transcribed by the French colonial engineer and draftsman, Alexandre de Batz. Courtesy of Archives d'Outre Mer, Aix-en-Provence. A reproduction can be found in Waselkov, "Indian Maps," p. 298.

The Captain of Pacana, an Alabama Indian war leader, obtained most of the information portrayed in this map from Mingo Ouma, a Chickasaw war leader. This map symbolically describes the political and military divisions among the southeastern Indians from the perspective of the Chickasaws, who were involved in a protracted war with the French and their Indian allies. The captions and key can be translated as follows:

A. The English B. The Cowetas C. The Kasihtas D. The Yuchis E. The Tugaloo Cherokees F. The Cherokees who speak a different language than E G. The Okfuskees Abekas H. The Alabamas I. Mobile or the French K. The Choctaws L. The whole Chickasaw Nation, which is white within, but the space surrounding it is of nothing but blood. It is white because they claim that only good words come from their villages, but those of the surrounding country lose their minds by not listening to them at all, and this stains their lands with blood. M. The Huron and Iroquois villages and nations and those they call Nantouague N. The villages and nations of the Tamaroas, Piankashaws, etc. O. The Arkansas or Quapaws P. The Chakchiumas, whom they are going to attack at once Q. These are warpaths that do not go as far as the villages, because they hope that they will become white when they make peace with those toward whom they lead. R. River of the Alabamas and the path from that nation to Mobile. It does not go as far as Mobile because they say they would not dare to go there, but in spite of that it is white for us S. White paths that lead to their friends T. War paths V. Hunting paths of the Alabamas, white.

7th of September, 1737, De Batz.

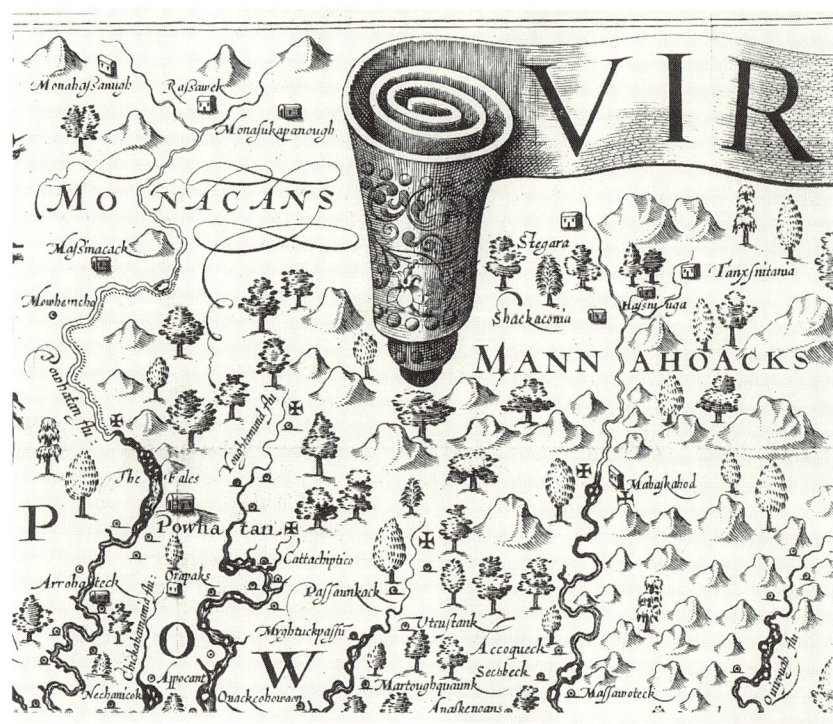

9.3 Detail of "Virginia... discovered and discribed by Captayn John Smith," from *A Map of Virginia with a Description of the Countrey, the Commodities, People, Government, and Religion* (Oxford: Joseph Barnes, 1612). A detail of Captain John Smith's map of Virginia showing the upper reaches of the James (Powhatan), Pamunkey (Youghtanund), Mattaponi (Mattapanient, next to the village of Passaunkack), and Rappahannock (Toppahanock, next to the village of Accoqueck) rivers. The Tuscan crosses indicate Smith's westernmost explorations, but the dotted line on the upper James River represents Christopher Newport's expedition into Monacan territory. Courtesy of the Edward E. Ayer Collection, the Newberry Library.

could not yet have been proficient in the Algonquian languages of Virginia. But we have no other source to call on. "Three days they used this Ceremonie, thereby to know (as they said) whether hee [Smith] intended them well or no. The circle of meale signified their Countrey, the two circles of Corne the Sea-bounds; and the stickes his Countrey. They imagined the World to be flat and round like a trencher, and themselves in the midst."[11] If this image in sticks and corn was meant, as Smith surmised, to represent Powhatan cosmography, then he saw a third type of Southeastern Indian map that is not represented by any surviving examples.

In his "Map of Virginia," first published in August 1612, Smith drew together all of the locational data and place-names garnered from personal explorations, discoveries of other Jamestown colonists, and facts solicited from Native American informants. That he reserved pride of place for his own discoveries cannot be doubted; he appended these lines to a description of his map:

> Thus have I walkt a wayless way, with uncouth pace,
> Which yet no Christian man did ever trace:
> But yet I know this not affects the minde,
> Which eares doth heare, as that which eyes doe finde.[12]

Smith was exceptionally forthright in his acknowledgment of Indian sources on lands yet unvisited by the English, which he delimited by placing Tuscan crosses at the farthest points reached by Europeans. According to the map key, "to the crosses hath been discovered—what beyond is by relation." Smith elaborated in the accompanying text "that as far as you see the little Crosses on rivers, mountaines, or other places have beene discovered; the rest was had by information of the Savages, and are set downe, according to their instructions."[13] For these regions at the periphery of Powhatan's domain, beyond the limits of the Jamestown colonists' personal experience, the village locations and place-names shown on Smith's map and unequivocally attributable to native cartography have been a valuable interpretive tool for archaeologists.

In western Virginia, Smith's crosses appear at the falls of the major rivers. Five Monacan villages are shown on the upper reaches of the James River, and a like number of Manahoac villages occupy the upper Rappahannock and Rapidan valleys, although other historical records suggest that the true number of piedmont towns was much greater. Smith's knowledge of the Manahoacs seems to have come from Amoroleck, a wounded captive, whom he interrogated through the Algonquian interpreter Mosco. Smith also had some brief direct contact with Monacans, including "one of their pettie Werowances [chiefs]" that the English briefly held hostage, and he probably obtained additional information from the Powhatans.[14] Since almost everything known about the Monacans and Manahoacs derives from Smith's map and several short written descriptions, archaeologists have long sought to correlate archaeological site locations with the village names shown on the map. In the 1920s and 1930s, David Bushnell intensively surveyed the region and thought he had identified many of these places on the ground.[15] Unfortunately, archaeologists of that era had not yet developed effective methods of dating prehistoric archaeological sites in eastern North America. Many of Bushnell's sites, we now know, predate by hundreds or even thousands of years the villages reported to Smith in 1607 and 1608.

More sophisticated recent surveys of the Monacan and Manahoac homelands have yielded only a few convincing matches between archaeological sites of the proper age and the village symbols shown on Smith's map. For example, a site on the north bank of the James River, near its confluence with the Rivanna River, may be the location of Rassawek village, "Chief habitation" of the Monacans that is depicted by Smith near a major confluence of the James.[16] Even after years of intensive searching, though, these identifications remain few and tentative for several reasons. Some of these sites have probably been lost to erosion over the last four centuries and others may lie under modern cities, inaccessible to archaeologists. The dispersed and shifting nature of seventeenth-century Indian villages may also be an important factor. If early historic piedmont villages consisted of widely separated houses scattered across the landscape, interspersed with gardens and fields as we know many Powhatan villages were, then these villages will appear to archaeologists as clusters of small, discrete sites (each consisting of one or two houses) rather than as large, unified town sites. A third problem is the imprecise nature of Smith's map, which precludes pinpoint identifications of most features on a map drawn to this scale.[17] The first potential problem, that of site loss through time, is a given for any archaeologist attempting to rediscover an extinct society's settlement pattern by systematically surveying the present landscape. But the other two problems relate directly to the nature of this map in particular, and to Native American maps of the Southeast in general, and so deserve more discussion.

John Smith and his fellow English colonists evidently found Virginia Indian settlements difficult to understand and map effectively. The Powhatans' houses, according to Smith, were "in the midst of their fields or gardens, which are small plots of ground. Some 20 acres, some 40, some 100. some 200. some more, some lesse. In some places from 2 to 50 of those houses together, or but a little separated by groves of trees."[18] On his map of 1612, Smith reduced this complexity to two kinds of settlements, as explained in his map key: 67 villages with "Kings howses," each represented by a drawing of an Algonquian longhouse, and 161 villages with only "Ordinary howses," each represented by an encircled dot. The so-called Zuñiga Map of 1608, probably a copy of a manuscript map drawn by John Smith and sent to England in that year, indicates how the English initially recorded the confusing diversity of Powhatan settlements. Instead of the individual encircled dots and longhouse symbols found on Smith's map of 1612, the earlier Zuñiga Map shows dot clusters, each containing from 3 to 35 dots, distributed along the rivers of the coastal plain.

While these dot clusters may have more accurately represented the various sizes of the internally dispersed communities actually observed by the English colonists, Smith opted, in his published version, for the dichotomous symbols—a circle or a longhouse—that conformed more closely to seventeenth-century English notions of settlement hierarchy.[19] But Smith's inspiration for this map symbolism may have been Native American instead of English. The use of circles to represent villages, whether dispersed or nucleated, and other levels of social organization is a common feature of all known Southeastern Indian maps from the early historic period. Smith had many opportunities to see such maps and presumably he adopted that convention in order to simplify his own very complex manuscript map, crowded with detail.

Most modern scholars of John Smith's map have assumed from the map legend, "to the crosses hath been discovered," that he and other English colonists personally rowed their exploratory barges to those limits indicated on the rivers and visited all of the villages shown. A careful reading of Smith's voluminous writings, however, leads to a different interpretation. Smith's explorations, and those of his compatriots, are well documented during the initial years of the Virginia colony. During Smith's slightly more than two years in Virginia, the colonists did make their way up most of the major rivers and streams, but their passage was frequently challenged and their landings blocked at numerous locations. On one occasion, rowing up the Rappahannock River the explorers "passed by Pisacack, Matchopeak, and Mecuppom" without stopping.[20] Their Algonquian interpreter, Mosco, was on board and must have supplied the place-names. During an earlier exploration of the eastern shore of the Chesapeake, the English were prevented by force from entering the Nanticoke River;[21] yet a considerable length of the river and three villages on its banks are depicted on Smith's map. The appearance of 62 unnamed villages on his map—most of which are found in the upper reaches of the James, York, and Potomac rivers—suggests that Smith's knowledge of a substantial number of Native American settlements was obtained secondhand. From the accounts of John Smith, George Percy, and Gabriel Archer,[22] there is strong evidence to believe or we can confidently infer that the English visited at least 58 of the 228 native villages shown on Smith's map. Other villages, such as many along the Potomac River, are simply listed without specifying how the explorers learned of them. While it seems likely that during the first two years the colonists did visit some additional native villages, particularly those near Jamestown, without leaving records of those visits, it seems equally likely that their personal observations may have extended to only about half of the 228 villages portrayed on the 1612 map. So, aside from the cartography of piedmont regions beyond the Tuscan crosses, openly acknowledged by Smith as derived from Indians, we can also attribute to Indian informants considerable data on village locations, place-names, and chiefs' habitations in the coastal plain.

Archaeologists have attempted for several centuries to use Smith's 1612 map of Virginia, with its substantial Native American contributions, to match place-names to specific sites in the coastal plain, just as they have in the piedmont, as already described.[23] Thomas Jefferson, in his 1785 book, *Notes on the State of Virginia*, apparently was the first to suggest locations for some of the village sites shown on Smith's map.[24] A recent, thorough review of all succeeding efforts concludes that only thirteen of these proposed associations, totaling "no more than 10% of all Powhatan settlements," are reasonably certain.[25] The reasons for such a poor showing are essentially the same ones that have frustrated archaeologists working in the piedmont, with one further complication—how to distinguish villages occupied by chiefs from those occupied by commoners.

Until recently, archaeologists studying the Powhatans assumed that chief's villages were larger (with more than ten longhouses) than villages occupied only by commoners (with ten or fewer longhouses).[26] There is, in fact, no archaeological or historical basis for this supposition. Chief's villages differed from commoners' villages only by the added presence of the chief's personal residence, mortuary temple, and storehouse, as well as the houses of close kin. Houses in both types of villages were typically dispersed over several hundred acres and are recognizable archaeologically as clusters of small, discrete sites.[27] If all of the 228 villages on Smith's map were ever to be discovered by archaeologists, they would very probably be represented by over 1,000 archaeological sites, most of them quite small (each the location of only two or three longhouses), but originally part of larger, widely scattered communities. Now that archaeologists understand the nature of Powhatan village organization, and the village symbols that John Smith derived from Indian maps and informants, the search for those historic Native American sites should progress much more rapidly than it has in the past.

A third interpretive problem facing archaeologists intent on using John Smith's map, with its substantial Indian content, or any of the surviving copies of Southeastern Indian maps from the early historic period, involves the effects of scale. Nearly all of these maps cover vast areas, territories encompassing thousands of square miles, scaled to fit a deerskin or a sheet of foolscap. Precise locations, at the level required for field identification of archaeological sites, are generally not retrievable from these maps.

A notable exception, however, is a plan of the Chickasaw villages drawn by the Captain of Pacana, an Alabama headman who acted as a spy for the French and scouted the Chickasaws' defenses in July 1737 in anticipation of a French military offensive. His map was redrafted by the French engineer Alexandre de Batz, and that copy still remains in the collections of the Archives d'outre Mer in Aix (fig. 9.4).[28] Thanks to extensive archaeological excavations from 1937 to 1940 and subsequent historical research, James R. Atkinson (National Park Service archaeologist stationed at the Natchez Trace Parkway in Tupelo, Mississippi) has been able to establish the archaeological locations of the villages depicted on the Captain of Pacana's map.[29] There are some discrepancies between the de Batz copy and Atkinson's modern interpretation, the most serious being a mix-up in the names of the southernmost Chickasaw villages of Tchoukafala, Apeony, and Ackia, presumably due either to a mistake by the copyist or to a misunderstanding on the part of the Alabama spy. We must assume that de Batz was responsible for appending a compass rose and a distance scale; the former agrees closely with Atkinson's map orientation, the latter exaggerates distances by about a factor of three. This error is probably the French engineer's, since the Captain of Pacana's map key accurately describes certain villages to be within musket range of each other, which contradicts the drawn scale.[30] Despite the prevalence of topological, non-Euclidean manipulation of space in most Indian maps of the Southeast, our only extant example of a local map treats distance and direction in an inflexible, Euclidean manner.

One might surmise that if other Native American maps of specific localities in the Southeast remain undiscovered or unrecognized in private or public archives, they might prove as useful as the Captain of Pacana's plan of the Chickasaw villages. In addition to providing village place-name identifications and locations, agricultural fields are also shown and can be correlated to several specific landforms and soil types.[31] Unfortunately, this map of the Chickasaw villages in 1737 remains the exception, uniquely apposite to archaeological purposes, construed in the narrow sense.

9.4 *Opposite*, *Plan et scituation des villages Tchikachas, mil sept cent trente sept* [Plan and Situation of the Chickasaw Villages, 1737], the Captain of Pacana's map of Chickasaw towns, 1737, as redrawn by Alexandre de Batz. Courtesy of Archives d'Outre Mer, Aix-en-Provence.

A translation of the map legend and key follows:

A. Ogoula Tchetoka, fort where M. d'Artaguiette attacked; there are 60 men. B. Etoukouma C. Achoukouma D. Amalata E. Taskaouilo F. Tchitchatala; the fort is the most important, the said village is of 60 men. G. Falatchao H. Tchoukafala I. Apeony, where the last party of Frenchmen attacked L. Aekya M. The Natchez, who still have forty men N. Bayous O. Paths between the villages P. Fields Q. Encampment of the last French party R. M. d'Artaguiette's Road S. Road of the last French party

The forts of the villages A, B, C, D, E, F, G are very near each other and almost within musket range. Likewise those designated H, I, L. Prepared and drawn up at Mobile, September 7, 1737. De Batz.

PLAN ET SCITUATION DES VILLAGES TCHIKACHAS.
MIL SEPT CENT TRENTE SEPT.

Les Ronds marques les Villages et dans chaque il y a un Fort à Trois Ronds de Pieux.

A. Ogoula-Tchetoka, Fort ou M. Dartaguiettes à Frappé il y a 60. Hommes,

B. Etoukouma, C. Achoukouma, D. Amalata, E. Taskaouilo, F. Tchitchatala, le Fort est le plus Considerable ledit Village est de 60. Hommes,

G. Falatchao, H. Tchoukafala, I. Apeony ou le dernier partis des François à Frappé

L. Ækya, M. Les Natchez qui Sont encore quarente Hommes N. Bayoües,

O. Chemins des Villages, P. Deserts, Q. Campement du dernier party François,

R. Chemin de M. Dartaguiettes, S. Chemin du dernier party François,

Les Forts des Villages A.B.C.D.E.F.G. Sont tres pres les uns des autres et presque à la portée du Fusil, Egalement Ceux H.I.L.

Fait et Redigé a la Mobille le Sept Septembre 1737. DeBatz

Archaeologists, however, are not entirely preoccupied with the particulars of the human past. This discussion may have thus far implied too narrow a view of the archaeological research endeavor. Modern archaeology is also a generalizing and comparative discipline, one to which the large-scale Indian maps of the Southeast can also contribute much.

Consider, for example, the preeminent Indian-drawn map of the Southeastern region of North America, now extant only as an English copy "from a Draught Drawn upon a Deer Skin by an Indian Cacique and presented to Francis Nicholson Esqr. Governour of [South] Carolina" around 1723 (see fig. 10.2, below). Enormous in scope, the map covers around 700,000 square miles from northeast Florida to southwestern Kansas, from southeast Texas to western New York. The map seems to have been drawn from a Chickasaw perspective (and a few notations are written in that western Muskogean language), so the "Indian Cacique" mentioned in the legend can be fairly confidently identified as a Chickasaw headman.[32] This and two other maps—including one dating to 1721 and attributed to a Catawba headman (fig. 10.1) and another drawn in 1737 by Mingo Ouma, a war leader of the Chickasaws (fig. 9.2)—are our only cartographic glimpses of the region from the viewpoint of Southeastern Indian elite. All are highly stylized, symbolically distinguishing allies from antagonists, and, in one case, Europeans from Indians. All employ topological manipulation of direction and distance in arranging that Southeastern mapmaking convention, the social circle. And all indicate an exceptional knowledge of distant places, far beyond the range of most individuals' personal travel in the early eighteenth century.

Archaeologists studying the late-prehistoric and early-historic eras of the Southeast have found abundant material evidence of long-distance trade or exchange. During the Mississippian Period (A.D. 1000–1550), marine shells from the Gulf and Atlantic coasts, translucent mica from the Appalachian Mountains, silvery cubes of galena from the central Mississippi River valley, copper from the Great Lakes and the Appalachians, and fine ceramics produced in the major population centers were routinely carried on an extensive communication system of foot trails and rivers navigable by dugout canoe. This prehistoric trade may have functioned either by means of down-the-line exchange, a series of transactions between neighboring villages or chiefdoms, or by long-distance traders who carried commodities from source to destination, or most likely by a combination of the two. By the late seventeenth century, written records confirm the existence of long-distance traders, including intrepid Apalachee middlemen who carried goods between the Spanish missions of Florida and Muskogean villages deep in the interior Southeast.[33]

While foreign artifacts, such as beads made of Gulf Coast whelk shell or gorgets formed from Spanish brass found far from the source of those materials, adequately document economic contact between distant societies, they cannot convey the social context of trade, the meanings of those exotic goods to the traders, or the motivations for their acquisition. However, historic and ethnographic records, including the three large Indian-drawn maps of the Southeast, do provide some of the missing contextual information so necessary for an understanding of early historic native Southeasterners.

Mary Helms, an ethnohistorian and ethnographer, has argued that the political and religious elite of most traditional societies valued esoteric knowledge and exotic materials obtained from distant lands and peoples. Their value derived from several factors. Long-distance travel was inherently difficult and dangerous; thus distant regions were relatively inaccessible to most members of every society, effectively beyond their personal experience. In a world where long-distance travel was limited, "geographical distance" often corresponded with "supernatural distance"; cosmological powers could be acquired and controlled by those so daring and so privileged as to visit far-flung locations, where they obtained exotic objects from their mythical sources. Possession of esoteric knowledge and extraordinary artifacts then tended to validate and legitimate the social status of political and religious elite by association with distant places perceived as sources of spiritual power.[34]

The Chickasaw and Catawba headmen that drafted our sample of highly stylized, social circle maps of the Southeast lived in the sort of world described by Helms. Political leaders, such as these headmen, often led long-distance expeditions for war, trade or diplomacy[35]—in fact, two of the maps were drawn for presentation to the colonial governor in Charleston. The stylistic similarity of these three maps, and their dissimilarity to the more practical landscape maps produced by lower-ranked Southeastern Indians, suggests that the elite did control and convey knowledge of distant places through a fog of arcane symbolism that we can only dimly penetrate today. If they did not all travel to the limits of their known world, clan and tribal leaders certainly obtained geographical esoterica acquired by earlier traveling elite who had preserved and transmitted their knowledge at least partly by means of circles painted on deerskins.

Despite interpretive obstacles, some archaeological implications can be derived from these maps. Focusing once again on the 1723 Chickasaw map (fig. 10.2), we can discern that the principal communication routes are portrayed either as continuous lines, representing open roads to allies and trading partners, or lines that end abruptly before entering the Chickasaw homeland, paths interrupted by war and unusable for trade or hunting.[36] Many of these routes can be identified as paths or rivers described in historical sources; some of them remained remarkably stable during the tumultuous eighteenth century. For instance, the Peoria, Cahokia, and Kaskaskia paths to French settlements and French-allied Illinois Indians (who are symbolically under attack, on the map, by the armed warrior shown leading a Chickasaw horse) were still in use thirty years later, according to a map drawn for the English around 1755 by Chegeree, who may have been a Miami or Twightwee Indian.

When trails remained open for trade, the ensuing exchanges of artifacts should, of course, have left "trails" of archaeological evidence. One sort of trade, the traffic in Indian slaves, is an understudied aspect of the colonial Southeast. Some of the societies shown along the far left edge of the 1723 Chickasaw map suffered severely at the hands of European-inspired slave raids by Southeastern tribes. Since a high proportion of those enslaved were female, and women were typically responsible for pottery production in this region, archaeologists should be able to recognize the presence of foreign slaves at sites located along the major communication routes. Material indicators of ethnic identity, particularly as reflected in ceramic decoration and form, are allowing archaeologists to follow the movements of refugee villages along many of these same routes. In fact, archaeologists are just beginning to devise methods to trace the courses of prehistoric and early historic paths across the Southeast,[37] and Indian maps of the region should be an important aid in that task.

Native American maps clearly have tremendous potential to inform us about past social landscapes. Archaeologists can contribute to their interpretation by identifying specific sites and communication routes depicted by Indian mapmakers, as they are in turn informed by the often unique knowledge and insights contained in those maps.

Notes

1. For a discussion of possible prehistoric maps from the Old World, see Catherine Delano Smith, "Prehistoric Maps and the History of Cartography: An Introduction," in J. B. Harley and David Woodward, eds., *The History of Cartography,* vol. 1, *Cartography in Prehistoric, Ancient, and Medieval Europe and the Mediterranean* (Chicago: University of Chicago Press, 1987), 45–49. For two examples of prehistoric shell art from southeastern North America with possible map symbolism, see Philip Phillips and James A. Brown, *Pre-Columbian Shell Engravings from the Craig Mound at Spiro Oklahoma* (Cambridge: Harvard University, Peabody Museum of Archaeology and Ethnology), pt. 1 (1978), pl. 122.3; pt. 2 (1984), pl. 167. One of these is discussed further by Robert H. Lafferty III, "Prehistoric Exchange in the Lower Mississippi Valley," in Timothy G. Baugh and Jonathon E. Ericson, eds., *Prehistoric Exchange Systems in North America* (New York: Plenum Press, 1994), 201–5.

2. Gregory A. Waselkov, "Indian Maps of the Colonial Southeast," in Peter H. Wood, Gregory A. Waselkov, and M. Thomas Hatley, eds., *Powhatan's Mantle: Indians in the Colonial Southeast* (Lincoln: University of Nebraska Press, 1989), 292–343. For a recently recognized original, see Morris S. Arnold, "Eighteenth-Century Arkansas Illustrated," *Arkansas Historical Quarterly* 53, 2 (1994): 119–36; and chapter 8 of this volume.

3. For a discussion of the general topic, see G. Malcolm Lewis, "Indicators of Unacknowledged Assimilations from Amerindian Maps on Euro-American Maps of North America: Some General Principles Arising from a Study of La Vérendrye's Composite Map, 1728–29," *Imago Mundi* 38 (1986): 9–34.

4. One such map must have been the basis for Thomas Kitchin's map, "A New Map of the Cherokee Nation," published in volume 29 of *London Magazine* in February 1760. He appended the attribution line, "Engrav'd from an Indian Draught." Unfortunately, nothing more is known about the original source.

5. In 1773, William Bartram, the Quaker naturalist, accompanied a party of colonial surveyors and Creek and Cherokee Indians that was establishing the boundaries of the New Purchase Cession in the colony of Georgia.

> We were detained at this place [a buffalo lick near the headwaters of Little River] one day, in adjusting and planning the several branches of the survey. A circumstance occurred during this time, which was a remarkable instance of Indian sagacity, and had nearly disconcerted all our plans, and put an end to the business. The surveyor having fixed his compass on the staff, and about to ascertain the course from our place of departure, which was to strike Savanna river at the confluence of a certain river, about seventy miles distance from us: just as he had deter-

mined upon the point, the Indian Chief came up, and observing the course he had fixed upon, spoke, and said it was not right; but that the course to the place was so and so, holding up his hand, and pointing. The surveyor replied, that he himself was certainly right, adding, that that little instrument (pointing to the compass) told him so, which, he said, could not err. The Indian answered, he knew better, and that the little wicked instrument was a liar; and he would not acquiesce in its decisions, since it would wrong the Indians out of their land. This mistake (the surveyor proving to be in the wrong) displeased the Indians; the dispute arose to that height, that the Chief and his party had determined to break up the business, and return the shortest way home, and forbad the surveyors to proceed any farther; however, after some delay, the complaisance and prudent conduct of the Colonel made them change their resolution; the Chief became reconciled, upon condition that the compass should be discarded, and rendered incapable of serving on this business; that the Chief himself should lead the survey; and, moreover, receive an order for a very considerable quantity of goods.

William Bartram, *Travels through North and South Carolina, Georgia, East and West Florida,* ... (Philadelphia: James and Johnson, 1791), 39–40. Also see Gregory A. Waselkov and Kathryn E. Holland Braund, eds., *William Bartram on the Southeastern Indians* (Lincoln: University of Nebraska Press, 1995), 39–40.

6. [Captain Gabriel Archer?], "A relatyon ... written ... by a gent. of yᵉ Colony [1607]," in Philip L. Barbour, ed., *The Jamestown Voyages under the First Charter, 1606–1609* (Cambridge: Cambridge University Press, 1969), 1:82. Cited by Louis De Vorsey, "Amerindian Contributions to the Mapping of North America: A Preliminary View," *Imago Mundi* 30 (1978): 72.

7. Archer, "A relatyon," 84; De Vorsey, "Amerindian Contributions to Mapping," 74.

8. John Smith, "The Generall Historie of Virginia, the Somer Iles, and New England," in *The Complete Works of Captain John Smith (1580–1631)*, edited by Philip L. Barbour (Chapel Hill: University of North Carolina Press, 1986), 2:183. Cited in G. Malcolm Lewis, "The Indigenous Maps and Mapping of North American Indians," *Map Collector* 9 (1979): 25.

9. Waselkov, "Indian Maps of the Colonial Southeast," 306–8. C. F. Feest, "Powhatan's Mantle," in Arthur MacGregor, ed., *Tradescant's Rarities* (Oxford: Clarendon Press, 1983), 130–35.

10. John Smith, "A Map of Virginia," in *The Complete Works,* 1:170–71.

11. Samuel Purchas, *Hakluytus Posthumus, or Purchas His Pilgrimes* (New York: AMS Press, 1965), 18:471. For another version, see John Smith, "A True Relation," in *The Complete Works,* 1:59.

Malcolm Lewis first noticed the significance of this account in Purchas, and published a sketch that attempts to reconstruct the Powhatan's world map. Based on this description, Lewis depicted England as "a pile of sticks" located between the two outer circles of corn kernels. G. Malcolm Lewis, "The Indigenous Maps and Mapping of North American Indians," *Map Collector* 9 (1979): 26. Judging from the account quoted here, though, the sticks seem to have been placed individually, interspersed between groups of kernels, in the two outer circles.

12. Smith, "The Generall Historie," 2:107. According to Barbour, Smith borrowed these two couplets from Lucretius and Horace, as adapted by Martin Fotherby in his work *Atheomastix: Clearing Foure Truthes, Against Atheists* (London: N. Okes, 1622).

13. John Smith, "A Map of Virginia," 1:151. The so-called Velasco Map, a copy of a 1610 English map of Virginia secretly obtained by the Spanish ambassador to England, Don Alonzo de Velasco, also contains an acknowledgment of native informants. "All the blue [i.e., the Great Lakes and the upper tributaries of the Chesapeake that are drawn in blue, rather than black] is done by the relations of the Indians." This note, and other internal evidence, suggests that Smith had a hand in compiling the Velasco Map. Philip L. Barbour, editor, *The Jamestown Voyages* (London: Cambridge University Press, 1969), Hakluyt Society, 2d ser., no. 187, 1:336; William P. Cumming, *The Southeast in Early Maps* (Princeton: Princeton University Press, 1958), 132; C. A. Weslager, *The English on the Delaware, 1610–1682* (New Brunswick, N.J.: Rutgers University Press, 1967), 10–13.

14. John Smith, "The Proceedings of the English Colonie in Virginia," in *The Complete Works,* 1:238. Jeffrey L. Hantman, "Between Powhatan and Quirank: Reconstructing Monacan Culture and History in the Context of Jamestown," *American Anthropologist* 92, 3 (1990): 678–79.

15. David I. Bushnell, Jr., "The Five Monacan Towns in Virginia, 1607," *Smithsonian Miscellaneous Collections* 82, 12 (1930): 3–5; "The Manahoac Tribes in Virginia, 1608," *Smithsonian Miscellaneous Collections* 94, 8 (1935): 8–10.

16. L. Daniel Mouer, "A Review of the Ethnohistory and Archaeology of the Monacans," in J. Mark Wittkofski and Lyle E. Browning, eds., *Piedmont Archaeology: Recent Research and Results* (Richmond: Archeological Society of Virginia, 1983), 21–39.

17. Jeffrey L. Hantman, "Powhatan's Relations with the Piedmont Monacans," in Helen C. Rountree, ed., *Powhatan Foreign Relations, 1500–1722* (Charlottesville: University Press of Virginia, 1993), 96.

18. Smith, "The Generall Historie," 2:116.

19. E. Randolph Turner and Antony F. Opperman, "Archaeological Manifestations of the Virginia Company Period: A Summary of Surviving Powhatan and English Settlements in Tidewater Virginia, circa 1607–1624," in Theodore R. Reinhart and Dennis J. Pogue, eds., *The Archaeology of Seventeenth-Century Virginia* (Richmond: Archeological Society of Virginia, 1993), 72–75.

20. Smith, "The Generall Historie," 2:174.

21. Ibid., 164–65.

22. Smith, *The Complete Works*; George Percy, *Observations Gathered out of "A Discourse of the Plantations of the Southern Colony in Virginia by the English, 1606,"* edited by David B. Quinn (Charlottesville: University Press of Virginia, 1967); Archer, "A relatyon."

23. Ben C. McCary, *John Smith's Map of Virginia, with a Brief Account of Its History* (Charlottesville: University Press of Virginia, 1957), 9; Philip L. Barbour, "The Earliest Reconnaissance of the Chesapeake Bay Area," *Virginia Magazine of History and Biography* 79, 3 (1971): 280–302; William P. Cumming, "Early Maps of the Chesapeake Bay Area: Their Relation to Settlement and Society," in David B. Quinn, ed., *Early Maryland in a Wider World* (Detroit: Wayne State University, 1982), 267.

24. Thomas Jefferson, *Notes on the State of Virginia* (London: John Stockdale, 1785), Query 11; Stephen R. Potter, *Commoners, Tribute, and Chiefs: The Development of Algonquian Culture in the Potomac Valley* (Charlottesville: University Press of Virginia, 1993), 43.

25. Turner and Opperman, "Archaeological Manifestations," 70–78.

26. Lewis R. Binford, "Archaeological and Ethnohistorical Investigations of Cultural Diversity and Progressive Development among Aboriginal Cultures of Coastal Virginia and North Carolina" (Ph.D. diss., University of Michigan, 1964), 85.

27. Potter, *Commoners, Tribute, and Chiefs*, 14–15, 27.

28. Waselkov, "Indian Maps," 299, 332–34.

29. James R. Atkinson, "The Ackia and Ogoula Tchetoka Chickasaw Village Locations in 1736 during the French-Chickasaw War," *Mississippi Archaeology* 20, 1 (1985): 53–72.

30. Ibid., 69.

31. Nine of the ten fields depicted on the Alabama Indian map of Chickasaw towns in 1737 were situated on upland prairie soils; the remaining field clearing was placed on a lower stream terrace. The historic Chickasaw's preference for farming upland prairie soils, which is well substantiated through archaeological survey, differs strikingly from late-prehistoric patterns of field placement in the bottomlands of major rivers, where annual flood-deposited silts replenished soil nutrients and permitted indefinite reuse of fields without the need of artificial fertilizers; Jay K. Johnson, Patricia K. Galloway, and Walter Belokon, "Historic Chickasaw Settlement Patterns in Lee County, Mississippi: A First Approximation," *Mississippi Archaeology* 24, 2 (1989): 45–52; Gregory A. Waselkov, "Changing Strategies of Indian Field Location in the Early Historic Southeast," in Kristen Gremillion, ed., *People, Plants, and Landscape: Case Studies in Paleoethnobotany* (Tuscaloosa: University of Alabama Press, in press).

32. Waselkov, "Indian Maps," 324–29.

33. Gregory A. Waselkov, "Seventeenth-Century Trade in the Colonial Southeast," *Southeastern Archaeology* 8, 2 (1989): 117–33.

34. Mary W. Helms, *Ulysses' Sail: An Ethnographic Odyssey of Power, Knowledge, and Geographical Space* (Princeton, N.J.: Princeton University Press, 1988), 3, 5, 13, 16–17, 119, 121, 132, 163, 266; Mary W. Helms, "Political Lords and Political Ideology in Southeastern Chiefdoms: Comments and Observations," in Alex W. Barker and Timothy R. Pauketat, eds., *Lords of the Southeast: Social Inequality and the Native Elites of Southeastern North America* (Washington, D.C.: American Anthropological Association, 1992), 187–88; Neal Salisbury, *Manitou and Providence: Indians, Europeans, and the Making of New England, 1500–1643* (Oxford: Oxford University Press, 1982), 49; Steadman Upham, *Polities and Power: An Economic and Political History of the Western Pueblo* (New York: Academic Press, 1982), 121–23.

35. Helen Hornbeck Tanner, "The Land and Water Communication Systems of the Southeastern Indians," in Wood, Waselkov, and Hatley, eds., *Powhatan's Mantle,* 6–20.

36. Waselkov, "Indian Maps," 303.

37. "Ancient Geographic Corridors: Conduits of Communication, Exploration, Conquest, Settlement, and Resource Distribution," Symposium Chaired by Carol Morrow and Mary McCorvie, Southeastern Archaeological Conference, November 10, 1994, Lexington, Kentucky.

10

Debriefing Explorers: Amerindian Information in the Delisles' Mapping of the Southeast

Patricia Galloway

Detailed European mapping of the lower Mississippi Valley and its environs east of the river did not begin until the French began to lay claim to the region in the late seventeenth century. It was not completed until the hydrography of the great river was finally understood at the beginning of the eighteenth century. This achievement was primarily that of the Delisle cartographic house, which bent its talents to the debriefing of the French explorers and diplomats who were finally learning about the interior by virtue of their relations with the Indians who controlled it.[1]

Indian maps and geographic information were so important a factor in the European development of knowledge of the North American continent that it is impossible to gain an adequate idea of the process without taking them into account.[2] The problem, of course, is that so very few Indian maps survive physically, even when documentary evidence describes specific instances when cartographic or geographic evidence was communicated to explorers by Indians. Furthermore, although some native cartographic records do survive and can be classified and analyzed in a limited way, it is very difficult if not impossible for us moderns, for whom European mapping conventions have established a hegemonic monopoly, to know what they meant as texts to their original users—how their conventions were "read" and mentally expanded.[3] Still it is vital for us to attempt to distinguish the Amerindian layers of the palimpsest of early European mapping of North America. By doing so we can begin to understand how Europeans manipulated and altered the reality of contact-period native geography to incorporate it into their mapping practice, often merely by making a new cartographic reality.

Several excellent studies of Amerindian mapping have helped to define the field of knowledge, so there is coming to be a fairly well known inventory of Indian cartographic documents, and important studies of Southeastern Amerindian maps have appeared.[4] For the Southeastern region of North America, a native culture area that included everything south of the Ohio valley and east of the Great Plains, the surviving evidence suggests that at least two kinds of maps were drawn, used, and preserved by Indians: what social scientists might term "sociogram"-like productions that render simultaneously the geography of physical and social space; and what I would call "event transcriptions" that render specific activities with some degree of geographical or social reference.[5] I believe it can be demonstrated that both kinds of information were incorporated into European maps, and that each of them made specific contributions.

The rendering of intangible aspects of the social environment along with topography in Amerindian maps seems less strange when we consider the degree to which the physical environment was (and indeed still is in many cases) continuous with the social environment for Amerindians. The tendency to find such a portrayal "naive" is a reflection of Euro-American ethnocentrism. The modern maps with which most people are most familiar take for granted that in some sense they are accurate representations of what will actually be found on the ground, yet we know perfectly well that the earth is round, so any map that portrays its surface naturalistically must use some conventional projection—and introduce some distortion—in order to be placed on a flat sheet of paper. We are familiar, too, with "unnatural" portrayals of our world, ranging from false-color infrared photographs of the Earth's surface showing vegetation to modern thematic maps that manipulate space to reflect—for example—travel time. The difference is that we have the key to this mapping convention, because it is contemporary, while to some degree the key to the equally thematic early Indian maps is either lost or is difficult to recover.

Amerindian "Sociograms"

The "sociogram"-style Southeastern maps have been discussed in useful detail by Gregory Waselkov, and I will summarize here.[6] The best-known examples are two maps on deerskins (now represented by paper copies) collected by governor Francis Nicholson of Virginia and sent to England. The first, apparently painted on deerskin by a Catawba Indian in 1721 (fig. 10.1), is especially striking because it shows the European space of Charles Towne as a rectilinear grid and Virginia as a rectangle, while it portrays the native interior as a network of circles, each portraying the "fire" of a native polity, a term by which Southeastern Indians referred to a community bound by political, genealogical, and ceremonial ties. The Chickasaw-drafted map of 1723 (fig. 10.2) is made up of a circle network alone, connected by both overland trails and recognizable rivers (with the Mississippi system portrayed in compressed form down one side). By contrast with the first, and perhaps with a different purpose, it incorporates the English and the French as similar "fire" circles, suggesting that the Chickasaws had begun to understand the European communities as similar to their own.

The term "sociogram" refers to a graphical convention sociologists have used to diagram social networks, where persons or groups are represented as nodes and the social connections between them make up the network connections.[7] I think it is applicable to these maps

DEBRIEFING EXPLORERS

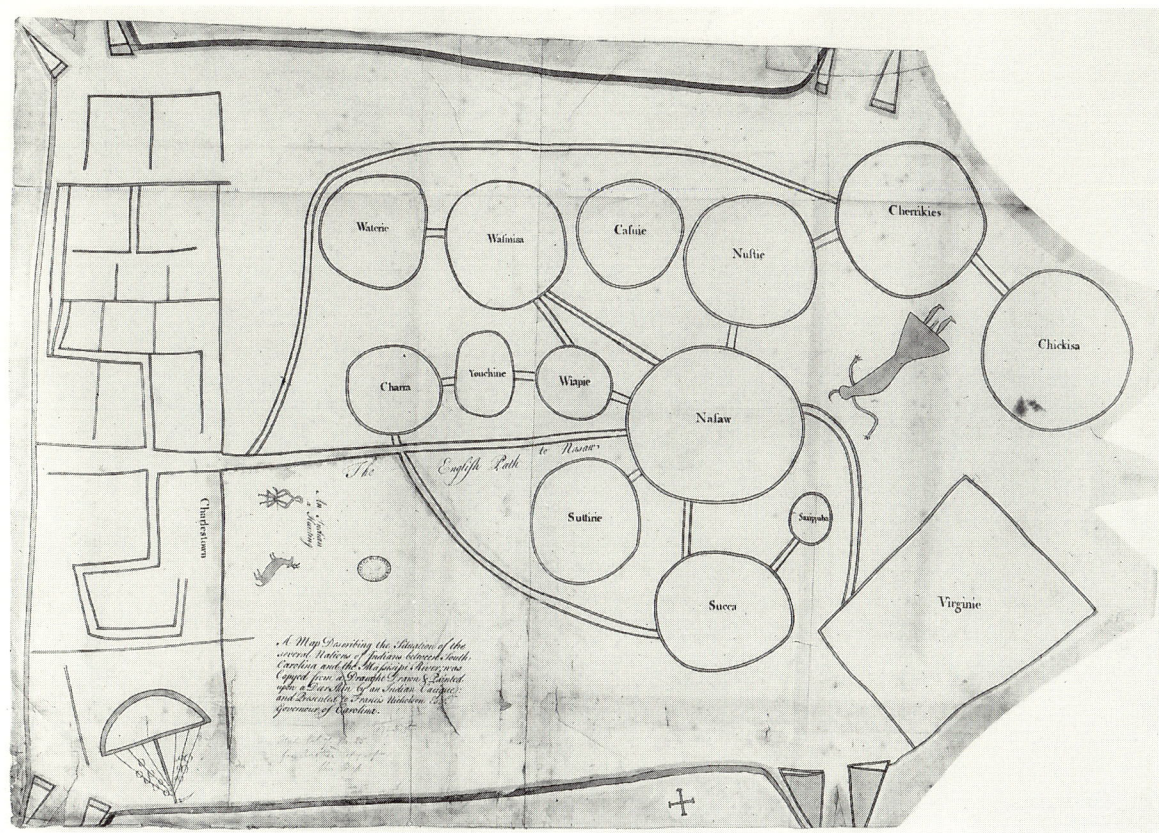

10.1 Catawba Indian map collected by Francis Nicholson, ca. 1721. Original in the Public Record Office, Kew, London. CO 700/6 (1).

because even to external observers it is obvious that they portray the social space their creators inhabited, although they make reference in a roughly relativistic way to topographical space. The connections between groups shown on these maps give overland paths and water routes equal status as connections, omitting such routes if they are not important to the maker. Thus these maps are not strictly speaking as "objective" as sociograms are intended to be, since the latter are usually constructed by observers of, rather than participants in, the groups they portray. Indeed Waselkov has shown that judged against what we know of the population and location of eighteenth-century Indian groups, they are clearly ethnocentric, in that each maker places his own polity at the center of the map and represents it by a disproportionally large symbol (much as Europeans map north as "up"). Yet the kind of political information such maps portray was crucial to European understanding of the balance of power in the native interior, and it would be odd if Euro-

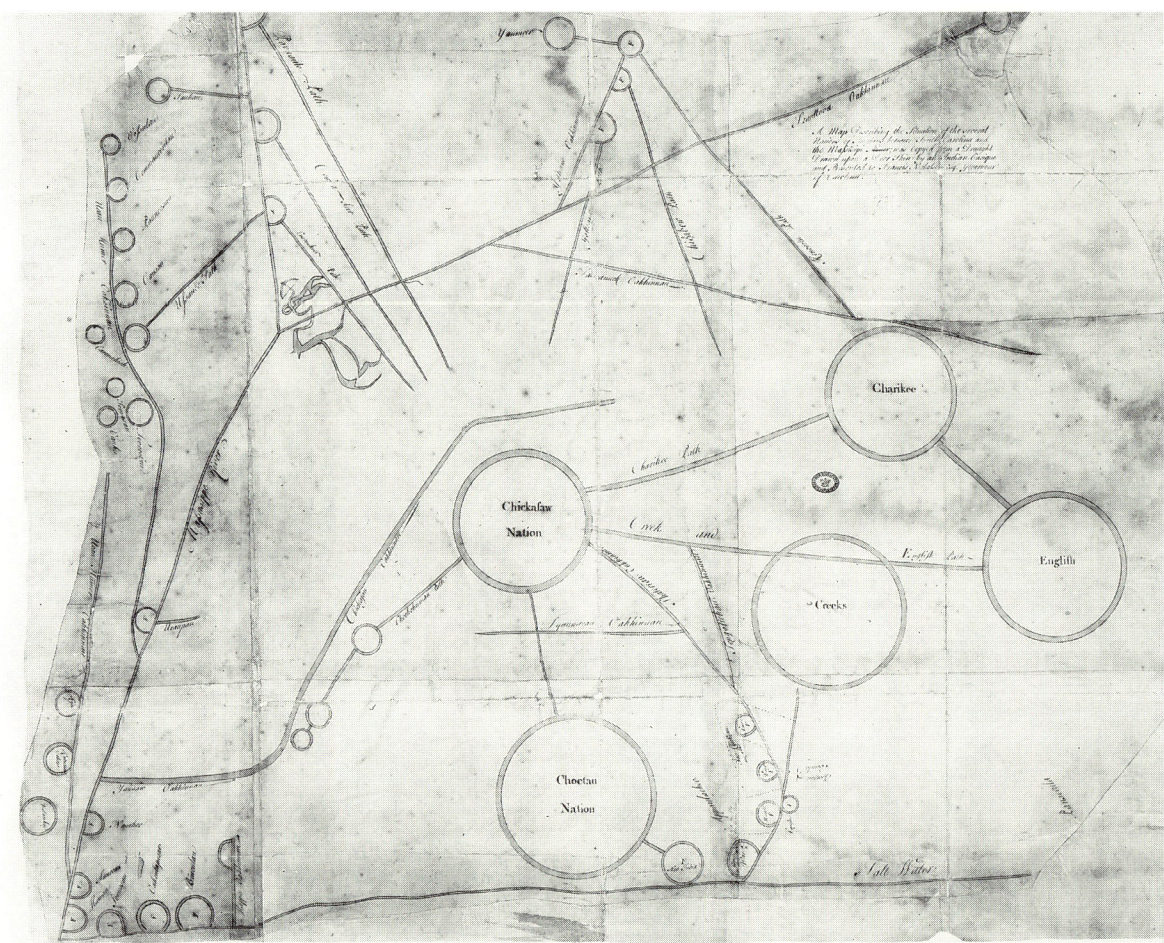

10.2 Chickasaw Indian map collected by Francis Nicholson, ca. 1723. Original in the Public Record Office, Kew, London. CO 700/6 (2).

pean mapmakers had not made use of the information.

That the sociogram style was a viable and ongoing native tradition is suggested by the appearance of two more examples from the same region fifteen years later. These two maps were copied by the French engineer Alexandre de Batz in 1737 on the basis of maps created by the Chickasaw war chief Mingo Ouma and the Alabama Indian Captain of Pacana. In the wake of the failure of the French war on the Chickasaws in 1736, the neutral Alabama chief Captain of Pacana acted as a go-between to trade a captured Chickasaw chief for a Frenchman held by the Chickasaws.[8] He dealt with Mingo Ouma during his visit, and upon his return he supplied the information for two maps known to us through versions created by Alexandre de Batz. One of them is a map of the Chickasaws' "friends and enemies" that constitutes Mingo Ouma's view of the political environment of the interior Southeast; the original was drawn by Mingo Ouma or by the Captain of Pacana at Mingo Ouma's instruction (see fig. 9.2). It is much like those collected by Nicholson, although the emphasis is much more single-mindedly upon the nature of the relationships between the Chickasaws and other tribes. The second map, drawn originally by the Captain of Pacana to summarize his observations in Chickasaw country, is drawn in the same style, but it is a map of the Chickasaw villages only, showing connecting paths, streams, and fields (see fig. 9.4). Contemporary French maps and archaeological studies have shown that it is topographically relatively accurate.[9] Since this last map is basically a kind of spy's report and was elicited from the Captain of Pacana for the purpose of obtaining accurate intelligence for military purposes, it is likely that more than any of the others in this style, it was influenced by European map conventions.

Event Transcriptions

What I am referring to as "event transcriptions" are only map*like* in the European sense,[10] but I am discussing them here because I believe they made a signal contribution to European mapping conventions. In fact they are related to pictographic representations of events like those recorded by Bernard Romans in the 1770s (fig. 10.3). Romans describes these "paintings" (he is not specific about the medium in which they were executed except to hint that they may have been marked onto

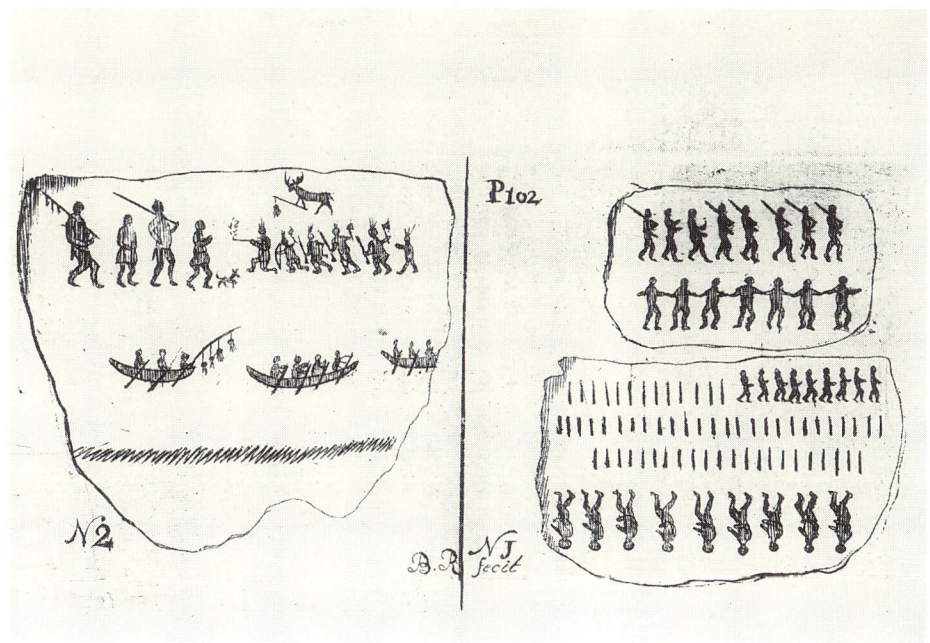

10.3 Creek (left) and Choctaw (right) "paintings" reproduced by Bernard Romans, ca. 1770. Courtesy of the Edward E. Ayer Collection, the Newberry Library.

trees; his drawings would suggest smooth bark, perhaps a peeled patch of tree trunk):

> [T]he first [painting is] Chactaw, and means that an expedition by seventy men, led by seven principal warriors, and eight of inferior rank, had in an action killed nine of their enemies, of which they brought the scalps, and that the place where it was marked was the first publick place in their territories where they arrived with the scalps.
>
> The second is a painting in the Creek taste, it means, that ten of the nation of the Stag family came in three canoes into their enemies country, that six of the party near this place, which was at *Oopah Ullah*, a brook so called on the road to the Chactaws, had met two men, and two women with a dog, that they lay in ambush for them, killed them, and that they all went home with the four scalps; the scalp in the stag's foot implies the honor of the action to the whole family.[11]

In the Choctaw version, the "seven principal warriors" are portrayed by seven figures shown frontally and holding hands; the "eight of inferior rank" are shown carrying weapons. The nine enemies killed are shown upside down, and the war party as a whole is shown simply as sixty-one marks and nine figures carrying scalps. The Creek-style drawing more closely portrays the event to which it refers, showing both the ten stag-clan warriors traveling in their three canoes and the actual altercation of the six stag-clan warriors with their four enemies.

The crucial point in both cases is the repetition of conventionalized elements to represent numbers. When we turn to a striking buffalo skin painting discussed at length by Morris Arnold in chapter 8 of this volume, apparently from a Quapaw Indian source and now in the Musée de l'Homme in Paris (see fig. 8.1),[12] we see these elements transformed into a map*like* setting, with the three major village names placed next to multiple dwelling and person symbols. As was the case with the Nicholson deerskin maps, the skin itself dictates something of the layout of the image. On one side is portrayed what seems to be a specific event, showing two opposing groups shooting at one another: in one group two men run away toward a palisade or group of buildings, while in the other, one of the attackers is armed with a bow. On the other side of the skin is what appears to be a conventionalized portrayal of a tribal group, consisting of nine figures alternating men and women, all holding hands (compare the seven leaders in the Choctaw pictograph cited above); below them appears the Illinois name given to the Quapaw, Ackansas. They are pictured above three groups of buildings labeled with the names of Quapaw villages, Ouzovtovovi (Osotouy; 6 buildings), Tovarimon (Touriman; 5 buildings), and Ovo8appi (Ukakhpakhti?; 8 buildings plus one larger building, perhaps the council house mentioned for that village in other sources[13]). Quapaw dwellings of the eighteenth century were multifamily longhouses with arched roofs,[14] and Father Vivier's estimation of Quapaw population in 1750 was 1,400,[15] so it would seem impossible for the symbols to be directly representational, though they may be proportionally representational. At the tail end of the skin is found a group of what appear to be European buildings, perhaps the rebuilt French post of the 1730s and 1740s.[16] The skin certainly must date after 1727, when a fourth major village Tongigua moved west of the Mississippi and merged with Touriman,[17] and it certainly reveals some French influence, if only in the portrayal of European buildings and in the Roman script of the village labels (although the presence of the "8" suggests an early date, since it was used to render the "ou" sound in Canadian missionary transcriptions of Indian languages but was not much used in the Louisiana colony, of which the Arkansas post was a part).[18] The point here is, as with the "heiroglyphicks" recorded by Romans, again the use of multiple conventional symbols to represent numbers.

Early Explorers and the Influence of Indian Maps

At the time of the exploration and colonization of the Americas, European cartographers had themselves developed a fairly conventionalized set of symbols for representing human settlement. The central fact of this symbology was the use of a single symbol for a population agglomeration constituting what Europeans perceived as a political unit, although the symbol could be varied in size and elaborateness to communicate the relative size of the population or its political importance[19] (fig. 10.4). When Europeans came to map the settlements of Amerindians, however, they were often without directly observed data from European explorers. Turning to less well understood Amerindian data in default of better knowledge, they also sometimes adopted a new symbology. Where cartographers used information derived from sociogram-style maps, with their single circle for a tribal group, the superficial similarity with European mapping conventions would obviously make this use difficult to detect—except perhaps by virtue of the portrayal of paths or of identifiably distorted spatial relationships—unless the actual native symbology was used also. Where the use of multiple symbols was adopted to

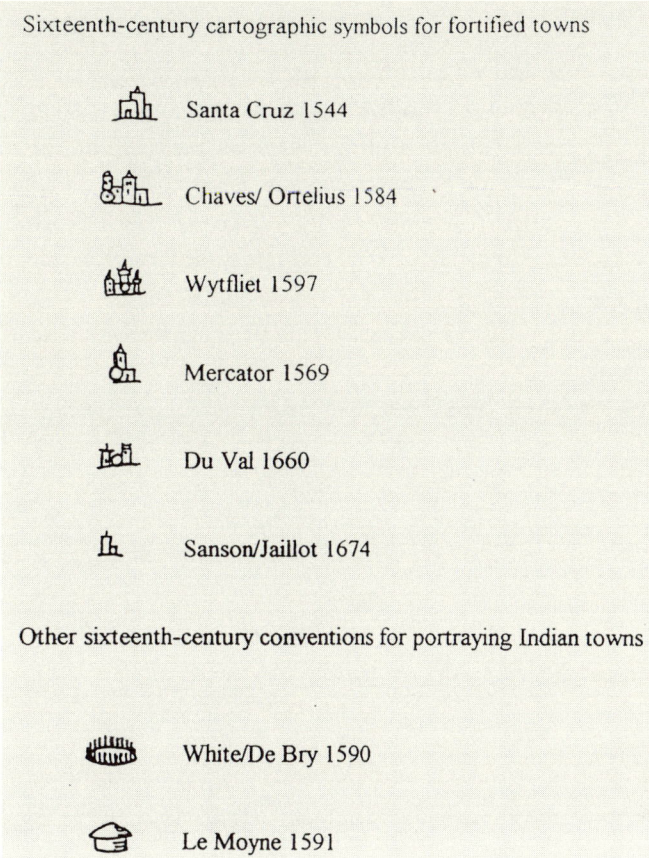

10.4 Some examples of Indian-house symbols from early maps of the Americas. By the author.

indicate population size, on the other hand, I think it is reasonable to suggest that cartographers were using direct Amerindian evidence from event-transcription maps, with their multiple-symbol representations.

In the Southeast the process of change in representational style is dramatically clear, since the sixteenth-century maps made on the basis of the accounts of the interior expedition of Hernando de Soto clung to the European single-symbol representational conventions, however unsuitable they may have been.[20] After the hiatus of a hundred and thirty years, when Marquette and Jolliet descended the Mississippi and returned to report their findings, we see a striking change in the style of representation. First, the stylized fortified-town symbols have been done away with in favor of a simple triangle (used on European maps of the period to indicate a peasant dwelling).[21] The Marquette map does not directly represent numbers using this symbol, but does use the symbol in arbitrary small multiples (two and three) to portray native groups; only in a few cases are these individual villages, most of which Marquette in any case did not see.[22] Jolliet's map goes further, in several cases using multiple symbols to represent number of villages (fig. 10.5). Interestingly, Jolliet's multiples appear precisely where the explorers did *not* go, and where native

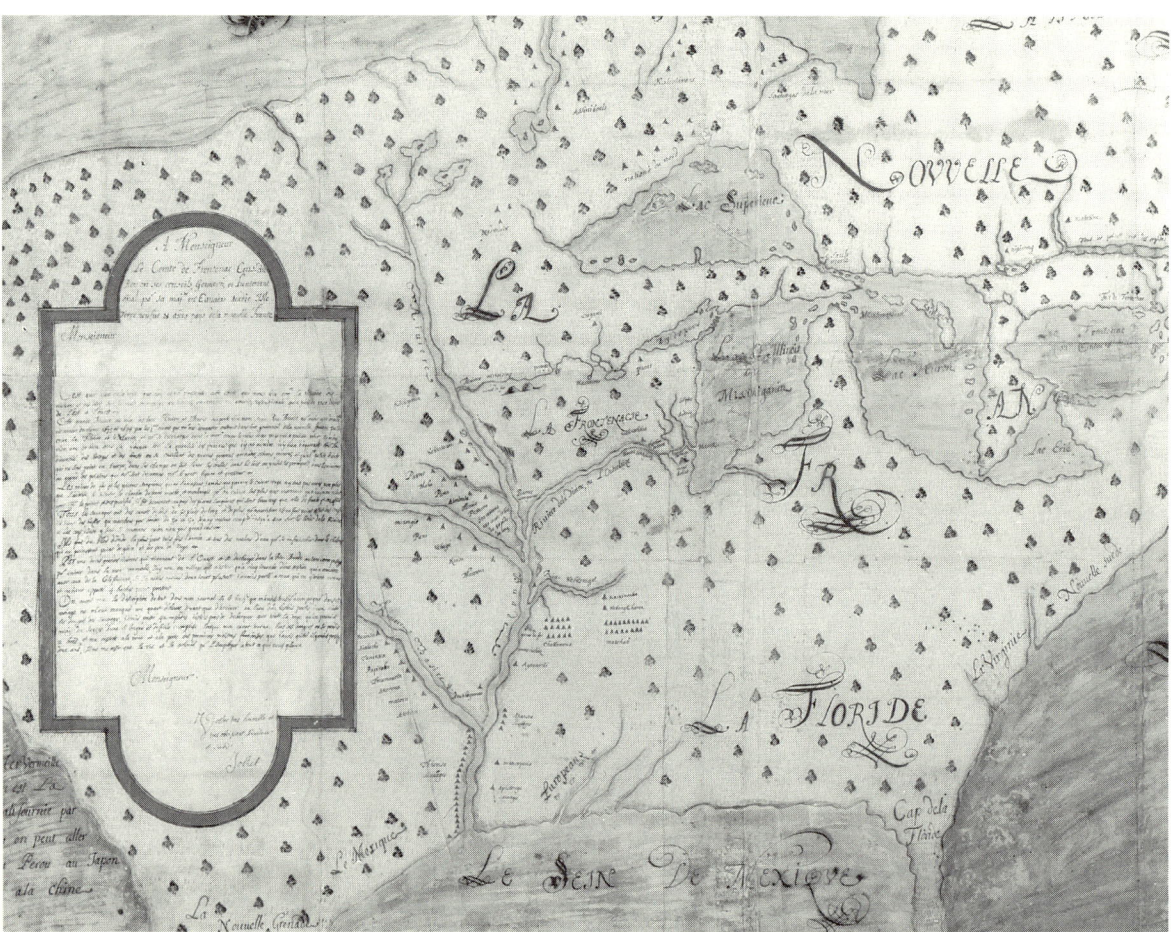

Fig. 10.5. Detail of lower Mississippi valley, Louis Jolliet, *Nouvelle decouverte de plusieurs nations dans la Nouvelle France* (ca. 1674). Courtesy of the John Carter Brown Library, at Brown University.

reports or maps must have been the only sources of information.

The same kind of information and the new convention was soon adopted by the Italian cartographer Vicenzo Coronelli in one of his manuscript sketches for his well-known maps and globe gores.[23] Where Coronelli uses multiple symbols, most obviously in the case of the Taensas, an accompanying legend confirms that each symbol is meant to represent a single village. This map is particularly interesting because it records a specific item of information available nowhere else and probably obtained from a specific native informant on a specific occasion. While searching for a lost man at the bluffs where Memphis now stands, some of La Salle's men captured two Chickasaw Indians, who guided a French party overland to visit their people (until La Salle decided the trip was too long) and one of whom eventually accompanied them on much of their journey downstream.[24] Although no other known document records the information these men must have given about their people, the Coronelli sketch provides the names of eight villages, of which later evidence confirms some six.[25] Coronelli's links to the French court were such that he was doubtless able to interview La Salle himself or others who had accompanied him.

The importance of Indian informants to the activities of French explorers in the interior remained a constant of French policy as the Louisiana colony was founded and settled. Readers of Iberville's accounts of his early visits to Southeastern North America will be well aware of his constant attention to the information of Indian informants. One Indian, a Taensa guide who accompanied him up the Mississippi in 1700, provided particularly rich information in the way of tribal names and locations along rivers, which Iberville recorded in his log.[26] This Taensa guide also apparently drew maps that are not preserved—except perhaps in a Delisle sketch of 1700. Like the Coronelli sketch, this map has written inscriptions that reinforce the meaning of the multiple-symbol convention (in this case, tiny gable-end rectangular buildings with flags flying

10.6 Detail of lower Mississippi valley, Guillaume Delisle, sketch of the Mississippi-Alabama region, ca. 1700. Courtesy of the Edward E. Ayer Collection, the Newberry Library.

from their roofs are used) by literally stating that they represent villages (fig. 10.6). At the same time, in many cases Soto-era polity names not reaffirmed by Amerindian evidence by 1700 are represented by single symbols.

The Choctaw Origin Problem

I became interested in the issue of Amerindian evidence in European maps because I had to solve a problem. In working on the early history of the Choctaw Indians I soon came to realize that the ethnogenesis of the historic Choctaw tribe had taken place not in prehistory but since European contact.[27] When I began the work of trying to uncover the history of this process, there was very little archaeological evidence available to elucidate the origins of the Choctaws. Scattered eighteenth-century observations by Europeans, read with a suspicious eye, seemed to suggest that they were multiethnic. If so, they might have in fact been a confederation of refugees from the upheavals of demographic collapse elsewhere than in the region of east-central Mississippi now defined as their homeland (fig. 10.7).

Although all of this took place when there were no Europeans on the scene, I hoped that some historical evidence might turn up on Euro-

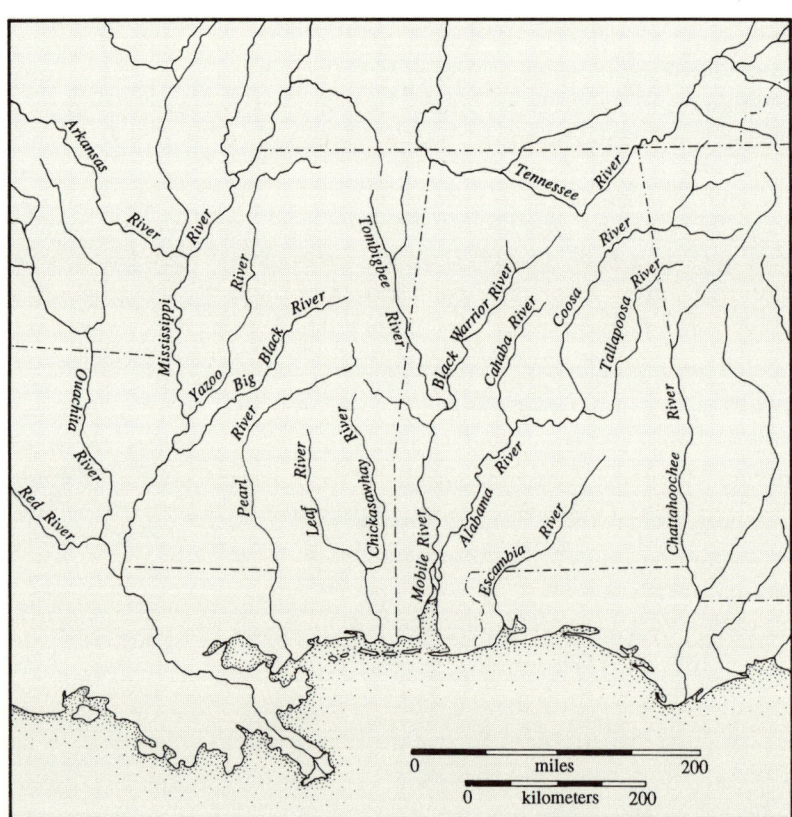

10.7 Reference map of Mississippi-Alabama region. By Julie Smith.

pean maps. With that in mind I undertook a close study of the large series of sketches and maps developed by the Delisle cartographic house in its remarkable project of the mapping of North America. The house, founded by the scholarly Claude Delisle and led by his eldest son Guillaume, was informed by the same scientific spirit that led to the mapping of France by Jean Dominique Cassini and his sons; indeed Cassini had instructed Guillaume Delisle.[28] Jean Delanglez has argued that the early work of the house, especially through 1703, was mostly that of Claude, and that his hand was also involved in the work—to an increasingly diminished degree—until his death in 1720.[29] But clearly there is a unity of critical spirit in all of the work of the house, leading to the recognition of Guillaume as *Premier Géographe du Roi* in 1718 and of his younger brother Joseph-Nicolas as Russian royal astronomer. The Delisle house dominated the mapping of North America from the end of the seventeenth century through the first quarter of the eighteenth, and its maps were copied by others for years afterward.[30] It is possible to reconstruct their practice and to recognize that critical spirit because their notes, sketches, proofs, and other materials were collected by Joseph-Nicolas and at least until recently were compactly preserved in the French archives in Paris.[31]

The first portrayal on a European map of the people we now call the Choctaws was one we have just seen: Delisle's 1700 sketch made on the basis of 1699 and 1700 data supplied to him by Iberville. Although Iberville's data for the tribes of the Mississippi River region must have come in turn from the redoubtable Taensa guide, the source of this information on the Choctaws—portraying them using forty-four house symbols located west of the Tombigbee River—is not as obvious. Iberville had met with many Indian leaders on the Gulf Coast, and it is likely that this information was obtained in that way. Iberville discussed the Choctaws with the Pascagoulas, their neighbors to the south, during his second visit to the Gulf Coast in April of 1700. He concluded that the Choctaws had "more than fifty villages" and "more than six thousand men."[32] On May 16–17 of 1700 Iberville actually spoke personally with two Choctaws aboard his ship, but did not record anything specific about population at that time. Since the Choctaws were in fact located where Delisle places them on the 1700 sketch,[33] there was no reason to become particularly excited about this map beyond remarking that it reflects a combination of new and old information and that new Indian information is often marked (where multiple villages were observed) by multiple symbols.

But in 1701 the Delisles produced at least three maps that discarded this information and portrayed the Choctaws in forty-four villages spread out along what can only be the Black Warrior River, an eastern tributary of the Tombigbee. It is difficult to decide which of these three 1701 maps is first, especially since one of them is now apparently lost,[34] but a close examination of the photostatic copies held in the Karpinski Collection by the Newberry Library reveals that there are indeed three separate maps that reflect this information. The first printed map, *Carte des environs du Mississipi par G. De l'Isle Geogr.* (fig. 10.8),[35] is clearly very closely related to the second map of the same title, except that the author's calling is spelled out as "Geographe."[36] A careful examination of both of these maps, however, reveals that the second has other affinities besides Choctaws on the Black Warrior with the small-scale summary manuscript map ASH 140–4: its portrayal of the Apalachicola river system is nearly identical with that map.[37]

I suspect that we are dealing here with a family of maps that never went into production or distribution because the information they bore was so quickly superseded—as the example of the Apalachicola River system shows.

What does the information mean? Was it just wrong information, using the multiple-symbol style simply because the Delisles had already ascertained the number of villages from Iberville? Or could it have reflected new data obtained from Indian informants, historical data perhaps derived from an "event transcription" or cartographic history, to use Elizabeth Boone's useful term, of the migration of the eastern Choctaws, rendering in that form the story told in their

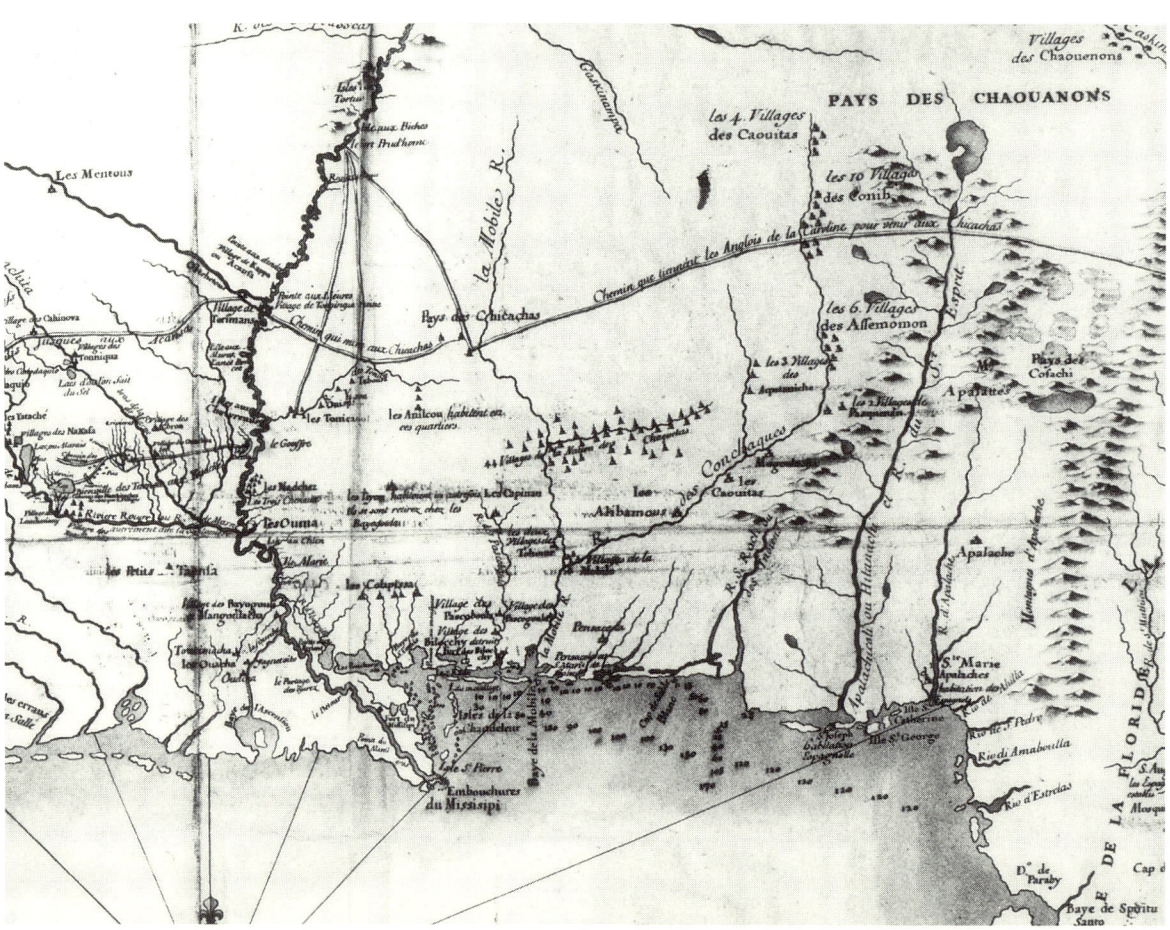

10.8 Detail, Guillaume Delisle, *Carte des environs du Mississipi par G. de l'Isle Geographe* (1701). Carte no. 4 of Recueil 69, Service historique de la Marine à Vincennes.

migration legend? If it can be so construed, and I believe the use of multiple symbols is very suggestive of genuine Indian evidence here, it would be a powerful assistance in confirming my suspicions, based upon similarities across the region in pottery decoration, mortuary ritual, and origin myths, that the eastern Choctaws' heritage stretched back to the impressive Moundville polity on the Black Warrior River.[38]

There were plenty of opportunities for such new "old" information to be gathered. One that certainly did generate information used by the Delisles was the Charles Levasseur expedition into the interior in 1700. Preserved in fact in the Delisle Papers, the journal records Levasseur's journey through the Mobile River delta to the confluence of the Alabama and Tombigbee rivers and recites a list of all the thirty-six nations living on the Alabama River, although Levasseur did not visit them.[39] Bienville himself was sent by his brother Iberville in 1702 to investigate an important sacred site in the Mobile River delta, and he must have had plenty of opportunity to discuss with his guide the political geography of the region.[40] That the Delisles had access to native maps now lost is suggested by the sequence of Delisle maps leading up to that of 1703. This map sequence shows a river system that starts as the Apalachicola, then marches west to become the Chattahoochee; but more significantly, this river has the linear-river-plus-"tadpole"-lake configuration that Lewis has shown is a strong indicator of Amerindian source.[41]

Later British evidence that adds strength to the connection of the eastern Choctaws with the Black Warrior River is recorded on maps as well as in texts. Two convenient examples are exemplary of others. Thomas Nairne's 1711 inset to Edward Crisp's map of North America labels the Black Warrior as "Pedegoe R.,"[42] while Henry Popple's 1733 map of the region terms it "Tascalasa or Pedoge R."[43] This terminology comes very close to reflecting one native name given to the eastern Choctaw people, "Ayepategoula."[44]

In 1702 the Delisles drafted a map—one closely related to the last, small-scale 1701 draft map—in which they corrected the "erroneous" information they had been using and returned the Choctaws to a location west of the Tombigbee at the same time that they began the rationalization of the Apalachicola river system.[45] By this time the Delisles knew that the people whom the French were calling Choctaws were living west of the Tombigbee River (now called Riviere de la Mobille), for they had the benefit of Henri de Tonti's first-person account of his lengthy journey among the Choctaws and Chickasaws, since the letters reporting on it were preserved among their papers.[46] They had also presumably debriefed Iberville on his important meeting with Chickasaw and Choctaw representatives brought back by Tonti to the newly founded town of Mobile; after meeting with four Choctaw chiefs, Iberville had written in his log that they had 1,090 houses, 3,800–4,000 men, and were divided into three "villages" or divisions.[47] The map indeed shows three fortified-town symbols for the Choctaws. The Delisles were thus able to adjust their map in the continuing effort to arrive at creating the completely contemporary representation of the political geography of North America that was so vital to the interests of France.

As the Delisles brought their work to a close they began to consolidate their knowledge of

10.9 Detail of lower Mississippi valley, Guillaume Delisle, Carte du Mexique et de la Floride (1703). Courtesy of the Mississippi Department of Archives and History.

native groups into fewer and fewer symbols, progressing toward a symbology that would again represent at least some native groups in the same way that European towns were represented (fig. 10.9), precisely because their maps had come to represent what they believed to be a well-understood geography with *political* as well as topographical aspects. It may also be suggested that this development reflected as well the assimilation of the synchronic "sociogram" representation that the Indians themselves used for portraying synchronic political geography. That the French did not always portray native groups and knowledge about them in this way is obvious when we consider maps of the region made at times of crises in Indian relations, when both French and British cartographers attempted painstaking portrayals of the kind of "ground truth" about Amerindian population distributions that would be of use to military strategists. The Baron de Crenay's famous map of 1733 attempted the detailed representation of precise locations of villages across the region after the violent Natchez Revolt of 1729 made the French aware of the necessity for such precise data.[48] A map detailing with even greater precision the Choctaw villages and the streams on which they were located, reflecting even the dispersed Choctaw settlement pattern around a rather amorphous "village" center, was made by the engineer Ignace Broutin around 1746–47 at the outbreak of the Choctaw civil war.[49] Similar efforts were made by the British as they came to take possession of the Louisiana colony east of the Mississippi in the 1760s and 1770s, when their interests turned to the acquisition of lands from the Indians and the registration of such acquisitions with legal precision. This attitude is well exemplified in the Stuart-Purcell map of 1775, which portrays an exhaustive inventory of Indian towns of the Southeast.[50]

It is interesting that these two European map forms, the political map representing polities and the analytical map portraying the spatial reflection of a historical moment, harmonized so well with the Amerindian "sociogram" and "event transcription" forms we have seen. It is only surprising if we assume that Amerindians were uncivilized and savage, for the human purposes being served are characteristic of complex societies, whether European or Amerindian.

Notes

1. The process is clearly revealed in the records they kept of their work; see for example Jean Delanglez, "The Sources of the Delisle Map of America, 1703," *Mid-America* 25, 4 (1943): 275–98.

2. See Elizabeth Boone, "Maps of Territory, History, and Community in Aztec Mexico," this volume, for a view of how this worked elsewhere in the Americas.

3. See Peter Nabokov, "Orientations from Their Side," this volume, for some hint of how rich such knowledge might possibly be.

4. Louis De Vorsey, Jr., "Amerindian Contributions to the Mapping of North America: A Preliminary View," *Imago Mundi* 30 (1978): 71–78; and "Silent Witnesses: Native American Maps," *Georgia Review* 46 (1992): 709–26; Gregory Waselkov, "Indian Maps of the Colonial Southeast," in Peter H. Wood, Gregory A. Waselkov, and M. Thomas Hatley, eds., *Powhatan's Mantle: Indians in the Colonial Southeast* (Lincoln: University of Nebraska Press, 1989), 292–343.

5. I would guess that these classes of documents may be compared very usefully with both types of Aztec maps, "organizers," and "records of movement" or "cartographic histories," discussed by Elizabeth Boone.

6. Waselkov, "Indian Maps."

7. See Mary L. Northway, *A Primer of Sociometry*, 2d ed. (Toronto: University of Toronto Press, 1967).

8. An account of the Alabama chief's trip is in Diron d'Artaguette to Maurepas, October 24, 1737, Archives des Colonies, ser. C13A, 22:233–43v, translation in Dunbar Rowland, Albert G. Sanders, and Patricia Galloway, trans. and eds., *Mississippi Provincial Archives: French Dominion*, vol. 4 (Baton Rouge: Louisiana State Univer-

sity Press, 1984), 147–51.

9. See Jesse D. Jennings, "Chickasaw and Earlier Indian Cultures of Northeast Mississippi," *Journal of Mississippi History* 3 (1941): 155–226; and James Atkinson, "The Ackia and Ougoula Tchetoka Chickasaw Village Locations in 1736 during the French-Chickasaw War," *Mississippi Archaeology* 20, 1 (1985): 53–72.

10. G. Malcolm Lewis, "Indicators of Unacknowledged Assimilations from Amerindian Maps on Euro-American Maps of North America: Some General Principles Arising from a Study of La Vérendrye's Composite Map, 1728–29," *Imago Mundi* 38 (1986): 9–34.

11. Bernard Romans, *A Concise Natural History of East and West Florida* (New Orleans: Pelican, 1961; transcription of 1772 edition), 71.

12. Musée de l'Homme MH.34.33.7. This painted skin is illustrated as no. 22 in *Robes of Splendor: Native American Painted Buffalo Hides* (New York: New Press, 1993). For a recent analysis, see Morris S. Arnold, "Eighteenth-Century Arkansas Illustrated," *Arkansas Historical Quarterly* 53 (1994): 119–36.

13. Arnold, "Arkansas Illustrated," 126.

14. George Sabo III, *Paths of Our Children*, Arkansas Archeological Survey Popular Series no. 3 (Fayetteville: Arkansas Archeological Survey, 1992), 33–34.

15. John R. Swanton, *Indians of the Southeastern United States*, Bureau of American Ethnology Bulletin 137 (Washington: Smithsonian Institution Press, 1979; reprint of 1946 edition), 176.

16. Arnold, "Arkansas Illustrated," 123–25.

17. Ibid., 120 n. 3.

18. Arnold's argument for significant French influence or perhaps a *métis* provenience is I think not justified in the context of the elements shared with other "event transcriptions" in the southeastern region and with the conventions of the buffalo robe genre.

19. François de Dainville, *Le langage des géographes: Termes, signes, couleurs des cartes anciennes, 1500–1800* (Paris: A. and J. Picard, 1964), figs. 27, 29, 30, 31, and 40.

20. For an example, see Heironymus Chiaves's map *La Florida* printed by Ortelius in 1584.

21. Dainville, *Langage*, 228, fig. 30.

22. Father Marquette's sketch map may be found in the collections of the Collège Sainte-Marie, Montreal.

23. This sketch of the Mississippi, which is untitled, is in the Map Collection, Yale University Library.

24. John Stubbs, "The Chickasaw Contact with the La Salle Expedition in 1682," in Patricia Galloway, ed., *La Salle and His Legacy: Frenchmen and Indians in the Lower Mississippi Valley* (Jackson: University Press of Mississippi, 1982), 41–48.

25. The village names are Fabarchaour, Malata, Achehophani, Torchinaske, Chichafalara, Onichapatafa, Pakaha, Chikouolika; at least three of these village names were still in use as late as the Stuart-Purcell map of 1775 (Malata = Mellataw, Achehophoni = Ashuk Hooma, Chikuouolika = Chicaluia), and others are confirmed between the two dates (e.g., Fabarchaour = Falatchao on the Pacana/de Batz map of 1737, fig. 9.4 of this volume).

26. See Pierre Le Moyne, sieur d'Iberville, *Iberville's Gulf Journals*, translated by Richebourg G. McWilliams (Tuscaloosa: University of Alabama Press, 1981). Iberville's logs twice refer to maps drawn by Indians: 60, 71.

27. Patricia Galloway, *Choctaw Genesis 1500–1700* (Lincoln: University of Nebraska Press, 1995).

28. For an outline of the work of the Delisles see Lloyd A. Brown, *The Story of Maps* (New York: Dover, 1979; reprint of 1949 edition), 241–43.

29. Delanglez, "Sources." Since it is not really possible without detailed study of all the notes and sketches to assign authorship precisely, I will simply speak here of "Delisle" maps, and by that mean a collaboration between Claude and Guillaume, although a few of the maps are attributed officially to the son.

30. R. V. Tooley, "French Mapping of the Americas: The Delisle, Buache, Dezauche Succession, 1700–1830," *Map Collector's Circle* 4 (1967): 7.

31. See A. Isnard, "Joseph-Nicolas Delisle: Sa biographie et sa collection de cartes géographiques à la Bibliothèque Nationale," *Bulletin du Comité des Travaux Historiques et Scientifiques, section de géographie* 30 (1915):

34–164; information on the Delisle Papers on pp. 59–62. I have not managed to unravel the logic behind the breakup of the Delisle Collection, which had been preserved as a unit from the eighteenth century until at least World War II; but although I have not visited to investigate for myself, researcher Donna Evleth in Paris informs me that the Delisle sketches are now distributed to several repositories in Paris and even in Aix-en-Provence, as the map references accompanying this paper demonstrate. Some, which were fortunately photographed for the Karpinski photostat series, are now apparently lost.

32. Iberville, *Iberville's Gulf Journals*, 141.

33. Patricia Galloway, "'So Many Little Republics': British Negotiations with the Choctaw Confederacy, 1765," *Ethnohistory* 41 (1994): 513–37.

34. I was informed in 1993 by the archive that the original ASH 140–4 could not be located.

35. This map, *Carte des environs du Mississipi par G. De l'Isle Geogr.*, was once known as BSH C4040–4 but is now known as Carte no. 4 of Recueil 69 and is kept at the Service Historique de la Marine, Vincennes—inexplicably shorn of most of its cartouche.

36. This map, which is very much larger than the previous one, is marked on its edge as AN, 193–12.

37. ASH 140–4 is an untitled drawing by Delisle dated 1701, photostat copy in the Karpinski Series, Newberry Library.

38. See my *Choctaw Genesis 1500–1700* for a discussion of the noncartographic evidence of this population movement. The history and archaeology of the Moundville polity is usefully summarized in Christopher Peebles, "Paradise Lost, Strayed, and Stolen: Prehistoric Social Devolution in the Southeast," in Miles Richardson and Malcolm C. Webb, eds., *The Burden of Being Civilized: An Anthropological Perspective on the Discontents of Civilization* (Athens: University of Georgia Press, 1986), 24–40.

39. The manuscript is numbered AM, 2JJ56, no. 16; it is translated in an article by Vernon J. Knight and Sherée Adams, "A Voyage to the Mobile and Tomeh in 1700, with Notes on the Interior of Alabama," *Ethnohistory* 27 (1981): 179–94.

40. The Bottle Creek mound site, which is undoubtedly the one Bienville visited, is now being investigated by Ian Brown. Bienville's discoveries are described in *Iberville's Gulf Journals*, 168–69.

41. Lewis, "Indicators of Unacknowledged Assimilations."

42. Thomas Nairne, *A Map of South Carolina Shewing the Settlements of the English, French, & Indian Nations from Charles Town to the River Missisipi*, inset to Crisp's *A Compleat Description of the Province of Carolina in Three Parts*, copy in the Henry E. Huntington Library and Art Gallery, San Marino, CA.

43. Henry Popple, *A Map of the British Empire in America with the French and Spanish Settlements adjacent thereto* ([London]: S. Harding, 1733), Library of Congress.

44. John R. Swanton, *Source Material for the Social and Ceremonial Life of the Choctaw Indians*, Bureau of American Ethnology Bulletin 103 (Washington: Government Printing Office, 1931), 57.

45. Guillaume Delisle, *Carte du Canada et du Mississipi*, originally among the holdings of the Ministère des Affaires Etrangères, Paris, but now missing; fortunately there is a photostatic copy in the Karpinski Collection at the Newberry Library.

46. Patricia Galloway, "Henri de Tonti du village des chactas: The Beginning of the French Alliance," in Galloway, ed., *La Salle and His Legacy: Frenchmen and Indians in the Lower Mississippi Valley* (Jackson: University Press of Mississippi, 1982), 146–75.

47. Iberville, *Iberville's Gulf Journals*, 174.

48. *Carte de Partie de la Louisiane* (1733), Dépôt des fortifications des colonies, Louisiane 1A, Archives Nationales, Centre des Archives d'Outre Mer, Aix-en-Provence.

49. This Broutin map, *Plan Figuré des Villages Tchactas*, must have been kept in New Orleans for purposes of military planning, because it ended in the Spanish archives and thus was presumably transferred when the Spanish took over the colony in 1763. It is dated to around

1746 on the basis of the villages named, since the Choctaw civil war that began in that year ended in 1750 with the destruction of several of the towns shown on the map.

50. John Stuart and Joseph Purcell, *A Concise General Map of East and West Florida*. Two copies of this map are in the British Public Record Office, while a third is held in the Ayer Collection of the Newberry Library; see Louis De Vorsey, "The Colonial Southeast on 'An Accurate General Map,'" *Southeastern Geographer* 17 (1966): 20–32.

11

Orientations from Their Side: Dimensions of Native American Cartographic Discourse

PETER NABOKOV

This essay explores three broad dimensions of American Indian cartographic discourse. When I characterize cartography as a discourse I mean the uses of material representations of space, whether depicting terrestrial or cosmological environments, as a mode for cross-cultural argument, a mirror for collective self-expression, a rhetorical device for staking out social or diplomatic positions, or a visualization technique often used in conjunction with oratory or storytelling for the charting of proper behavior or spiritual development. While there are glimpses of such a discourse within American Indian traditions which suggest that these dimensions of cartographic thought are not exclusively a product of Indian and white relations, it is generally when natives and Euro-Americans come into contact over issues of land, political power, and cultural authority that they acquire higher profile.

In those historical contexts the rhetoric of claims and counterclaims often seeks documentary support through graphic representations of territory—both when those territories are known, measurable landscapes and when they are imagined, encompassing cosmologies. It is also in those charged circumstances when cultures are jockeying for the political or psychological upper hand that underlying sentiments regarding the relative value of Indian and white worldviews and ethical orientations not uncommonly find their analogic expression in representations of space and human movements through it.

In this discourse the arrangements and mappings of territories and places—whether literal or idealized—are produced and reproduced, challenged and renegotiated in a process that sees cartographic expression as a modality for the establishment of cultural and political authority. Specifically, this essay reviews some of the ways that depictions of space have played their part in Indian-and-white contestations, provided glimpses of tribal notions of the wider universe, and offered guidelines for right living and historical revision.

The Domain of Power

For many American Indian peoples, the land was often its own best map and demanded knowing first on its own terms, almost as if the topography itself possessed some sort of volitional authority. Before representing it, for instance, some native traditions expected you first to listen to its stories and learn its names, to follow it with your feet or to find a way to dream at its most propitious locations. Only after practicing a range of such knowledge-engendering practices *with* the landscape might you be able to truly depict it on a flat surface. This was often the reverse of the non-Indian process of appropriating space by first naming and drawing it, and only then by striding over or settling what was thereby already your own(ed) conception.

Some of this contrast in cultural approaches to what we might call the *map-as-experience* is illustrated by what happened in the late summer of 1879, when the Narragansett Indian council of Rhode Island found itself forced to defend its unified existence before state politicians who wanted to extinguish their "tribal relations." In the opinion of the non-Indian legislators, those "relations" exercised "a pernicious influence" among the Narragansett remnant population by encouraging "pauperism and vagabondism."

Responding on behalf of the 119 Narragansetts who still resided on their sixty-four-square mile reservation were members of their tribal council. Particularly at issue was the fate of 3,020 acres of common lands, which "lie idle and unproductive." By purchasing them, the state would in effect eliminate their claim to tribal unity. When defending his people's rights to this "public land," one Narragansett Indian councilman, a Mr. Ammons, was challenged by Rep. Dwight R. Adams:

"You keep records?"
"Yes, sir; a kind of record of what we do."
"How long back have you got that?"
"1849"
"And what were the records before that?"
"We didn't have any then."
"They were only legends handed down?"
"Yes, sir . . ."
"Have you a map of the tribal lands?"
"No, sir. He [Dr. Griffin, a consultant] said he never had seen that map, and he supposed the map would give us a clue to all our lands outside of this one; but he said he never had seen it, and he supposed it must have been destroyed." . . .
"How do you Indians hold the fee [to the tribal lands]?"
"In the first place, when it was taken up as a reservation, the sachems took what they wanted, and most of them pitched on to the sea coast; and then it was left for the families to go around as they wanted, where the other members of the tribe hadn't gone. They went around a piece and marked it, and that has been subdivided among their heirs right down, and that is the way we came by it; and what was not taken out by these families is what we call public land. It is for the advantage of the tribe."[1]

A few motifs that are probably recurrent throughout American Indian cartographic history arise from this tense conversation between different cultural points of view. First is the public Indian deference shown to the authority of Western cartographic tradition in their respect for the "lost map," that arbiter of tribal claims whose mysterious disappearance often carries the suggestion of skullduggery in the archives. Among the Nez Percé of Idaho was told a similar story of a fabled lost map. It was drawn in pencil by either Chief Joseph or his brother, Ollokot ("the Frog"), in response to another map, drafted by the nemesis of Nez Percé freedom, Gen. Oliver O. Howard, that the Indians felt unfairly minimized their original holdings on the Grande Ronde, Wallowa, and Imnaha rivers.

The Nez Percé map, according to Charles Erskine Wood, was a "crude but quite accurate sketch, drawn on paper with pencil . . . covered with the usual Indian pictures representing whites, Indians, and Indian women, animals, guns, etc., intended to depict the scene and actors in the killing of the Indian . . . a kind of historical painting [of the infamous McNall-Findley crime]."[2] Either irretrievably misplaced in Washington's Indian Office archives, or conceivably interred with Chief Joseph when he was reburied at Nespelem in 1905, the story of the lost map more than the actual document itself has served to legitimize for the Nez Percé both the rightness of their unrequited claims and the bias in the white man's ways of recording history.

Corollary to this "missing map" legend is a second theme, underpinning the Nez Percé anecdote as well, which is how the Indian acceptance of white man's maps as argument constituted an implicit devaluing and replacing of their own legends, which once dignified oral tradition with the ability to transmit an accurate record of hard facts such as numbers of acres and legal claims. But even more debilitating to Indian territorial interests than the disparaging of their own cartographic discourse may have been the accompanying demotion of their cartographic practices.

In a nonliterate society the vernacular recitation we hear from Mr. Ammons concerning how single-family homesteads and a common for the "advantage of the tribe" were established, especially coming from a reputable orator, would normally constitute all the proof one needed of their legal validity. As anthropologist Bruce R. Caron enumerates the practices that are often part of "mapping the sacred" in small-scale, preindustrial societies, "These are composed of various genres and forms: poems, songs, epic narratives, sculptures, mandalas, drawings or paintings. However they are never . . . *maps* in the modernist sense."[3]

A powerful illustration of this difference between what G. Malcolm Lewis has called the "precise boundary delimitation" associated with the modernist enterprise of Euro-American mapping and the "absence/weakness of a linear boundary concept" in Native American cartographic thought,[4] is found in what happened when the Hidatsa of Fort Berthold along the middle Missouri River were first approached by government emissaries about advancing their territorial claims at the all-Plains Indian treaty conference that was planned for Fort Laramie in summer 1851. But what boundaries should they claim, puzzled their leading chief, Four Bears? So he sought out tribal members with stories and rights to sacred sites that figured in the tribe's Earthnaming ceremonies, and then drew them with charcoal on a piece of white hide. What were essentially proven locations for especially potent vision-questing which had been ritualistically linked under the term "Earthnaming," were abruptly converted into boundaries of political territory.[5] It is the same conversion of which the novelist Thomas Pynchon has recently written, as he devotes an entire novel to dramatizing the way in which Euro-American culture "slowly triangulates its Way into the Continent, changing all from subjunctive to declarative, reducing Possibilities to Simplicities that serve the ends of Governments,—winning away from the realm of the Sacred, its Borderlands one by one, and assuming them unto the bare mortal World that is our Home, and our Despair."[6]

But as a reminder of the diversity of tribal responses within this cartographic discourse one might turn to the Lakota reaction to the very same treaty. When instructed to bring their boundary claims to Fort Laramie in 1851, the natives of Dakota Territory found the entire process "ludicrous," according to Indian historian Joseph Marshall III. "Where, they wondered, did the whites get the power to say where the land should begin and end simply on a map?" To his people, writes Marshall, it was like telling someone "to suddenly stop being who and what they were."[7]

Even worse than this ominous delimitation of their homeland was the intrusion of the 2,100-mile-long Oregon Trail, part of which soon sliced across it. And when the white man claimed his Great Father could safeguard the stream of families, wagons, and cattle on this pioneer highway, the Lakota had new cause for alarm. In Marshall's view the Lakota turned to their own cultural categories to make sense of this invasion's deeper significance,

> And how, they wondered, could a group of people be under the protection of a "Great Father" who was so far away that he could not see his people on the road they must stay on to be protected by him? Perhaps, came the suggestion, the road is sacred or holy in some fashion. Thereafter the White Man's Road (*Wasicu Tacunku kin*), known in other circles as the Oregon Trail, the Mormon Trail, and the Platte River Road, became known to the Lakota as "The Holy Road."[8]

For them the twenty-year reign of this major artery represented a time of invasion. "We should not forget the Holy Road," they told themselves, "and we should not let it happen again."[9]

Most likely such testimony as vision quest accounts and the other expressive modes cited above by Caron would have been inadmissible in the forum of late-nineteenth-century Rhode Island, where Mr. Ammons's efforts to protect his people's holdings were considered but one isolated piece of personal evidence, and without supporting evidence really little more than hearsay, to be assessed by non-Indian adjudicators with little sympathy for any remnant Narragansett sense of tribalhood. Which is not to deny the possibility that some element of the Indians' moral claim might still not acquire almost mystical persuasiveness, in disproportion to its conformity to Rhode Island's legal procedures. For in Indian-white cartographic discourse, as with many other debates between the two cultures, it is so often those implicit, psychological claims, strengthened by the abiding hold that Indians have on America's imagination and conscience, on which so much of Indian survival in popular culture, and even a portion of Indian success in the legal sphere, is still founded.

In the case of Euro-American maps that purport to depict American Indian homelands, other sorts of hidden agendas often lie beneath the surface. The imported practice of drawing geographical outlines and landscape features to scale on two-dimensional surfaces was never very sympathetic to Native American economic interests or territorial claims. For a good example of what one might characterize as their often dueling cartographies—in which mapping came to be employed by both Indians and whites to reinforce cultural hegemony—one need only go back to early July 1874, when a wagon train accompanied by geologists, botanists, and paleontologists led by George Armstrong Custer wound its way into the Black Hills of western South Dakota, ostensibly only to get the lay of the land.

To the Sioux, these were sacred hills, surrounded by the mythic racetrack around which the world's first animals raced to decide their hierarchy in the finished world. In a map produced by the Oglala Lakota artist Amos Bad Heart Bull sometime between 1890 and 1913 the artist emphasized the density of mythic sites and sacred mountains such as Bear Butte and Harney Peak just within and without the sacred racetrack (fig. 11.1). That this was his intent is suggested by the fact that Bad Heart Bull even altered the topographic reality altogether, plucking *Mato Tipi*—meaning "Grizzly Bear Lodge" but called "Devil's Tower" by whites—out of its location above Wyoming's Belle Fourche River and dropping it within the protective orbit of the Black Hills' front ridge in South Dakota. Another significant place whose location he adjusted so it would join the cluster of sites he wanted to stress was Ghost Hill *(Inyan Kara)*, which Bad Heart Bull represented by a buffalo head standing for a spirit animal who kept mysteriously eluding hunters by disappearing into this butte, which certified the site as possessed by great power.[10]

While Army engineer William Ludlow's official map and accompanying report of the Custer reconnaissance the following year contains a drawing of the craggy spires at the southern end

11.1 Amos Bad Heart Bull, Lakota artist, map depicting Black Hills region of South Dakota, with hills surrounded by oval of the mythic racetrack and featuring sacred sites in and around the region. Reproduced from Amos Bad Heart Bull, *A Pictographic History of the Oglala Sioux*, by permission of the University of Nebraska Press. © 1967, 1995 University of Nebraska Press.

of *Inyan Kara* mountain—bearing a rough likeness to buffalo horns which may have underscored its association to the Lakota—Ludlow was less concerned with underlying beliefs than with underground geology—which soon meant gold.[11] This Custer expedition, which violated the Sioux Treaty of 1868, triggered the last series of Plains Indian wars, whose heavy-handed American military response, as the cliche goes, "changed the map" of the American West and initiated the extractive economy of mineral resources that remains contentious from the Black Hills to the Bighorns still today.

While the earliest maps that delimited tribal locations sometimes included a pictorial hint of indigenous land tenure, such as Theodore De Bry's famous 1607 engraving of the great Powhatan presiding over his Virginia domain, that changed with later representations as native territories were generally dismissed as *incognita* until they received their first official names from their European discoverers. On George Best's 1578 map of the seaward entrance to Hudson Bay, for example, this egotistical practice shows the narrows virtually clogged with the names of noble European claimants.

To be sure, Indians often collaborated in this cartographic process, yielding the sort of culturally hybrid depictions whose decodings have been pioneered by G. Malcolm Lewis. In 1817, for instance, Lord Selkirk accompanied his land cession treaty with local Chippewa and Cree with a representation of the Red and Assiniboine River systems, adding native pictographs to indicate where particular tributaries were claimed by specific clans.[12]

However, as Selkirk himself was made aware, underlying any culture's geographic consciousness was a system for the cultural constitution of space which could differ widely from that of Euro-Americans. When some Chippewa asked Selkirk's surveyors what two miles was, the men pulled out their measuring chains. But the Indians disliked this, preferring, according to anthropologist A. Irving Hallowell, measurement in terms of a "unit of activity." Hence, when their chief noticed some horses grazing where the surveyors' chains now rested, he promptly declared that two miles was "as far as a man could see daylight under the belly of a horse," whereupon his companions squinted westward and "were satisfied that they knew what two miles meant."[13] Amusing as this anecdote might seem today, the upshot of this contrast in spatial conceptualizations was that these Chippewa clans lost their western lands.

Not that the pinpointing of locations, the cartographic delimitation of farmlands, or the notation of coastal contours were out of line with one traditional function of native American geographical consciousness, even when that information was generally kept in oral tradition rather than committed to paper. From Sonora, Mexico, to San Cristobal, New Mexico, as well as points west, one can find examples of rock art which some argue are the representations of irrigation systems, routings of trail systems, or contours of wider landscapes, often appearing as if they had been envisioned from a bird's eye. Perhaps more convincing evidence of the native Indian ability to graphically reproduce their connecting of the dots, so to speak, created by their successive camp moves, comes from the work of Hugh Brody among the British Columbia Athabaskans,[14] Michel Audet among the eastern Hudson Bay Inuit,[15] and Edmund Carpenter among the Aivilik Inuit of Southampton Island.[16] These examples of native recordings of seminomadic activities over space and time leave no doubt about the profound impact of the *experienced* environment upon Native American cartographic memory.

It was the decisive next step, the creation of cadastral boundaries between actual land areas, which

Indians often found confusing—especially when it was done not on foot or for obvious purposes of remembering how to reach favored hunting and foraging spots, but through imaginary lines and for highly suspicious reasons. Indians might have instant evidence of the power of the white man's firearms, but the destructiveness of invisible alignments drawn through the surveyor's transit took a while to hit home, as William Faulkner conveyed in his short story, "The Courtship," concerning Mississippi's early Choctaws:

> Issetibbeha and General Jackson met and burned sticks and signed a paper, and now a line ran through the woods, although you could not see it. It ran straight as a bee's flight among the woods, with the Plantation on one side of it, where Issetibbeha was the Man, and American on the other side, where General Jackson was the Man. So now when something happened on one side of the line, it was bad fortune for some and good fortune for others, depending on what the white man happened to possess, as it had always been. But merely occurring on the other side of that line which you couldn't even see, it became what the white men called a crime punishable by death . . . Which seemed foolish to us.[17]

When it came to that third step, the enclosing of space, cutting Indians off from hunting lands and isolating them from each other, that went beyond "foolishness," as author Will Campbell describes in *Providence*, his fictionalized chronicle of one square mile in central Mississippi. Here Campbell dramatizes the cartographic reincarnation of this land in the form of Section Thirteen, Township Sixteen, Holmes County—a piece of ground that was sliced from the one-time heartland of the Choctaw Indians:

> "We're surveyors," Mr. Payten said, although the boys had not asked what they were doing. "We're making maps, plats, measuring all this land, dividing it into townships, sections, all that." He was leaning against a black oak tree, writing as he talked.
> "Why?," Luther asked. He moved close enough to see what was being written.
> "So people will know who owns which land. The government will be giving patents."
> "What's a patent?" one of the men heard Jesse whisper to Luther.
> "Decree of the king," the man laughed.
> "The Great Spirit owns the land," Luther said. "The Great Spirit provided it for the children of Nanih Waiya to live forever."[18]

Like the Choctaw people, Indian tribes of the western Plateau were also subject to the bloodless process of territorial diminishment through repeated remapping of their aboriginal estate. Sometimes it was ostensible friends who oversaw the surveyors' tools, as when Smithsonian Institution ethnographer Alice Fletcher facilitated the government's policy of subdividing tribal lands into single-family parcels (the General Allotment Act of 1887) among the Nez Percé in the early 1890s. Their immediate neighbors, the Flathead, also underwent land loss through the imposition of allotment cartography, as described by the Newberry Library's own D'Arcy McNickle, a Flathead Indian writer, in his last novel, *Wind from an Enemy Sky*,

> Men from a far country, from somewhere east of the mountains, came and said: "You must build fences. Four strands of barbed wire, stretched and stapled to cedar posts set sixteen feet apart. That's a legal fence and you can prosecute for trespass if anybody lets his stock in on you . . ." . . . the new men, coming from across the mountains, set family against family, telling them to build legal fences, tear up the sod, build little houses.[19]

Not surprisingly, the rhetoric of Indian resistance to the United States government's late-nineteenth-century project of cultural assimilation often focused upon this issue of what one might call *territorial reinscription*. In the Columbia Plateau, under the influence of the revitalization prophet of the Wanapum, Smoholla, the white man's treatment of the land was likened to corporeal dismemberment. For Smoholla himself, in perhaps the most reproduced of all Indian quotations, the plowing of furrows near his Priest Rapids community was compared to an earthly mutilation. "Shall I take a knife and tear my mother's bosom?" he demanded of Maj. J. W. MacMurray in July 1884.[20] Under Smoholla's influence, the nativistic faction of the nearby Nez Percé took this land-body association even closer to heart. "The earth is part of my body," said Toolhulhulsote, the spiritual adviser to the tribe's famous Chief Joseph, in 1877, "and I never gave up the earth . . . What person pretended to divide the land and put me on it?"[21]

The Realm of Cosmology

We should not forget that while some of the native cartography we research was also implicated in the imperialistic enterprises discussed above, behind the native spatial concepts that insinuated their way into much historic-period cartography often lurked traces of an indigenous cosmology, which could run at cross-purposes to non-Indian territorial depictions and which was frequently informed by a geographical consciousness that was grounded in myth and religion. Thus in the very term "indigenous maps" we can detect the hint of a contradiction in terms, a messy discourse rather than a tidy artifact, a problematic cover term for the often opposed vested interests of cartography and cosmology. And as for the real-world consequences of that opposition, they are clearly visible on any maps of the shrunken Indian land base today.

It is interesting, however, that for some Indian spokespeople the mutual and, they might argue, "natural," interactions between those realms of power and cosmology, of politics and worldview, were most persuasively expressed through spatial representation. After one elderly Kiowa medicine man named White Bird heard U.S. commissioners in November 1866 complain about Kiowa raiding along the southern Plains, he had his helpers lay out "two circular pieces on the floor; one blue and one white." After praying, the medicine man actually addressed himself to Lone Wolf, the head chief and official spokesman for the tribe, who translated his words for Col. J. H. Leavenworth and the other visitors: "That piece of paper (pointing to the white) represents the earth. There is a big water all around the earth. The circular blue paper is the sky. The sun goes around the earth. The sun is our father. All the red men in this country, all the Buffalo are his."[22] Having thus established *the way things really were*, White Bird's logic built into his argument that, as a medicine man who had been trained in controlling the weather, his access to the all-powerful sun, to his "Great Father" and therefore to his power, was closer and stronger than that of the white man's "Great Chief" in Washington. Hence, in this Kiowa's eyes, as anthropologist Raymond J. Demallie has pointed out, his people's resistance to the power of Washington rested upon firm cosmological grounds and moral precepts.

For a more explicit example of a Plains Indian leader making a cartographic case for his tribe's sacred destiny as a chosen people in a divinely ordained land, we turn to an interaction between a major Crow Indian chief, Iron Bull, and a Jesuit priest named Father Pierpaolo Prando. Meeting in 1883, Iron Bull first illustrated with some sort of drawn circle the tribe's claimed homeland in present-day Wyoming-Montana. "This is the earth the Great Spirit made [for] them," Father Prando wrote a fellow priest, describing the scene. "The Piegans [Blackfeet] he [the Great Spirit] put them there, indicating a point in the line of the circle he made, then the Great Spirit made the Sioux, the Snakes [Shoshones], Flat-heads and many others [and he] located them all around the earth." Finally, we can imagine Iron Bull pointing to the circle's center—the Big Horn valley—as he added with pride, "The Great Spirit put us right in the middle of the earth, because we are the best people in the world."[23]

Since my own introduction to the workings of Indian geographical consciousness was prompted by my study of Native American built environments, I will first review some generalized relationships between cosmology, sacred geography, and architecture, and then examine a few cosmographies and cosmograms that Native Americans produced in order to orient themselves in worlds that were and are just as real to them as those that Rand McNally interprets for non-Indians today. To do this I shall adopt the heuristic conceit of a generic American Indian cosmology with its generalized, frequently encountered motifs, such as Center, Four Directions, Sky Dome and Vertical Axis, all the while flashing a warning reminder about the great diversity of tribal variants that can render those motifs fairly useless whenever one starts to conduct more controlled regional comparisons.

Centers

As the historian of religions Mircea Eliade fervently proclaimed, and the cultural geographer Yi-Fu Tuan then reiterated a bit more temperately, "The claim to cosmic centrality is widespread: this is the habit of structuring the world with one's own group at the center."[24] Whether an Indian people emphasized creation in place, and considered themselves an original or first people in an autochthonous homeland, or whether they emphasized epic migrations, and considered themselves a destined or chosen people in an envisioned or promised homeland, the notion of center-places, the pivots of a sacred geography, or of multiple "earth navels," as they are sometimes described for Southwestern Pueblos, and the importance of ritual sites that might temporarily be consecrated for ceremonial activities, are not uncommon throughout Native America.

For the Choctaw of Mississippi, who seem to have vacillated between creation and migration scenarios in their own accounts of ethnogenesis—perhaps dependent on what momentary advantage either might provide—the *Nanih Waiya* "mother mound," which is still visitable today in southern Winston County, was the pivotal site in both scenarios. For the northwestern California Yurok people, their attention to this notion of a center was associated with the sacred post of the sweat house, the structure originally said to have been built where the world's first redwood tree grew.

In the Pueblo Southwest the exceptionally strong sense of centrality seems most accessible to the outside researcher. In many Keresan communities, such as San Felipe Pueblo just north of present-day Albuquerque, the concept is focused around prayer deposits planted beneath their plazas, which function almost as pivots for a spatial scheme that organizes concentric zones that extend to a perimeter customarily demarcated by four sacred mountains. For the Zuni Pueblo of far western New Mexico their center is commemorated by a modest sandstone shrine in their present village, Zuni Pueblo, also known as *Itiwana*, or the "Middle Place." For the Hopi of Arizona one of their principle center shrines is a travertine dome near the Little Colorado River, out of which they claim to have emerged. While outside of Taos Pueblo, northernmost of the Rio Grande pueblos, perhaps the foremost of their center shrines is Blue Lake in the Wheeler Peak area, site of the community's great August pilgrimage.

Directions

As one stands in the center, faces the sun, and spreads out one's arms—addressing, as did tribespeople as diverse as the Acoma and the Cherokee, that light-giving being who is considered crucial to space and time coordinations in most Native American cosmologies and their calendric religious systems—then the four directions become apparent: right (south), left (north), in front (east) and behind (west). Among the Southwestern Pueblo traditions, especially, one finds these directions also associated with specific fauna and flora spirit-intermediaries; in altarpiece cosmograms of the Hopi they are portrayed as the color-coded snakes that can bring moisture from those corners of the universe. This emphasis makes us wonder whether it was not early Pueblo influence that made four-directional coding so ubiquitous throughout the range of Navajo dry paintings associated with their distinctive curative "chantways."

Yet one finds an equally strong sense of four-directional symbolism within Southeastern traditions. Among the Creek, Cherokee, and Yuchi, for instance, the centerpiece of their Green Corn "square ground" ceremonies is the four-log fire, kindled at dawn as a piece of the living sun, which becomes a central place for afternoon oratory and the hub for late-night stomp-dancing (fig. 11.2). But the four directions are also prominently linked with Plains-dwelling peoples. The origin myth for grass houses built by the Wichita of the southern Plains calls for only four of the structure's split red cedar ribs, standing for the sacred directions, to jut beyond the twined-grass steeple. In the prototypical structure that the culture hero Red Bean Man teaches the early Wichita, they are aligned directly above four directionally positioned doorways.

Sky Dome

But this flat orientation, like a floor, must be given shelter, like a roof—as well as light, movement, and time. That happens as the east-born sun describes its arc across the sky, producing not only warmth, illumination, and the dome of heaven but also the calendric seasons. The synchronization of solar movement with human endeavors was, for many Pueblo communities, the concern of designated sun-watchers for whom the horizon became an orienting map that timed all phases of the agricultural process and the rituals that assured its success. With varying degrees of salience, this "sky dome" or "celestial vault" symbolism also appears across native North America. When an early government agricultural agent

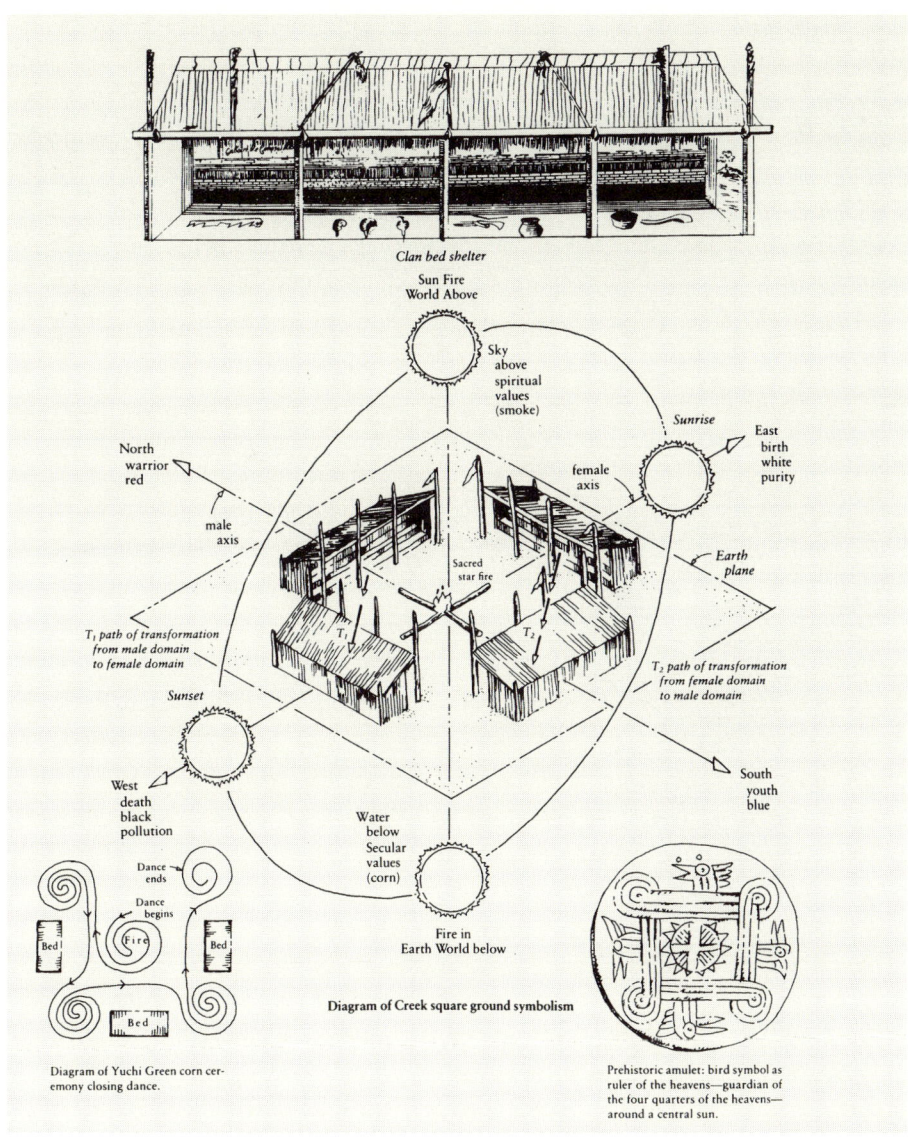

11.2 Symbolism of Southeastern Indian ceremonial ground, organized around four-directional sacred fire. Drawn by Robert Easton, based upon Yuchi and Creek sources including work by anthropologist Amelia Bell Walker. From *Native American Architecture*, by Peter Nabokov and Robert Easton, p. 110. Copyright © 1989 by Peter Nabokov and Robert Easton. Used by permission of Oxford University Press, Inc.

named Alfred Reagan worked among the Jemez Indians he saw the mural of such a sky dome in one of their kivas (fig. 11.3). Clearly emphasized were multiple sky and underground levels, the sun and moon, and the water-connected serpents that linked realms on behalf of the people and their agricultural cycle. And when the Oglala Lakota seer, Black Elk, depicted the famous vision quest he experienced at the age of eight, the event transpired beneath such a cosmic arch, which is often equated with the rainbow. When contemporary Indian painters make an effort to illustrate tribal creation myths, such as Albert Johnson has done for the Navajo, or when they more freely invent scenes using traditional elements such as one Tewa artist has done in his painting of Pueblo clowns climbing a cosmic ladder, this curving sky dome motif, like an inverted bowl, often serves as an internal frame for the workings of an entire universe.

11.3 Jemez Pueblo depiction of the tribal universe from a kiva wall, Jemez Pueblo, New Mexico. It shows a rainbow sky dome over sun, moon, stars, and multilayered earth. Courtesy of the Edward E. Ayer Collection, the Newberry Library.

Sometimes native cosmography was as astronomical as it was terrestrial, producing the cartographic representations of heavenly activity which are analyzed by the relatively new subfield known as archeoastronomy. Portrayals of stars, exploding nebulae, planets, and constellations are plentiful among the rock art sites, from the southwestern Anasazi to the Chumash of south-central California. Perhaps the most controversial of such celestial representations is a buckskin map of the heavens which the Skidi Pawnee of Nebraska safeguarded in their Big Black Meteoric Star Bundle. On closer inspection, however, its characterization as a star chart may be a misnomer, since, as Pawnee scholar Douglas R. Parks has shown, nearly half of its features actually reflect earthly, cultural aspects of Pawnee life, such as a Pawnee camp circle and important Pawnee medicine bundles. Rather than an objective depiction of the night sky at a particular moment in time, then, its point is more likely to draw the crucial connection in Pawnee cosmology between cosmic forces—identifiable constellations such as the Big Dipper, the Little Dipper, the Pleiades, plus sunrise and sunset respectively—and the tribe's social and religious organizations that reflect and interact with those heavenly forces. To Parks the chart is best appreciated as a complex mnemonic device that is referenced as much to Pawnee mythology as to their astronomy.[25]

Vertical Axis

Yet the sky dome is pierced, as we know, by the sun, which is often represented by the smokehole of a lodge that is considered a model of the universe. While the smokehole thus allows for entry to cosmic worlds above, the topography beneath one's feet can also open to nether levels, subterranean underworlds, which for the Pawnee were represented by those celebrated natural mounds known as "animal homes" where the tribe's shamans received their healing powers. Thus a number of American Indian cultures add Zenith and Nadir to their cosmologies, making six directions, with that center spot on which one stands sometimes represented as a seventh direction. Linking these tiered realms is the cosmic axis, a motif which assumes diverse manifestations in different Indian worldviews. Among the Iroquois of upper New York State, whose origin narratives feature the creation of human beings by means of descent from an above sky-world rather than emergence from out of the earth, a domestic dispute in that above-world uprooted an all-illuminating tree. Through the resulting hole dropped a pregnant woman, falling onto the back of a turtle that floated in the primordial sea. Once an earth-diver animal helped to create land, it was up to her descendants to nurture the earthly counterpart of that cosmic tree: the sacred "white" tree of peace.

But Iroquois tradition and ritual are not alone in featuring such a world tree or sacred pole connecting cosmic realms, which Eliade has dubbed the *axis mundi* motif. In the Northwest Coast such a cosmic tree also links levels of the universe, and is even portrayed with mother-of-pearl buttons on trade cloth blankets. In the wood-plank lineage houses built by the Kwakiutl this axis is represented by a ritual pole, stripped of its branches, which juts through the gable roof during the great winter ceremonials and is climbed by shamans as they dramatize the acquisition of their superhuman powers from above-world spirits. For nearly all the Southwestern Pueblo societies access to nether worlds is provided by the long-poled ladders by which one descends into the semisubterranean kivas, their religious chambers. It is these kivas that, in turn, bring human beings as close as they can come symbolically to the underworlds described in most Pueblo mythic narratives of origin.

The architectonic importance of a central pivot for ritual dramas and healing ceremonies is especially keen in Plains Indian sun dances. A cottonwood tree is carefully selected and harvested; for the four-day duration of the ritual, within the brush-walled lodge that becomes a microcosm of the universe, that tree functions as a conduit for the sacred powers that are then transmitted to the fasting supplicants surrounding it or even tied by their flesh to it. Even when such a ritual structure is permanently constructed, as is the case with the wood-shingled dance houses still built and used by native traditionalists in central California, the hardwood center post becomes the revered conveyor of curative powers.

Whole Cosmologies

Portraying these cosmic elements of center, cardinal directions, sky dome, and zenith/nadir in two-dimensional graphics, as symbolically inclined anthropologists are inclined to do, can produce intriguing cosmograms. When ethnographer Leslie White asked his Santo Domingo informants in New Mexico how to map their cosmos, he was told their world was round, with four color-coded sacred directions. In the Santo Domingan view, history and cosmology were not necessarily antagonistic, so locations where earlier Santo Domingans had formerly lived were collapsed within the cosmic perimeter; Santo Domingo Pueblo was positioned at the center of it all, and historical ruins such as "White House" were also noted (fig. 11.4).[26]

An attempt by the ethnogeographer T. T. Waterman to depict what he learned from north western California Yuroks of their view of geographic reality produced a much different cosmogram. Known as "that which exists," they conceived of this world as an island floating beneath a sky that had been fashioned by "World-Maker" from a fishnet. Anchored in the ocean all around the horizon, this net was pierced by a sky hole through which the geese could fly. Since to the Yurok all directions were oriented to the Klamath River, they considered our north their "downstream." Throughout this mythologized landscape, and in the "hard land" beyond the sky net itself, a host of supernatural beings had their homes.[27]

In other instances Indian ritualists themselves fashioned models of the universe for their own religious reasons that remain obscure to us today. A 1987 exhibit of masterpieces from Oklahoma's Philbrook Museum featured what its catalog described as a portable mobile presumably crafted by Tewa Pueblo ritualists. Constructed of two intersecting disks of hide stretched on bent saplings, its white dots on a dark field, and its male and female figures, seemed to depict "the interdependent dualities of night and day, the gods and man . . . with fertility symbols and celestial lights."[28] Possibly used for cosmological orientation during divination rites, it hung from the roof of a cave in the Gobernador territory of New Mexico, and was accompanied by natural crystals and other ceremonial objects.

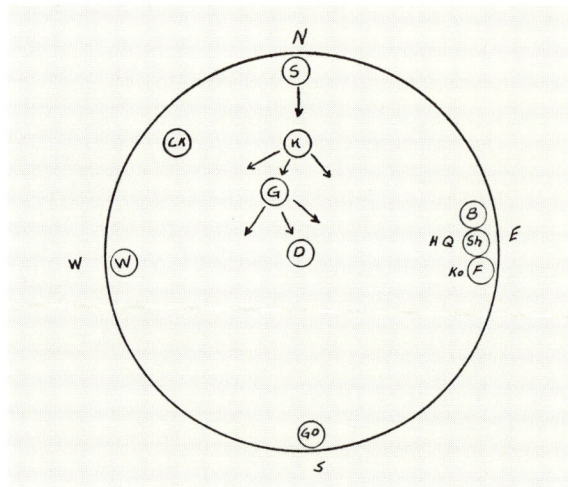

11.4 Drawing of the "mythological landscape" of Santo Domingo Pueblo, New Mexico. Place of origin is *Shipap* (S), with *Gykyatsagya* (G) indicating a former historical village of this tribe and also denoting the current site of Santo Domingo village (D). Other locations include the mythic origin "homes" for the tribe's Boyakya Society (B), the Cikame Society (Sh), the Flint Society (F), and the Koshairi, or sacred clown, Society (Ko). Reproduced by permission of the American Anthropological Association from Memoirs of the American Anthropological Association, no. 43, 1935. Not for further reproduction.

No one knows who painted what appears to be a Lakota cosmos on a tipi cover that was destroyed in a German museum by Allied bombs in World War II.[29] Its "mural in the round" seems to reveal an entire Sioux bestiary, and includes other cosmic forces that are still central to Lakota traditionalists. But we can affirm that the traditional Blackfoot tipi allowed individual visionaries to portray their personal spirit guardian animals within a stereotypical cosmic frame: the stylized horizon and mountains below, stars and rainbows above.[30]

The Navajo conception of the cosmos, with its emphases upon directional mountains, gender divisions, and solar movement, was writ small in their male hogan—with its directional posts, consecrated during construction and its gender-specific seating arrangement during ceremonies. In its function during curing ceremonies as a shelter for the chantway dry-paintings, the patient was positioned directly on these stylized cosmograms in order to regain congruence with Navajo notions of balance and beauty.

The Conduct of Life

Our confidence that we are uncovering native worldviews by synthesizing and decoding cosmograms should be tempered by the realization that native cosmologies generally have two sides. One is these neat arrangements of cultural categories, often hierarchical, along with key elements of a culture's abstracted universe. Passive and orderly, such diagrams are pliable to cross-cultural comparison. But the other side includes the messier human aspect related to moral meanings, living beliefs, and the proper relationships by which Indian peoples contributed to what has been termed their "participatory maintenance" of their universes. "Cosmologies have more than aesthetic or intellectual interest," the scholar Yi-Fu Tuan wisely reminds us. "They are not simply fantasies of the mind floating above the hard realities of day-to-day living. They derive from lived experiences and at the same time give them meaning and *direction*."[31]

Here we are interested in how Indian peoples attempted to "live their maps," so to speak. Their cosmologies, and even their own pictorializations of them, were not only *models of* how the world looked, but in the phrase of anthropologist Clifford Geertz, they become *models for* cultural practice as well. Giving such "direction" to behavior could be an everyday endeavor, as some notion of "right action," or avoiding what the Berens River Chippewa called *madjiijiwe baziwin*, "bad conduct," involved conforming to proper behavior within what we might call a moral cartography.[32]

Not surprisingly, it was often at key life-crisis moments in a person's developmental and educational cycle when Indians were exposed to these dynamic world portrayals. Initiation, whether as part of puberty ritualism or not, was a prime opportunity to impress upon the young these notions of their roles and responsibilities in the broader scheme of things. Among a half-dozen southern California native societies, each "ground map" or dry painting that was created on the floors of ceremonial enclosures depicted the heavens, the land of the dead, the topography of this world, and supernatural animals and spirits.[33] These cosmograms were the centerpieces of rite-of-passage rituals, when pubescent girls and boys were given lengthy sermons on how to behave throughout their lives (fig. 11.5). As elders hectored the young, referencing their admonitions to drawings at their feet, their descriptions traveled from the symbols of earthly creatures to the world of the sky and the otherworld of life after death,

> See these, these are alive, this is the bear-mountain lion; these are going to catch you if you are not good . . . The earth hears you, the sky and wood mountain see you. If you will believe this you will grow old . . . Do not forget this that I am telling you; pay heed to this speech, and when you are old

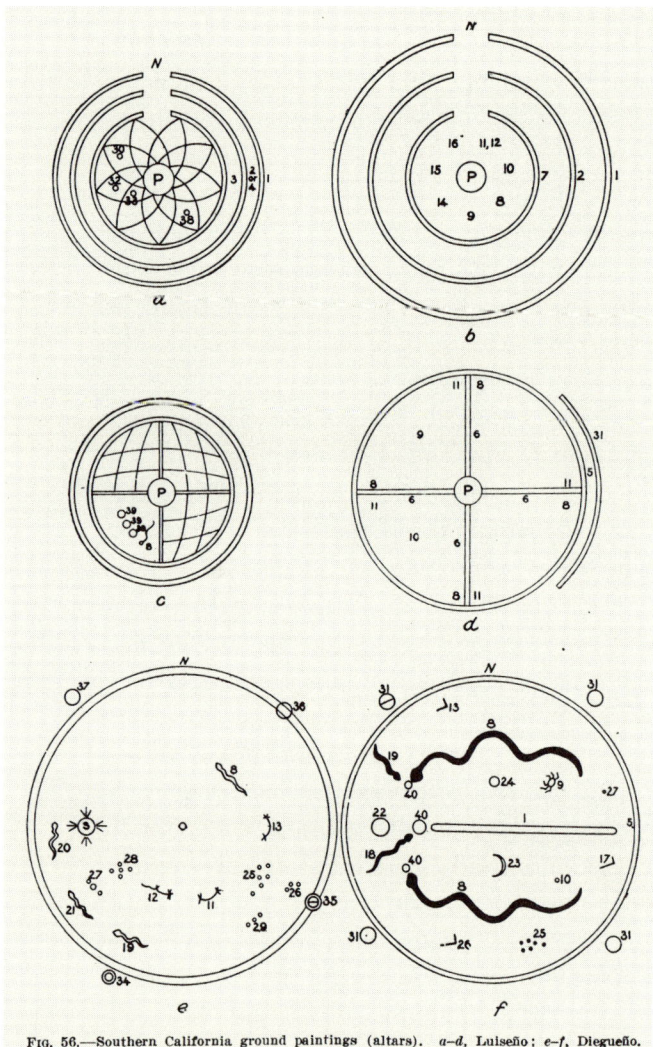

FIG. 56.—Southern California ground paintings (altars). *a–d*, Luiseño; *e–f*, Diegueño.

11.5 Cosmology and topography depicted on floors of ceremonial enclosures by Southern California Indians. Used to educate the young during puberty rituals, they illustrate such features as, (1) the Milky Way, (2) Night (or sky), (5) the world, (22) the Sun, (23–24) New and Full Moon, (25) the Pleiades, the constellations of (26) Orion, (27) Altair, (28–29) the "Cross" and "Shooting," (30) the sea, (31) the mountains, (34) Coronado Island, (35) the "Mountains of Creation," (36) San Bernardino, and (37) Santa Catalina Island. This image also appears in A. L. Kroeber, *Handbook of the Indians of California*, Smithsonian Institution, Bureau of American Ethnology Bulletin 78 (Washington, D.C.: Government Printing Office, 1925), 663.

like these old people, you will counsel your sons and daughters in like manner, and you will die old. And your spirit will rise northwards to the sky, like the stars, moon, and sun."[34]

Another way such cosmograms guided human action was through their incorporation of what seems to be one of the most fertile, widespread tropes in American Indian consciousness: the road, the trail, the path or the journey. Once again, people often "lived their maps" by presaging the intended endeavors to come—which might literally include a journey—through ritual actions such as pilgrimages through mythologized landscapes, or through ceremonial enactments for which, as anthropologist Caron writes, "The performance of the practice is itself a map."[35] In the Southwest the ritual practices of "road-making" and "road-closing," which customarily entailed sprinkling lines of blessed cornmeal so that benevolent spirits such as the rain-bringing *katsinas*, the Sun, the Sky or other mythological beings could safely travel into and out of the communities, was so ubiquitous among both the eastern and western pueblos that it seems almost certainly of precontact origins.[36] During the Hopi Snake Dance, the cosmogram of a four-quartered circle is drawn with cornmeal on the plaza floor so that the snakes, who then are serving as the messengers of the Hopi, may follow those four "roads" as they carry the Indian's prayers for rain to the weather-producing powers of the sacred directions.

At the height of the annual Shalako ceremony, held every December at Zuni Pueblo for the blessing of houses and well-being of inhabitants, the road metaphor is heard repeatedly, as with this closing invocation from the Sayatacha kachina at the House of the Gods, "May your roads be fulfilled; May you grow old; May you be blessed in the chase; To where the life-giving road of your sun father comes out May your roads reach; May your roads all be fulfilled."[37]

The same procedure also blessed human movements. For roads were "opened" and sanctified with cornmeal when religious leaders headed for sacred springs, when woodcutters returned home with fuel to burn during special ceremonies, and when processions of marriage parties walked back to their matrilineal homes. And Zuni elders taught their youngsters about the *onanne*, the "road of life," which has been established for each individual by the Sun Father, and they described how it was the job of medicine men's societies to remove any obstacles so as to prevent one's incomplete passage on that road.[38] But this practice of road-making can sacralize events of a more secular, historical nature as well: in August 1980 at Taos Pueblo, at the outset of the 375-mile Tricentennial Run, when all the Southwestern Pueblos joined in commemoration of their successful uprising against the Spanish in 1680, I witnessed a young Hopi priest similarly open and bless the asphalt roads ahead with cornmeal just before the relay runners departed before dawn.[39]

For the Papago of the southern Arizona desert, this "road of life" was expressed as a single-path labyrinth that was often woven into their basket designs. It showed the tribe's culture hero, I'itoi, at the mouth of the circular labyrinth, with directional lines indicating the four winds, which are his messengers. While the Papago say that human existence yearns to complete the road leading to the earth center, or I'itoi's home on Babiquivri Mountain, this cannot be achieved without surmounting the tribulations thrown up along the labyrinth's path.[40]

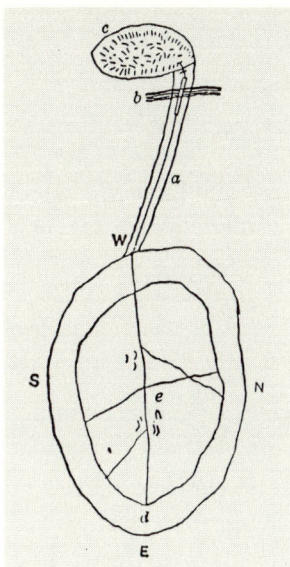

11.6 A map of the world drawn by a Thompson River Indian, British Columbia, showing the the route taken by the dead from the land of the living or "the middle place" (shown as an oval-shaped earth lodge at bottom) as they head across (a) the trail to the underworld, crossing (b) a mythic river, until they arrive at (c) the land of the dead, which is shaped like a great, mound-like earth lodge containing double rows of fires through it, the sweet smell of flowers, and welcoming, happy people. Courtesy of the Edward E. Ayer Collection, the Newberry Library.

In the Columbia Plateau region a Thompson Indian of interior British Columbia drew for James Teit a map of life's road which depicted how the soul traveled out of this earthly existence, which they envisaged in the form of a typical plateau semisubterranean pit-house (fig. 11.6). Then the soul moved toward the setting sun to cross a river, which indicated the transition of death, before entering a second pit-house structure made of granite, the house of spirits, where bygone friends welcomed you with feasting and celebration.[41]

To illustrate how in Indian geographic consciousness maps *for* proper action can possibly overlap with maps *of* landscape, one might examine some of the foregoing concepts within the vast sub-Arctic culture area, which was the focus of fieldwork by A. Irving Hallowell, a University of Pennsylvania anthropologist who was particularly interested in native notions of space and time. Based upon his work in the 1920s among the Berens River Chippewa of Lake Winnipeg, Hallowell learned that when these seminomadic hunters were on the move their mental maps focused upon movement between time-honored points, whose topographical cues were lodged in many different aspects of their culture. Hallowell could almost be writing of the Apache, Navajo, Penobscot, or numerous other Indian peoples when he says, "Place naming, star naming, maps, myth and tale, the orientation of buildings, the spatial implications in dances and ceremonies, all facilitated the construction and maintenance of the spatial patterns of the world in which the individual must live and act."[42]

For the Chippewa with whom Hallowell canoed and whose "behavioral environment" he sought to understand, the narratives that told of North Star and the different personalities of those four brothers, the directional winds, provided orienteering guidelines for responding to the behavior of additional entities, like the winds, Thunder Birds, and moving rocks, whose categorical identity Hallowell explained through his memorable gloss, "Other-Than-Human-Persons." Hallowell argued that we can decipher native maps only if we appreciate how their directional orientation is seen to reflect not mechanical instrumentalization but the lived experience of pragmatic travelers, constantly adjusting their guiding compass not to magnetic north but to culturally constituted landmarks, cosmically mediated observations, and the behavior of those "Other-Than-Human-Persons."[43]

To reduce the impact of the unexpected, these boreal forest travelers also utilized ingenious strategies to reassure themselves during ventures through unfamiliar landscapes. From the stories he heard, Hallowell gathered that a traveling band's step-by-step rather than goal-oriented progression through their landscape might be reinforced by their tales of a series of mysterious old women who were magically positioned at intervals along the route. As if hand-delivering the travelers from wise woman to wise woman, together with their advice at each stage about the intervening countryside, these stories psychologically protected the nomadic Chippewa with a sequence of familiar comfort stations along the way.

A similar concern with maximizing social stability in the face of often unpredictable environmental circumstances seems to inform what ethnographer Adrian Tanner learned from Mistassini Cree storytellers along James Bay to the east. They told him about Pukwat Skwes, a mythic woman, who was traveling with a hunting band but urged her camp-mates to leave her behind. Upon reaching the next campsite, however, they were surprised to find her there with their tents already erected. Moreover, the tent covers stretched around the very same tent poles from the previous location. Perplexed, the next time she lagged behind her friends stayed to spy on her. As they watched she sang a song whereupon the tent magically flew into the air, poles and all. But when she caught sight of the onlookers, the spell was broken and the tents tumbled to the ground. Hence, as a Cree named Bally Husky explained to Tanner, children today are admonished from ever counting the poles in their tent, lest their scrutiny impair the smooth moves from camp to camp, and jeopardize the job of making campsites identical.

More to my point, however, Bally added that this way, "people say that no matter where they move their camp to, they are always really staying in the same place."[44]

Within the different types of dwellings that Tanner watched the Mistassini Cree employ in different locales at different times of year, he noted a similar concern over reproducing social stability amidst geographic mobility and seasonal diversity. Since his Indian friends observed strict rules based upon gender, kinship, and age to establish consistent sleeping positions of each family member no matter the season or the size of the building, it was as if they were always occupying the same social location, and hence not really moving at all.

This concern with securing a preordained, auspicious future produced other uniquely Great Lakes and sub-Arctic techniques for transcending time and space. In the divination rite popularly known as the Shaking Tent, a shaman stepped within a barrel-shaped enclosure fashioned according to precise instructions. Once swathed in birchbark, the shaman dispatched his magical Turtle helper to places distant in terms of both dimensions. Onlookers learned of the whereabouts of game and the well-being of distant relatives, but also about calamities or good fortune to come. As the interior of the lodge was described to Hallowell, he realized that, with the Thunders and Stars inside and the floor envisaged as the upper shell of a mythic turtle, the little body box was a model of the generic Algonquian universe as it existed in the time of creation, a time when such transcendence was the norm.[45]

While the shamans hidden within the shaking tent provided their audiences with verbal maps of distant circumstances, a second sub-Arctic practice for casting auguries produced a graphic representation of the unknown future and the unseen faraway. This was scapulimancy, a technique by which the flat shoulder bones of caribou, hare, and even dog were marked with charcoal to indicate the territorial confines of specific interest, then held by a cleft stick in a fire. In a session recorded by writer Alika Webber and her photographer husband in 1962, the diviner, a Naskapi from Davis Inlet in northern Labrador, then called upon the caribou spirits to produce a crude map which, in its rudimentary way, collapsed those concepts of place and action as we have discussed earlier.

The ritual involved withdrawing the shoulder blade from the flames and interpreting its cracks and marks. The diviner said that a blotch in the center indicated starvation—"there are no paths to follow." Pointing to one split in the bone, he said that it was their present location, and that if they headed for a second crack they would find caribou, many in fact, but they might be too far away for easy procurement. A third mark indicated the whereabouts of other Indians, perhaps five of them, who might have extra meat for them.[46]

After the arrival of Europeans in North America came the alien cartographic concepts and their new and frightening political applications discussed in the first section of this essay. With regard to American Indian religious traditions whose incorporations of geographical consciousness into liturgies and rituals were documented well after contact with whites, however, the degree of Euro-American influence on Indian doctrine is a matter of debate. The question remains open, for instance, whether the *Midewewin*, known as the

Great Medicine Lodge ceremony and practiced by the numerous Chippewa groups around the Great Lakes, was a pre- or postcolonial synthesis, an aboriginal religion or a revitalization movement. One of its key symbols, however, expressed in story, enacted in ritual, and depicted in the birchbark scrolls associated with Medicine Lodge rites, is the concept of a sacred route. For celebrants this highly generative symbol simultaneously evokes the historical migrations of the Chippewa, offers a key scenario for the ritual procedure that sees participants ascending through a series of degrees of ritual knowledge and purity, and also serves as a metaphor for the trajectory of one's life and for the proper conduct in between—the "Road to Life and Death" which the Winnebago Jasper Blowsnake described to anthropologist Paul Radin.[47] Quite likely this key metaphor built upon still older cosmological notions of the road that the Chippewa told A. Irving Hallowell led to the southerly Land of the Dead.

In this way the song literature, the pictographic paraphernalia, and the ritual practices of the Great Medicine Lodge so well summarized elsewhere by Selwyn Dewdney were suffused with cartographic awareness.[48] As it related to reconciling migration and creation stories and with chartering brand-new beginnings for immigrants, the ceremony sacralized both history and geography. As shown by ethnomusicologist Thomas Vennum, Jr., it was the tribal histogeographer, known in the Chippewa tongue as the "preserve man," who deciphered the mnemonics marked onto birchbark scrolls. According to Vennum, his rich song repertoire relating the migration legend contained detailed geographic placenames, to such an extent that Vennum was able to correlate most of twenty-one places with (1) song references, (2) bark scroll pictographs and (3) locations on a contemporary map of the Great Lakes.[49]

One of these Chippewa cosmograms copied by early ethnographer Walter J. Hoffman from a birchbark scroll seems to illustrate a fusion of such an origin story with the sort of moral map we have mentioned earlier. The Great Spirit appears as a large central figure on the right of the scroll, who is flanked by the world's first two men and two women. Read from the right to the left, down the major central road of the scroll are shown four supernatural beings who assisted in the creation. From this middle portion to the left the scroll is concerned with following the road of life and death of the Great Medicine ceremony itself. As celebrants move through the four degrees of ever-deeper membership in this religious association, they learn to live on earth in a sacred manner, as signified by a half rectangle at the very far left of the chart.[50]

A woodland Indian ceremony whose non-Indian influences seem less debatable was the final rendition of the *Nkamwin*, or Big House ceremony, practiced by the Delaware Indians in Oklahoma until around 1924. A world-renewal ritual that was intended to bring good health, prosperity, and blessings to all Delawares, for twelve nights its participants followed the oval-shaped "white path" inside the special, gable-roofed log house built expressly for the ceremony. As each night they figuratively ascended ever closer to the uppermost realm in the twelve-layered Delaware cosmos, dancing around the center post which stood for the "creator's staff," it was said they "pushed their existence in front of them." On the final night their movement on this sacred road had brought them as close as human beings could ever come to their heaven. As this concluded the ceremony, they exited out the building's eastern doorway, and the Delaware universe was restored for another year.⁵¹

While the white man's political uses of cartography were strange and intimidating to Indians, the maps his missionaries drew to promote Christianity may have been more acceptable; indeed, there are even some hints that Indians borrowed some of what we might call their rhetorical cartography from Euro-Americans, as if reterritorializing the newcomer's cosmology. In the mid-eighteenth century a Delaware religious prophet named Neolin, "The Enlightened," began encouraging his followers south of Lake Erie to reject European ways, to abandon rum and the use of firearms and gunpowder. According to Anthony F. C. Wallace, the prophet "made a chart of the path to heaven, showing obstacles placed by white men, by their rum, and by the Indians' own forsaking of primitive virtue, in the way of entry."⁵² Thereafter Neolin's deerskin map "was reproduced and distributed among Indian people to serve as reminders of his spiritual teachings."⁵³

The zeal of anti-white charismatics like Neolin was often directly proportional to the intensity of indoctrination they had suffered under European missionaries. While avenging that crusade to change them, however, they may also have picked up a few pointers. Although we cannot be sure how exposure to Christian proselytizing marked Neolin's pedagogical technique, we know that a century later both Protestant and Catholic missionaries were using so-called ladder

11.7 Eliza Spalding, Protestant ladder chart, ca. 1842. Reproduced by permission of the Oregon Historical Society, #OrHi 627.

charts as visual aids for converting Indians in the American Plateau region (fig. 11.7). Long paper scrolls busy with biblical scenes, they mapped the rungs by which one climbed upward "on the one true road to salvation" toward heaven. On one Catholic chart the Protestants were depicted headed on a dead-end side road, while the Protestant ladder showed the denominations following two parallel ladders, with the Catholics going straight to a fiery hell.[54]

The metaphor of roads to dramatize one's desired progress through Christian cosmologies seems to have been appropriated by Indian prophets in imaginative ways. An exceptionally rich study of such syncretism was conducted by Robin Ridington among the Dunne-za, or Beaver Indians, of British Columbia in the 1960s (fig. 11.8). From the "Dreamer" Charlie Yahey he learned of an earlier prophet, Makenunatane, whose name meant "His Tracks, Earth, Trail." Ridington understood that the name evoked both the Beaver culture hero known as Saya, whose trail runs around the rim of the sky, and swans, which migrate between the seasons.[55] This same Makenunatane had prophetic dreams of the coming of whites, of Jesus, and especially of Yagatunne, the all-important "Trail to Heaven," by which the good might join their relatives after death. In this way Indian logic was able to incorporate and even preempt, as it were, the invasive cosmology.

The interdenominational struggle between Catholics and Protestants may have been irrelevant to Indians, but they seem to have found the Christian mappings inspiring where they depicted the crucial issue of moral choice. The late eighteenth-century Iroquois prophet named Hand-

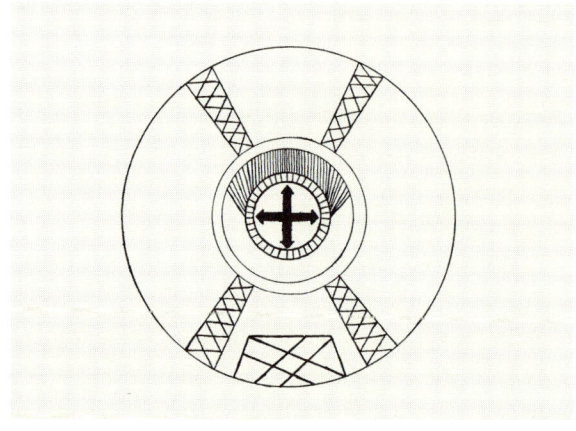

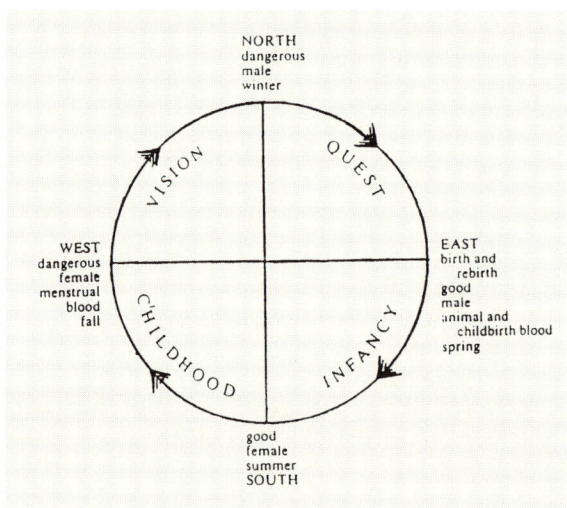

11.8 Drum head drawn by Beaver Indian shaman from northeastern British Columbia, showing the four-quarter symbol that anchors the earth in the midst of the primeval waters, and which also becomes the link between cosmic realms above and below. In addition, the four directions each have their associated colors, times of day, seasons, genders, and emotional qualities toward human beings.

some Lake, who had been exposed to Quaker teachings in upper New York State, preached of the *deyohthaho:gen*, or "the forked path" on which people go "when they leave Mother Earth."[56] While the wide path "led to the place of punishment," the preferable narrow road led "to the land of the Creator."

Another rendition of this probable Christian concept was recorded by ethnographer Donald Cadzow among the Plains Cree of Saskatchewan in 1924. Visiting a tribal elder named Rock Thunder on the Piapot reserve, Cadzow heard him decipher the cosmogram painted on his sacred tipi. According to Rock Thunder, the structure's easternmost tripod pole should be placed "in the path of the sun," while its complex diagram depicted the road we travel through life, and then the thunderbird, lightning, and sun, respectively. At the end of life's road, we are met by a creator spirit (fig. 11.9). Here Christian notions seemed to enter the picture, as Rock Thunder explained the choice between "the wide road to the real home" and the unclean path into the ground, where "what happens there we do not know."[57]

The case of the Lakota holy man Black Elk is one of the best examples of how a Christian teaching tool could transmit this visual metaphor for the new ingredient of *life choice* and "Indianize" it, as it were. For in the image of the split in the road, or the two roads, morality and culture, in the doctrines of both white missionaries and Indian prophets alike, could fuse. Half a century after his life story, *Black Elk Speaks* (1932), brought him renown as a spokesperson for traditional Lakota culture, Black Elk was revealed as having been a staunch Catholic catechist at the same time that he was an eloquent synthesizer of Indian beliefs. The mass-produced, polychrome chart that Black Elk was photographed using with Lakota children in 1935 when he was teaching the Christian cosmology probably derived from those older religious charts, and was popularly known as "The Two

11.9. Tipi painting by Rock Thunder, Plains Cree elder from central Canada, documented by Donald Cadzow, showing (1) the "road we travel through life," which is guarded by (2) the thunderbird, (3) the lightning, and (4) the sun. At the end of this road one finds (5A) day or (5B) night and is received by the Great Spirit, or "Manito." According to Rock Thunder, the moon and stars watch out for us day and night (6A, 6B). If a person has lived a clean life the (7) "wide road to the real home of Manito" is opened, and they live forever in (8) the place where the Manito sends the four winds. But those who have lived unclean lives take (9) another road, and according to Rock Thunder, "what happens there we do not know." Courtesy of the National Museum of the American Indian, Smithsonian Institution.

Roads Map." Anthropologist Michael F. Steltenkamp has itemized the similarities between this printed map and those oral teachings which, in *Black Elk Speaks*, are treated as timeless and pristine Lakota lore. These common motifs include "thunder beings, a daybreak star, flying men, tree imagery, circled villages, a black [evil] road, a red [good] road, friendly wings, an evil man living in flames, a place where people moaned and groaned."⁵⁸

One faction of Hopi spiritualists have put great stock in a "life plan" map which is carved on a large boulder just east of their Oraibi Pueblo, in northern Arizona. Known today as Prophecy Rock (*Wutakyawvi*), when it was first noticed by John Wesley Powell in the early 1870s his Hopi contacts interpreted it as a petroglyphic record of early Spanish assault on a Hopi village. But according to historian of religions Armin W. Geertz, that reading had changed dramatically by the mid-1950s when "traditionalist" spokespeople told U.S. Government officials that its stick figures, circles, and lines charted the original Hopi emergence, the creation of their clans, and the diverging roads leading into humanity's future that illustrated moral choice. Pointing to the movement of human images along two scratched lines, they depicted how those following the wider "white man's road" eventually faced hardship, decapitation, and catastrophe. But Hopis who pursued the narrower "right road" would enjoy old age and "plenty of corn and the good life."⁵⁹

Rather than the domination or dilution of tribal traditions, however, these interpenetrations of belief-systems can also be viewed as Indian strategies for mitigating unavoidable innovations through selective adaptation. Nowhere is this more evident than in the ritual "Road" that was introduced to North America a little over a century ago, and which was a central symbol of the pan-Indian church which was incorporated in Oklahoma in 1918. The peyote ceremony of the Native American Church uses the hallucinogenic cactus *Lophophora Williamsii,* generally known as peyote, as its sacramental food. For peyotists, Jesus Christ is a minor male savior alongside Peyote Woman, who is believed to have first brought this plant to the Indians of North America.

Peyotism elevated the root metaphor of the road to a new status, as it now embodied the proper way to live. The ritual's presiding functionary is called a Road Man. Curving along the ridge of the crescent-moon altar made of earth is a groove that stands for the path of one's life, until one meets Father Peyote. Thereafter you set out upon the Peyote Road, remaining helpful to your fellow peyotists, and living a right life. As one member told anthropologist Warren L. d'Azevedo. "You can't see nothing better than all them Indian people working in their meetings. They're all on the Road together."⁶⁰ And in the rhetorical move common to many revitalization ideologies, this consultant also saw the invaded as coopting the invader's theology: "The Creator put this Herb on Earth for all the people," he explained to d'Azevedo. "Jesus tried to tell the white people how to use It. They forgot, I guess . . . If we keep on the Road It shows us we will have good life."⁶¹

Expressing collective and individual movement through a moral universe, the road and its journey remain dominant metaphors in Indian thought. Books by native authors continue to reference it: as authors Russel Lawrence Barsh and James Youngblood Henderson explain their title *The Road: Indian Tribes and Political Liberty*, "Tribalism has always been the Road, that is, the heart and spirit of the Indian people." ⁶² To Kenneth Lincoln and Al Logan Slagle, who tell how "the sacred Road, the red road" (of *Black Elk Speaks*) inspired their chronicle of "focused wandering" around Indian country, *The Good Red Road: Passages into North America*. While for Acoma poet Simon J. Ortiz, the title for his 1977 poetry collection *A Good Journey* evoked "the vision by which we see out and in and around."⁶³ The health of this key symbol for the persistent moral and spiritual movement of Indian people through space and time also serves contemporary Indian graphic artists. A painting by Virgil Marchand of Santa Fe's Institute of Indian Arts has updated the two-roads map to depict future choices for Indian peoples as a whole; and catalogs for the most recent exhibitions by Washington's new National Museum of the American Indian feature such titles as *Creation's Journey, The Path We Travel,* and *All Roads Are Good*.⁶⁴

For a summarizing example of dramatized geographic history, I return to a deeper reading of the interplay of Crow Indian history and religion to which Chief Iron Bull alluded as quoted earlier in this essay. Although Iron Bull's cartographical prop argued that Crows were created in situ in their current landscape, scholars believe that this Siouan-speaking people arrived in their homeland of the Big Horn Valley region in two waves from present-day North Dakota. Yet other Crow narratives offer an even more complex and historical picture of Crow origins than the one portrayed by Iron Bull.

These stories explain the Crow ethnogenesis as a distinct, buffalo-hunting Plains people out of an older, prehorse horticultural tradition as originally enabled and still commemorated through the oral traditions and ritual activities of their Crow Tobacco Society. Agreeing with white scholars that they split off from the Hidatsa Indians of present-day North Dakota, these narratives describe two men who went on a vision quest. One received a vision of corn and stayed put, but the other received a vision of sacred tobacco seeds. He was also instructed to lead his followers on an epic search for these seeds, and a proper place to plant the tobacco, which would hereafter stand as a key symbol for Crow identity. Once his quest climaxed alongside Montana's Big Horn Mountains, the religious leaders of the Tobacco Society began the annual planting and harvesting rituals for their sacred plant, and commenced the adoption ceremonies that would bring new members into its fold. While the sacred tobacco gardening rites have been discontinued since the 1970s, and although it requires a considerable investment of time and money, Crows still manage to stage their Tobacco Society adoption rituals, known as *bassucuaa,* or "the soaking," from a metaphoric phrase for the special preparation given to the sacred tobacco seeds prior to planting.⁶⁵

The formal sequence of this adoption ritual features a solemn procession in which adoptees are escorted from a preparatory lodge, a wall tent, about fifty yards to an adoption lodge, an oversize, squat tipi framed with ten poles and shrouded with two canvas covers. A woman leads this procession, and there are four points en route when all the celebrants stop for singing and prayer. When they finally reach the sacred adoption lodge, with its altar of ground cedar symbolizing the sacred tobacco, the ritual reaches its climax.

But to interpret the symbolic meanings of this ceremony we must backtrack a moment. Contrary to the scholarly accounts on the tribe proposed by outsiders, most Crow folklore says that their original migration from the middle Missouri region, the one they prefer to sacramentalize, was not a simple, straight route as the crow flies. Instead, Crow oral traditions describe their people's original quest to find and plant the first tobacco as a circulatory journey, during which time they tasted all the corners of the Great Plains until they discovered their "promised land," as the tribe's own bilingual literature for primary schools characterizes this Big Horn Mountain country today. Thereafter, their great nineteenth-century chief, Sits in the Middle of the Land, staked out his people's claim during a treaty conference with this visual map: "When we set up our tipi poles, one reaches to the Yellowstone, the other is on the White River, another goes to the Wind River, and the others are on the Bridger Mountains. This is our Land."[66]

It is this four-tipi-pole metaphor, plus Crow folk versions of their epic migration, which are evoked during the tobacco adoption ceremony. As the procession halts four times, it imitates the four stops made by their legendary wandering forebears at the four-directional extremities of the vast Plains. As they reach the special adoption tent and cluster around a miniature altar garden where ground cedar is said to stand for their sacred tobacco plants, they are symbolizing the biological process of rebirth, as the initiates who have been recategorized as "babies" are both invested into the society and given a new set of societal "parents." But simultaneously they are symbolizing their historical arrival at their promised land, as well as the cultural process through which their transformation as a new ethnic group on the Plains took place. With their feet they have retraced their mythic map, collapsing past and present, biology and culture, historical cartography and ritual sacrament.

By now this essay has strayed far from the flat surfaces of native-drawn maps discovered in archives and the ways that they can reveal historical knowledge, in an effort to explore some metaphysical dimensions of Indian cartographic consciousness and to reflect upon the roles that such conceptualizations of space and human behaviors can play in cultural and historical processes. This trespassing beyond the customary boundaries of a discussion on native cartography seems to suggest that the linking of imagined or actual locations to urgent human concerns on the ground is an integral part of that geographical awareness that we might consider distinctively American Indian. As a native from interior British Columbia told Hugh Brody, "Oh yes, Indians made maps. You would not take any notice of them. You might say such maps are crazy. But maybe the Indians would say that is what your maps are . . . Different maps from different people—different *ways* . . . You may laugh at these maps of the trails to heaven, but they were done by the good men who had the heaven dream, who wanted to tell the truth. They worked hard on their truth."[67]

Notes

1. *Narragansett Tribe of Indians: Report of the Committee of Investigation; A Historical Sketch and Evidence Taken, Made to the House of Representatives at Its January Session, A.D.1880,* State of Rhode Island and Providence Plantations (Providence: E. L. Freeman and Co., 1880), 30–32.

2. L. V. McWhorter, *Hear Me, My Chiefs!: Nez Perce*

History and Legend (Caldwell, Idaho: Caxton Printers, 1952), 157.

3. Bruce R. Caron, "Magic Kingdoms: Towards a Post-Modern Ethnography of Sacred Places," in *Kyoto Journal* 25 (1993): 126.

4. Personal communication, October 9, 1994.

5. Alfred C. Bowers, personal communication, July 1984..

6. Thomas Pynchon, *Mason and Dixon* (New York: Henry Holt and Co., 1997), 345.

7. Joseph Marshall III, *On Behalf of the Wolf and the First Peoples* (Santa Fe, N.M.: Red Crane Books, 1995), 84.

8. Ibid., 85.

9. Ibid., 86.

10. Amos Bad Heart Bull, *A Pictographic History of the Oglala Sioux* (Lincoln: University of Nebraska Press, 1967), 289–90.

11. William Ludlow, *Report of a Reconnaissance of the Black Hills of Dakota, Made in the Summer of 1874* (Washington, D.C.: Government Printing Office, 1875).

12. Mary Jane Schneider, *North Dakota's Indian Heritage* (Grand Forks: University of North Dakota Press, 1990), 52.

13. A. Irving Hallowell, "Some Psychological Aspects of Measurement among the Saulteaux," *American Anthropologist* 94 (1942): 62–77.

14. Hugh Brody, *Maps and Dreams* (New York: Pantheon Books, 1981).

15. Michel Audet, "Le reseau spatial des Qikirtajuarmiut, reflexions thoriques," *Recherches Amerindiennes au Quebec* 5, 3 (1975): 40–47.

16. Edmund Carpenter, "Space Concepts of the Aivilik Eskimos," *Explorations* 5:131–45, 1955.

17. William Faulkner, *Collected Stories* (New York: Random House, 1950), 361.

18. Will D. Campbell, *Providence* (Atlanta: Longstreet Press, 1992), 53.

19. D'Arcy McNickle, *Wind from an Enemy Sky* (Albuquerque: University of New Mexico Press, 1988), 30–31.

20. James Mooney, *The Ghost Dance Religion and the Sioux Outbreak of 1890*, Fourteenth Annual Report of the Bureau of American Ethnology (Washington, D.C.: Government Printing Office, 1892–93), 721.

21. Oliver O. Howard, *Nez Percé Joseph* (Boston: 1881; reprint ed., New York: Da Capo Press, 1972), 64–66.

22. Raymond J. Demallie, "Touching the Pen: Plains Indian Treaty Councils in Ethnohistorical Perspective," in Frederick C. Luebke, ed., *Ethnicity on the Great Plains* (Lincoln: University of Nebraska Press, 1980), 49.

23. Father Prando to Cataldo, September 26, 1883, Gonzaga College Jesuit Archives, p. 5 of translation from the Italian by Paul Gehl, Newberry Library, Chicago, courtesy of Frederick E. Hoxie, director, McNickle Center for Indian History, Newberry Library, Chicago.

24. Yi-Fu Tuan, *Man and Nature,* Resource Paper no. 10, Association of American Geographers (Washington D.C.: Commission on College Geography, 1971), 32.

25. Douglas R. Parks, "Interpreting Pawnee Star Lore: Science or Myth," review essay of *When Stars Came Down to Earth: Cosmology of the Skidi Pawnee Indians of North America,* by Von Del Chamberlain, *American Indian Culture and Research Journal* 9, 1 (1985): 63–64.

26. Leslie White, *The Pueblo of Santo Domingo, New Mexico,* Memoirs of the American Anthropological Association, no. 43 (Menasha, Wisc.: American Anthropological Association, 1935), 29-32.

27. T. T. Waterman, "Yurok Geography," *University of California Publications in American Archeology and Ethnology* 1, 5 (1920): 189-95.

28. William C. Sturtevant, "The Meanings of Native American Art," in Edwin L. Wade, ed., *The Arts of the North American Indian: Native Traditions in Evolution* (New York: Hudson Hills Press, 1986), 31.

29. Charles R. Corum, "A Teton Tipi Cover Depiction of the Sacred Pipe Myth," *South Dakota History* 5 (1975).

30. Ted J. Brasser, "Tipi Paintings, Blackfoot Style," in *Contextual Studies of Material Culture,* National Museum of Man, Canadian Ethnology Service, Mercury

Series, Paper 43 (Ottawa: National Museums of Canada, 1978).

31. Tuan, *Man and Nature,* 24; emphasis mine.

32. A. Irving Hallowell, *Culture and Experience* (Philadelphia: University of Pennsylvania, 1955), 269.

33. Alfred L. Kroeber, *Handbook of the Indians of California*, Smithsonian Institution, Bureau of American Ethnology Bulletin 78 (Washington, D.C.: Government Printing Office, 1925), 661–65.

34. Kroeber, *Handbook,* 684–86.

35. Caron, "Magic Kingdoms," 126.

36. E. C. Parsons, *Pueblo Indian Religion* (Chicago: University of Chicago Press, 1939), 1:360–65.

37. Ruth Bunzel, *Zuni Ceremonialism* (Albuquerque: University of New Mexico Press, 1992), 721.

38. Dennis Tedlock, "An American Indian View of Death," in Dennis Tedlock and Barbara Tedlock, eds., *Teachings from the American Earth: Indian Religion and Philosophy* (New York: Liveright, 1975), 259–60.

39. Peter Nabokov, *Indian Running* (Santa Fe: Ancient City Press, 1987).

40. Sam D. Gill, *Native American Religions: An Introduction* (Belmont, Calif.: Wadworth Publishing Co., 1982), 83–84.

41. James A. Teit, *The Thompson Indians of British Columbia,* Memoirs of the American Museum of Natural History (New York: American Museum of Natural History, 1900), 2:343.

42. A. Irving Hallowell, "Cultural Factors in Spatial Orientation," in Janet L. Dolgin, David S. Kemnitzer, and David M. Schneider, eds., *Symbolic Anthropology: A Reader in the Study of Symbols and Meanings* (New York: Columbia University Press, 1971), 133.

43. A. Irving Hallowell, "Ojibwa Ontology, Behavior, and World View" in Stanley Diamond, ed., *Culture in History: Essays in Honor of Paul Radin* (New York: Columbia University Press, 1960).

44. Tanner, Adrian, *Bringing Home Animals: Religious Ideology and Mode of Production among the Mistassini Cree Hunters* (New York: St. Martin's Press, 1979), 73.

45. A. Irving Hallowell, *The Role of Conjuring in Saulteaux Society*, Publications of the Philadelphia Anthropological Society 2 (Philadelphia: Philadelphia Anthropological Society, 1942).

46. Webber, Alika and Ray, "Divinition Rites," *Beaver* (Summer 1964, Outfit 295): 40–41.

47. Paul Radin, *The Road of Life and Death: A Ritual Drama of the American Indian* (Princeton: Princeton University Press, 1945).

48. Selwyn Dewdney, *The Sacred Scrolls of the Southern Ojibway* (Toronto: University of Toronto Press, 1975).

49. Thomas Vennum, Jr., "Ojibwa Origin-Migration Songs of the *Mitewiwin*," *Journal of American Folklore* 91 (1978).

50. W. J. Hoffman, *The Midewiwin; or 'Great Medicine Society' of the Ojibwa*, U.S. Bureau of American Ethnology, Seventh Annual Report, 1885–86 (Washington, D.C.: Smithsonian Institution, 1891).

51. Speck, Frank G., *A Study of the Delaware Big House Ceremonies,* Publications of the Pennsylvania Historical Commission, no. 2 (Harrisburg: Pennsylvania Historical Commission, 1931).

52. Anthony F. C. Wallace, "New Religions among the Delaware Indians, 1600–1900," in Deward E. Walker, ed., *The Emergent Native Americans: A Reader in Culture Contact* (Boston: Little, Brown and Co., 1972), 351.

53. Arlene Hirschfelder and Paulette Molin, *The Encyclopedia of Native American Religion: An Introduction* (New York: Facts on File, 1992), 66.

54. Jacqueline Peterson, *Sacred Encounters: Father De Smet and the Indians of the Rocky Mountain West* (Norman: University of Oklahoma, 1993), 110–11.

55. Robin Ridington, *Trail to Heaven: Knowledge and Narrative in a Northern Native Community* (Iowa City: University of Iowa Press, 1988), 77–78.

56. Chief Jacob Thomas, *Teachings from the Longhouse* (Toronto: Stoddart Publishing Co., 1994), l00-l01.

57. , Donald A. Cadzow, "The Prairie Cree Tipi," *Indian Notes* 3, 1 (January 1926): 25–27.

58. Michael F. Steltenkamp, *Black Elk: Holy Man of the Oglala* (Norman: University of Oklahoma Press, 1993), 95.

59. Armin W. Geertz, *The Invention of Prophecy: Continuity and Meaning in Hopi Indian Religion* (Berkeley: University of California Press, 1994).

60. Warren d'Azevedo, *Straight with the Medicine* (Berkeley: Heyday Books, 1985), 7.

61. Ibid., 1–2.

62. Russel Lawrence Barsh and James Youngblood Henderson, *The Road: Indian Tribes and Political Liberty* (Berkeley: University of California Press, 1980), viii.

63. Simon J. Ortiz, *A Good Journey* (Berkeley: Turtle Island, 1977), 11.

64. National Museum of the American Indian, *Creation's Journey: Native American Identity and Belief* (Washington: Smithsonian Institution Press, 1994); National Museum of the American Indian, *The Path We Travel: Celebrations of Contemporary Native American Creativity* (Washington: Smithsonian Press, 1994); National Museum of the American Indian, *All Roads Are Good: Native Voices on Life and Culture* (Washington: Smithsonian Institution Press, 1994).

65. Peter Nabokov, "Cultivating Themselves: The Interplay of Crow Indian Religion and History" (Ph.D. diss., University of California, Berkeley, 1988).

66. Edwin J. Stanley, *Rambles in Wonderland; or a Trip through the Great Yellowstone National Park* (Nashville: Southern Methodist Publishing House, 1885), 30.

67. Brody, *Maps and Dreams*, 45–46.

Part 3

Future Encounters in New Contexts

12

Future Encounters in New Contexts

G. MALCOLM LEWIS

Predicting new contexts for any kind of investigation is a difficult and dangerous but important and, from time to time, necessary task. If the attempt is not made, currently vigorous lines of research risk running their courses with diminishing impact; harmless hobbies for those involved, but with little or no significance for others. Yet anyone attempting to predict new contexts must consider the possible risks of so doing against benefits likely to be achieved. Promoting a particular context can result in wasteful use of resources, including the careers of those who are persuaded to become involved. It can also damage the status of the wider field of which it is but a part. These and other risks are not to be taken lightly by those who dare to make the predictions. Nevertheless, I am going to outline briefly five future encounter contexts in the firm belief that it would be negligent not to do so.

At this point readers may wish to remind themselves of my attempt in the Preface to present myself as subject. It is in writing this chapter that I have been most aware of and least inhibited about presenting my own views. Forgive me if I appear to have preached too much. Forgive me also if, in doing so, I appear not to value sufficiently the contributions of my coauthors in chapters 5–11. That is not so. Their contributions are current. Each will undoubtedly stimulate further work. What I am attempting to do in this chapter is not to predict the future courses of current encounters but to anticipate and foster new encounter contexts likely to emerge in the fairly near future.

The five contexts are not mutually exclusive. Nor are they presented as the only ones in which Indian and Inuit maps will be studied and used. There may well be others that I have failed to anticipate. What is certain is that the needs and directions of work in the new contexts will draw on the work reviewed and presented in this volume. Conversely, the new contexts will motivate more and better work along traditional lines. In so doing, they will lead to the rediscovery of lost sources, involve the application of new research methodologies, and demand greater procedural rigor. Above all, new contexts will actively involve new people. Interest in Indian and Inuit maps will no longer be dominated by geographers, cartographers, and others with an interest in maps per se. The focus of interest, as the chapters in part 2 of this volume have heralded, will swing more and more toward what the maps reveal. Increasingly, and by an ever widening range of people, they will be valued as evidence.

The Legal Context

Throughout North America, native/non-native land negotiations and agreements sometimes involved the use of maps and plans. Usually they were produced by the whites, sometimes by the natives, and occasionally by both. The following examples indicate some of the roles of these maps.

In late July and early August 1805, Col. William Claus, deputy superintendent of Indian affairs for Upper Canada, negotiated with the Mississauga Indians to the west of Toronto in order to establish the western limit of the so-called Toronto Purchase of 1787, which had "not been so perfected as to ascertain the exact limits," and at the same time to purchase further lands immediately to the west.[1] In making the request, Claus presented the Indians with at least one map, presumably on paper, and, probably, in part at least, based on survey. At first the Mississauga chiefs were unwilling to sell all that was being requested, their spokesman, Queneponen, supporting his reply with "a sheet of Bark with Lines representing the Tract they were willing to let their Father [George III] have."[2] Claus immediately rejected this offer and, on the following day, Queneponen spoke again "with a flat stone in his hand [it was a region of well-bedded sedimentary rocks] on which was represented the lines within which they had on reconsideration agreed to give their Father."[3] Many words were spoken, some of them describing supposed, proposed, or agreed boundaries. Used in conjunction with at least three maps, made on three different media, by representatives of two vastly different cultures, they do not, however, seem to have led to significant misunderstandings, either at the time or thereafter. The map artifacts have not, of course, survived. Nor is it known what roles they had in the negotiating process; in particular that intended by the Indians and that as interpreted by the whites. One suspects that the role as intended by the Indians was somewhere between empirical at one extreme and symbolic at the other. The accounts do not make it clear how the map artifacts were used. In particular, there is no indication of their relative importance to the whites vis-à-vis the Indians' speeches as received in translation. If there were misunderstandings, they were and have remained unrecognized.

In contrast to the 1805 negotiations with the Mississaugas, the Mississagi River band of Ojibwa on the north shore of Lake Huron used maps of their own to dispute the northern boundary of Reserve #8; not as they believed they had agreed to it in the Robinson Treaty of 1850, but as it was surveyed in the field two years later. Intermittently, and until very recently at least, the dispute rumbled on. A series of single-authored research reports in the early 1980s traced its development and revealed the essential factors behind it. It is not known whether the Mississagi River Band used maps to dispute the survey as it was being undertaken in the field. At Shawanaga to the east the local band of Ojibwa most certainly did. There, they "had their own 'Indian plan' on birchbark ready to indicate their desire for other areas than those identified in the 'Schedule of Reservations.'"[4] What the "plan" was like and how exactly it was used is not clear, but the Mississagi River Band used a similar map, albeit on paper, to make a retrospective protest about the northern boundary of Reserve #8. With a letter of June 25, 1892, sixteen members of the band sent a map of the boundary as it had been surveyed. In the letter, they disputed what had been done, asked for honest surveyors to be sent to resurvey the line, promised to accompany them in the field, and enclosed a "true map" drawn by the chief as evidence of how they had been "wronged." The map is extant and, though on paper, has most of the characteristics of a birchbark map.[5] A well-considered recent opinion concluded that "[a]dmittedly, the scale and orientation are shaky, but the features are all there . . . no 'better description of the points in which they considered themselves wronged' could have been provided."[6]

Such well-reported accounts and subsequently well-researched examples of Indian maps being used in negotiations for and disputes about land are rare or little known. In chapter 7 Margaret Pearce has demonstrated that, in negotiating southern New England land deeds at least, they were the exception rather than the norm. Even so, adopting a continent-wide perspective, it seems very likely that searches in appropriate archives and unpublished reports would reveal more examples than have hitherto been suspected of Native Americans using maps in treaty negotiations and land sales. Even though such use may have been exceptional, many questions arise.

Pearce has considered some of these. There are, however, many more. Why did Native Americans use their own maps in only a minority of cases and what circumstances motivated them to do so? How did white negotiators perceive native maps and what influence did these perceptions have on their negotiating strategies? Did the submission of maps by native negotiators increase or decrease the chances that boundaries supposedly agreed to by both parties would be disputed by natives once pegged on the land? What were the relationships between treaty- and deed-boundaries as defined in words and graphically delimited on native and white maps? It would not be difficult to pose further questions. The essential points are that they are important to a theoretical understanding of the negotiating process and to the legal resolution of some disputes.

Maps made by Indians and Inuit in negotiations relating to land should not be considered as still further evidence in support of such truisms that native North Americans made maps and that these were different in almost every respect from those made by whites. These generalities have long been recognized. Instead, such maps should be considered in the contexts in and for which they were made: not as artifacts in isolation but as symbols and information; in relation to both the proceedings in which they were used and the agreements reached; and with reference both to ground referents and to maps used by white negotiators or subsequently made by other white officials. In-depth studies of these maps, their roles, and the consequences of their having been used will need to draw on a wide range of concepts. Of these, intercultural differences in such things as semantic categories, traditions of graphical representation, and metrical systems will be fundamental. Less fundamental, more case specific, but equally important will be intercultural differences in intention, in geopolitical systems, and in knowledge and evaluation of both terrain and resources. Individual researchers will rarely be capable of dealing with each of these. Future studies of Indian and Inuit maps in legal contexts will need to be group encounters if they are to penetrate much below the level of what is immediately obvious. This will be in keeping with recent developments in both the sciences and humanities, in which once significant problems have been posed, solutions are sought by multidisciplinary teams.

Language, Linguistics, and Semantics in Translational Contexts

The central issues involved in the Mississagi River Band's dispute concerning the northern boundary of their reserve centered on the position of two rapids located on adjacent rivers and a straight line that was intended to connect them. Yet straight lines (or, at least, the means by which they were defined), whether or not drawn on maps, would seem to have been alien to traditional Indian understanding. In 1736, Cadwallader Colden, surveyor general of the lands of the Province of New York, was sent to investigate complaints made by Mohawk Indians that lands they believed they had sold to white settlers were not those that were being pegged out by surveyors in the forests. Sympathetic to the complaints, Colden wrote to the lieutenant governor of the province that in the deeds of purchase "the Boundaries of the land said to be purchased are in several cases expressed in points or Degrees of the Compass [and] by English Measures [presumably yards, chains, furlongs, and miles] which are absolutely unknown to the Indians." His proposed solution was that "in order to prevent such Impositions or Deceits on the Indians of the future . . . all lands to be purchased of the Indians be survey'd in their presence before any Deed for the same be sign'd."[7]

With time, some Indians did acquire an understanding of absolute linear metrics, but confusions persisted. Some of these arose from changes in colonial regimes. For example, the Robinson Treaty made in 1850 with the Ojibwa Indians on and near the north shore of Lake Huron, conveying lands to the crown, used miles in the schedule of reservations. Once the ground surveys began, this caused confusion and dissatisfaction over the extent of their reserves among some of the bands: they were considerably smaller than the leaders thought they had agreed to.[8] This was because through their long contact with the French Canadian voyageurs the Ojibwa had developed an experiential understanding of the French league, for which apparently, they assumed that "mile" was merely the equivalent English noun.[9]

Occasions in which native North Americans were confused by conceptual and linguistic differences between European regimes are easier to demonstrate and were far less frequent than those involving differences between their own and white concepts and ways of expressing them. Many examples could be cited: far more have still to be detected. While only some of these confusions are specific to the understanding of particular maps,[10] most have pervading significance.

In his letter of 1736 to the lieutenant governor, Cadwallader Colden gave another reason for the Mohawks' protestations: "the Deeds . . . obtained from the Indians are in the English Language & your Mem[orialist] has been inform'd that the Indians have been perswaded to sign these Deeds without having them interpreted by persons sufficiently skill'd in the English & Indian languages." Colden recommended that every "Deed be truely & fully explain'd to the Indians before the same be executed."[11] In doing so, he was probably thinking exclusively in terms of oral discourse heavily dependent on nouns. There were, of course, already English and French word lists for some Indian languages. But there were no grammars.

It is now apparent that much more fundamental issues arise when differences in syntax begin to be recognized; particularly in those embodying the expression of location and spatial deictic systems. Of the fifty-nine Indian language families recognized by Joel Sherzer, each has its distinctive way of expressing location in space (locatives) and relations between two or more different locations in space (spatial deixis).[12] The variables are prefixes, suffixes, prepositions, and postpositions. Arising from this, several important questions emerge. Were there regional differences in the native languages of North America in ways for expressing location and spatial relations? If so, where were the major geographical discontinuities in the pattern of differences, did these coincide with discontinuities in the cultural conception of space, and did these in turn act as barriers or distorters in the native geographical information systems? If so, what were the consequences for the areal extent, geometrical structure, and information content of indigenous maps of areas too large to have been based on the direct experiences of any one group? Quite clearly, there are big, complex, and important questions to be posed. Even posing them in meaningful ways will be difficult. Pertinent concepts such as language, translation, communication, and representation have concerned philosophers in recent years. The precision and consistency with which these and other concepts are used will have consequences for the findings. The involvement of linguists will be central but other disciplines will have to be involved. Geographical information passed furthest, but with inevitable distortions, along regular multistage trade routes. Hence, the nature of trade languages will be particularly significant; especially their syntactical limitations. In addition to linguists, archaeologists and ethnohistorians will have a great deal to contribute in explaining what determined the limits and information content of specific large-area maps. In chapter 9, Gregory Waselkov has already given some useful leads.

Categories, hierarchies, and classifications are clearly important influences on map content. For modern maps made in the Euro-American tradition these are usually precisely defined and consistently applied. They determine what is represented, how elements are differentiated and grouped, and to some extent the manner in which they are visually emphasized vis-à-vis each other. Especially in traditional societies, categories, and to a lesser extent hierarchies and classifications, are rooted in language. Only recently, however, have there been serious attempts to explore the linguistic roots of these organizing systems. The best known of these is a book by linguist George

Lakoff subtitled *What Categories Reveal about the Mind*. Its lead title is more opaque but much more exciting: *Women, Fire, and Dangerous Things*. Derived from Dyirbal, an Australian aboriginal language, these are only three of several Euro-American categories included in one linguistic category denoted by the Dyirbal classifier *balan*. In addition to women, fire, and many different dangerous things, this category includes platypus, most varieties of birds, fireflies, the hairy mary grub, some types of trees, the sun and stars, warriors' shields, and some types of spears.[13] From a Euro-American perspective this grouping may seem ludicrous. One wonders how a cooperative Dyirbal speaker would respond to a request to map *balan*. The extent to which categorizations in Indian and Inuit languages had an influence on map content has still to be explored. When it is, the results could be very surprising. The roles of conventional students of Indian and Inuit maps will then be to suggest suitable maps for case studies and then to find and work with appropriate linguists or, better still, with native speakers. Indeed, in all future encounters it is wherever possible important to involve appropriate descendants of the peoples who made the maps.

The above are but a few of the ways in which an understanding of spoken language, linguistics, and semantics might, if rigorously tested, lead to a deeper understanding of Indian and Inuit maps. Far more doubtless await speculative recognition. Many may be untestable, and others may fail rigorous testing, but some will be very revealing. After all, oral discourse and maps are each capable of conveying information about real and imagined differences from place to place on the surface of the earth, in the sky as perceived from it, and in imagined worlds. It would be very surprising if in a given culture they operated quite independently. Perhaps they share some of the deeper structures of Chomskyan transformational grammar. Though not yet applied to the oral languages and graphical maps of traditional cultures, there is already a considerable literature linking natural language and cartography.[14]

Cognitive Science Contexts

This context for future encounters is related to the previous one, but is even larger and will lead in directions where many with conventional interests in Indian and Inuit maps will feel even less comfortable. The 1980s and 1990s witnessed spectacular growth in the neurological and cognitive sciences. The processes involved in cognition are far better known than even a few years ago. In particular, there is now a vast literature on the human cognition of space and the ways in which brains process sensory information spatially. Although some of this focuses on minority groups (e.g., young children, the elderly, and the brain damaged) and some on groups subjected to extreme physical or emotional stress (e.g., mountaineers), very little research has been done on spatial cognition in traditional cultures, or indeed on changes therein consequent upon acculturation. Hence, because the structure and content of maps made by Indians, Inuit, and, of course, other traditional peoples, reflect aspects of the spatial cognitions of those who made them, those with conventional interests therein should do their utmost to bring the existence of maps to the attention of cognitive scientists who might use them as evidence, or even as sources of hypotheses. They should go further by focusing particular attention on the least acculturated examples and posing what seem to them to be the important questions, particularly about map structure, the answers to which might involve deeper structures of the mind.

From time to time, historians of cartography have drawn on the literature of psychology in an attempt to place maps and mapmaking in an evolutionary context. This implied that there were hierarchical cultural stages in development. My own attempt was criticized by Denis Wood.[15] Although the debate is to be welcomed, this is not the place in which to pursue it. It involves the use of evidence far beyond that of Indian and Inuit maps. Wood, however, cleared some of the intellectual confusion surrounding this issue with the important observation that "relationships among *spatial cognition*, the *ability* to *make* maps, and their actual *production* are not straightforward, and the failure of the latter cannot be taken to indicate an absence of the former."[16] Achieving further understanding of these and related issues will involve some hard scientists. Once they become involved, those with more conventional interests must not allow themselves to be frightened away.[17] The work will involve using as evidence maps from many cultures: past as well as present; traditional as well as Western. Those with a conventional knowledge of Indian and Inuit maps will have much to offer because these are better known than the maps of other traditional cultures.

Social Science Contexts

There is a small but growing literature on the social structuring of space. Robert Sack made the first synthesis and included a brief section on "primitive" conceptions of space.[18] Notwithstanding the volume's subtitle, *A Geographical Perspective*, the section seems to have been written without a realization that "primitive" people made maps that merit attention for what they reveal about societal conceptions of space.[19] We now have a Russian perspective on the same topic, although this embraces time as well as space and, in dealing with traditional societies at least, places the emphasis on time.[20] Like Sack, it neither mentions nor draws evidence from maps made within these societies. In part, at last, these lacunae reflect the limited academic awareness of maps and map use in traditional societies in the late 1970s when the two books were being written. Now is the time to make social scientists aware of such maps, as well as the contexts in which and the purposes for which they were made, the past and present societies for which they exist, where to find and consult them, and how to recognize different survival states. The last point is important, since some states afford more reliable evidence than others.

Artistic, Literary, and Performance Contexts

In every society, one of the roles of art, literature (in preliterate societies, folk narratives), and ceremonial performance is to represent. This is equally true of indigenous and commercial forms. It would be surprising, therefore, if evidence of maps, mapping, or at least space was not to be found in examples of some of these. The search, however, has barely begun.

Chapter 4 briefly mentioned two examples of maps being incorporated in native North American commercial art and another in which map drawing was part of a creative activity.[21] A search for further examples ought to be rewarding and could be very revealing, though it would be necessary to filter out Western-style maps of the kind engraved as mementos for tourists by Alaska coast Inuit after the end of the nineteenth century.[22]

Whether or not acculturated, the examples discussed in chapter 4 are undoubtedly maps. Equally revealing, indeed perhaps even more so, would be a search for art that, while not obviously incorporating maps, nevertheless represents the arrangement of things and events in space. Especially when working at the level of the subconscious, artists sometimes draw on deep traditions not generally evident in modern native societies. Even among commercial artists of course, such traditions can wane. In a study of drawings made by Baker Lake Inuit, Northwest Territories, Marion Jackson found it necessary to distinguish between first- and second-generation artists. First-generation artists were those who had spent their first fifty or more years living more or less traditionally on the tundra, with little or no contact with the outside world. Their drawings often fused spatial perspective to achieve planlike assemblages. In contrast, the second generation, although they had grown up in the tundra environment, had experienced sustained contact with the outside world and were more aware of themselves as artists.[23] One study suggests that, compared with the first generation, among "second generation artists the panoramic perspective becomes less well understood and more confused in implementation."[24] Quite clearly, attempts to use art as a means of furthering the understanding of maps must proceed cautiously and, as in all other future encounters, involve appropriate specialists.

Usually in the role of metaphors, in recent years maps have appeared increasingly in fiction. Not so, however, in fiction about or by native North Americans. Rudyard Kipling's *Just So Story*, "How the First Letter Was Written," may have been the first, although, notwithstanding the beaver swamp, carp, deer, and birchbark, it is not entirely clear on which continent the fictitious Neolithic characters with Edwardian English speech forms were supposed to have lived. But, given Kipling's brief residence in rural Vermont a few years before writing it, North America seems most likely. The letter, on birchbark, was certainly made in part to serve as a map for "Stranger-man" who was to take a message to little Taffy's mother:

> You go along till you come to two trees (those are trees), and then you go over a hill (that's a hill), and then you come into a beaver-swamp full of beavers. I haven't put in all the beavers, because I can't draw beavers, but I've drawn their heads, and that's all you'll see of them when you cross the swamp. Mind you don't fall in! Then our Cave is just beyond the beaver-swamp. It isn't as high as the hills really, but I can't draw things very small. That's my Mummy outside. She is beautiful.[25]

Kipling's birchbark map lacks the authenticity of the Blackfeet and Gros Ventre novelist James Welch. In *Fools Crow*, White Man's Dog makes love in the grass to Kills-close-to-the-lake. Afterward, "he felt her fingers tracing worlds on his back, and then he slept."[26] This has the ring of authenticity lacking in Kipling, raising questions, some of which verge on being hypotheses. If the "worlds" were traced, were they spatial? If so, they were maps, and mapmaking was as instinctive as lovemaking. Indeed, it must have been, for White Man's Dog to recognize it for what it was when he was on the verge of sleep.

If mapmaking was indeed the natural and intimate activity James Welch would have us believe, does it or did it ever occur in comparably private realms? The answer to that question is, probably yes—in dreams.

In 1639, the Jesuit François du Peron wrote that the Indians "throughout the Huron country" (Algonquian- and Iroquoian-speakers including Ottawa-Ojibwa, Nipissing, Huron, and Petun) "consider the dream as the master of their lives, it is the God of the country; it is this which dictates to them their feasts, their hunting, their fishing, their war, their trade with the French."[27] With the possible exception of feasts, these activities and pertinent predictions relating to them must, in part at least, have been spatial. Though Peron does not explain the nature of the dreams, to have been effective they must have contained maplike components.

More than three hundred years later, something very similar was reported by Robin Ridington among the Beaver Indians (Dunne-za) of northeastern British Columbia. According to folk memory:

> The Prophet tradition, as many people described it to me, began with a Dreamer named Makenunatane, also called the Sikanni Chief. Stories about this man tell that he dreamed ahead to predict the coming of the white men. The stories say that he began his prophetic career as a leader of communal hunts just before the white men came to Dunne-za country. His ability to "dream ahead for everybody" allowed him to visualize the pattern of a perfect communal surround of an animal, during the last days in which the technique was practiced, before the introduction of firearms. He could tell people where every one of them should be, in relation both to one another and to the game.[28]

Although Makenunatane was a nineteenth-century dreamer, Beaver Indians were dreaming trails to heaven until relatively recent times; and making and preserving map artifacts of them. Hugh Brody's account of such a map on moosehide which he saw at a meeting of Beaver Indians in 1979 has already been referred to.[29] Ridington mentions having seen a similar example a few years before:

> Prophet River [northeast British Columbia] is named for Deculta the [Beaver Indian] Dreamer, who painted his dream of the Trail to Heaven on a moosehide that Jumbie kept and showed to people on special occasions. As Charlie Yahey, the Dreamer whose practice I knew best, explained to me, "A person who is Prophet doesn't just dream for himself. He dreams for everybody."[30]

Dreaming a map, though harder for the Western scientific mind to understand, was evidently not as private as tracing worlds on one's lover's back. But as with commercial art, we should be cautious about drawing ideas from commercial fiction. A close reading of Indian and Inuit folklore might be more revealing and would certainly reduce the chance of deriving ideas from acculturated sources.

More public than lovemaking and dreaming but prone to changes and loss of meaning through time are certain forms of traditional performances. Dances and many ceremonies involved movements in two—sometimes three—dimensional space. Yet the possible existence of traditional action maps seems not to have been systematically considered. Much potential evidence certainly exists; particularly for the Pueblo Indians of the Southwest.

Necessary Conditions for Achieving Future Encounters

Whether or not I prove to have been correct in isolating law, language, the cognitive and social sciences, and the traditional and commercial arts as future contexts in which Indian and Inuit maps will be investigated and utilized, certain conditions will have to operate if significant progress is to be made. Most of these have been anticipated already but are so vital as to require a final systematic restatement. In order to achieve maximum impact they are listed here, but the sequence should not be misconstrued to imply an order of significance. With the exception of the all-important last one, they will be equally important.

A generally accepted operational definition of "map" is required. This must be wider than that normally admitted by cartographers but sufficiently precise to exclude graphics, behavior patterns, language, and mental constructs that lack a spatial component relatable to either real or imagined worlds.

Those with developed and, for the most part, conventional interests in Indian and Inuit maps must share their understanding and problems with students of traditional cartography on other continents.[31] They must also communicate their findings to a much wider community of scholars and scientists than hitherto, be honest about the problems of evidence, formulate carefully stated hypotheses, and seek to involve appropriate specialists in refining and testing them. Once a diversity of appropriate others has been involved it is vital that the conventional hosts do not take fright and abandon the field.

In every context, serious attempts must be made to inform, obtain permission from, and engage the involvement of native peoples. In many cases they are the rightful custodians of artifactual evidence, and inheritors of invaluable knowledge. A better understanding of their cartographic tradition will further many of their own causes.

Arising from the last condition, arguably more important than any of the others, and as a corrective to the gross bias implicit in almost all studies to date, it is essential that native peoples be encouraged to present their own perspectives on the maps and mapmaking activities of their forebears.[32] In short, to reveal their own mapworlds. As well as being the most important of the necessary conditions for achieving future encounters, this is the most urgent.

Notwithstanding recent developments in the understanding of Indian and Inuit maps, mapmaking, and map use, if progress is not made toward achieving the previously stated conditions, future encounters will be mere reencounters, leading to a hiatus comparable to that of the 1910 to 1970 period. From that, a resurgence of interest might never be achieved. If, on the other hand progress is made toward achieving the conditions, this book will have been in part responsible. As contributors, we are aware that our joint achievement may prove only to have placed a cairn on a low summit before descending again to the lowlands. I hope that that will not be so and what we will be seen to have created is a staging camp, some of us there to be joined by others before attempting more demanding ascents to ever more exciting peaks of achievement. One thing is certain. We will not have long to wait for the verdict: either a gentle and genteel descent with comfortable friends, or a renewed upward struggle, sharing both the dangers and the successes with new and lively comrades.

Notes

1. William Claus, in "Proceedings of a Meeting with the Mississaugues at the River Credit 31st July 1805," Indian Affairs, Lieutenant-Governor's Office, Upper Canada, Correspondence, 1796–1806, RG 10, 1:290, National Archives of Canada, Ottawa.

2. Chief Queneponen, "At a Meeting with the Mississagues at the River Credit, 1st August 1805," Indian Affairs, Lieutenant-Governor's Office, Upper Canada, Correspondence, 1796–1806, RG 10, 1:296, National Archives of Canada, Ottawa.

3. Chief Queneponen, in "Proceedings of a Meeting with the Mississagues at the River Credit on the 2nd August, 1805," Indian Affairs, Lieutenant-Governor's Office, Upper Canada, Correspondence, 1796–1806, RG 10, 1:298, National Archives of Canada, Ottawa.

4. David T. McNab, "Research Report: The Location of the Northern Boundary, Mississagi River Indian Reserve #8, at Blind River," typed report, Office of Indian Resource Policy, Ministry of Natural Resources, Province of Ontario, Toronto, November 17, 1980, revised March 8, 1984, 11.

5. Untitled manuscript map in ink by Chief Michael Sahgatchewaikezhis of the lower Mississaga and Blind Rivers, Ontario, accompanying a letter of June 25, 1892, from the chief and fifteen other members of his band, addressed to R. Sinclair, Acting Deputy Minister of Indian Affairs, Dominion of Canada, RG 10, vol. 7751, file 27013–6, pt. 1, National Archives of Canada, Ottawa.

6. David W. Lambden, "Mississaga Band of Ojibways: Location of the Northern Boundary," typed report to the Indian Commission of Ontario, dated April 6, 1986, Guelph, Ontario, 38.

7. Cadwallader Colden, "To the Honourable George Clarke Esqr. Lieutenant Governor of the Province of New York Ec... New York," November 3, 1736, in *The Letters and Papers of Cadwallader Colden*, vol. 2, *1730–1742*, Collections of the New York Historical Society, vol. 51 (New York: Printed for the Society, 1918), 158–59.

8. McNab, "Research Report," 22 n. 27.

9. Lambden, "Mississaga Band of Ojibways," 20.

10. For example, the explanation as Indian in origin of several strange features and patterns that characterized many eighteenth-century European maps of the northern interior of North America. See G. Malcolm Lewis, "La Grand Rivière et Fleuve de l'Ouest: The Realities and Reasons behind a Major Mistake in the Eighteenth Century Geography of North America," *Cartographica* 28, 1 (1991): 54–87.

11. Colden, "To the Honourable George Clarke," 158.

12. Joel Sherzer, *An Areal-Typological Study of American Indian Languages North of Mexico*, North-Holland Linguistic Series, no. 20 (Amsterdam: North-Holland Publishing Co., 1976).

13. George Lakoff, *Women, Fire, and Dangerous Things: What Categories Reveal about the Mind* (Chicago, University of Chicago Press, 1987), 92–93.

14. See for example, Alan M. MacEachren, *How Maps Work* (New York: Guilford Press, 1995), 152, 213–14, 325.

15. G. Malcolm Lewis, "The Origins of Cartography," in J. B. Harley and David Woodward, eds., *The History of Cartography*, vol. 1, *Cartography in Prehistoric, Ancient, and Medieval Europe and the Mediterranean* (Chicago: University of Chicago Press, 1987), 50–53; Denis Wood, "*The History of Cartography,* volume 1: Review Article," *Cartographica* 24, 4 (1987): 71–72.

16. Denis Wood, "Maps and Mapmaking," in Robert A. Rundstrom, ed., *Introducing Cultural and Social Cartography*, Cartographica Monograph no. 44, *Cartographica* 30, 1 (1993): 2.

17. For a recent contribution by two fellows of the Mapping Sciences Institute, Australia, see M. (John) Balodis and Gita Pupedis, "Geographic Information in Pre-literate Human Communities," *Cartography: Journal of the Mapping Sciences Institute, Australia* 25, 2 (1996): 13–22. They reexamine the functional role of several well-known maps made by aboriginal peoples in several parts of the world "in terms of contemporary geographic information concepts" (p. 13). Because, in their opinion, "pre-literate spatial orientation and model making has practically disappeared," they end with a plea for "a summary reserach program in pre-literate spatial communications [that] could not only be of some anthropological and geographical value, but [should also] clarify a range of psychophysical aspects of human spatial perception" (p. 21).

18. Robert D. Sack, *Conceptions of Space in Social Thought: A Geographic Perspective* (London: Macmillan, 1980), 170–77.

19. Throughout this book we have avoided using the term "primitive" because of its unfortunate connotation. Sack uses it to mean "primary or original in time or even in rank." Ibid., 170.

20. Murad D. Akhundov, *Conceptions of Space and Time: Sources, Evolution, Directions* (Cambridge: MIT Press, 1986), 19, 35, 38–42. The original Russian edition was published in 1982.

21. See chapter 4, nn. 114, 115, 117, 119.

22. See chapter 4, nn. 112, 113.

23. Marion E. Jackson, "Baker Lake Inuit Drawings: A Study in the Evolution of Artistic Self-Consciousness" (Ph.D. diss., University of Michigan, 1985).

24. Pete Osmers, "Inuit Perspective in Drawings," term paper in Inuit Art, submitted April 13, 1992, Carleton University, Ottawa.

25. Rudyard Kipling, "How the First Letter Was Written," in *Just So Stories for Little Children* (London: McMillan, 1902); I have quoted from *Just So Stories* (Ware, Herts.: Wordsworth Children's Classics, 1993), 129–30.

26. James Welch, *Fools Crow* (New York: Viking Penguin, 1986), 119.

27. François du Peron, Ossossané, April 27, 1639. Reuben G. Thwaites, ed., *The Jesuit Relations and Allied Documents* (Cleveland: Burrows Bros., 1898), 15:177–79.

28. Robin Ridington, *Trail to Heaven: Knowledge and Narrative in a Northern Native Community* (Vancouver: Douglas and McIntyre, 1988), 77–78.

29. Hugh Brody, *Maps and Dreams: Indians and the British Columbia Frontier* (London: Jill Norman and Hobhouse, 1981), 266–67.

30. Ridington, *Trail to Heaven*, 77.

31. This task will become easier with the publication of David Woodward and G. Malcolm Lewis, eds., *Cartography in the Traditional African, American, Arctic, Australian, and Pacific Societies*, vol. 2, bk. 3 of *The History of Cartography* (Chicago: University of Chicago Press, forthcoming 1998).

32. This will involve far more than correcting and revising earlier studies by whites. In trying to understand Beaver Indian dreamings, Robin Ridington came to recognize the fundamental difference between white and Indian thinking. "In our [educated white] thoughtworld, myth and reality are opposites. Unless we can find some way to understand the reality of mythic thinking, we remain prisoners of our own language, our own thoughtworld. . . . The language of Western social science assumes an object world independent of individual experience," e.g., the object world as supposedly represented on a topographic map. "The language of Indian stories [and likewise maps] assumes that objectivity can *only* be approached through experience." Ridington, *Trail to Heaven,* 71. These and related ideas are difficult for nonnatives to grasp, and much understanding is lost, diluted, or misconstrued when the experiences of one culture become objects for study by another.

About the Contributors

G. Malcolm Lewis taught geography at the University of Sheffield, England, throughout his career. His early work on the American Great Plains and different communities' perceptions of the region led him to the study of manuscript maps. He soon realized that little serious attention was being given to native maps of this and other regions of North America, and has devoted the last twenty years to the study of North American Indian and Inuit maps. Recently this interest has expanded to embrace maps in traditional cultures elsewhere in the world. He is currently coediting with David Woodward *Cartography in the Traditional African, American, Arctic, Australian, and Pacific Societies,* which is volume 2, book 3 of *The History of Cartography.*

Elizabeth Boone has been interested in the painted books of preconquest and early colonial Mexico since she was a graduate student at the University of Texas at Austin, where she took her Ph.D. in art history. Her publications include *The Codex Magliabechiano, Incarnations of the Aztec Supernatural, The Aztec World,* and (edited with Walter Mignolo) *Writing without Words: Alternative Literacies in Mesoamerica and the Andes.* A new book on Mexican pictorial histories is forthcoming. Mexico has recognized her contributions to Aztec studies by awarding her the Order of the Aztec Eagle. Formerly director of pre-Columbian studies at Dumbarton Oaks, she holds the Martha and Donald Robertson Chair in Latin American Art at Tulane University.

Barbara Belyea has lived, studied, and worked in western Canada, Quebec, Britain, and Switzerland. Most of her publications are in the fields of Canadian literature and Canadian history; they include two editions, E. D. Blodgett's *Driving Home,* and David Thompson's *Columbia Journals.* Currently she is editing another fur-trade journal and the diaries of an Ontario pioneer. She teaches literature at the University of Calgary.

Margaret Wickens Pearce is assistant professor of geography at Humboldt State University. Her dissertation, "Native and Colonial Mapping in Western Connecticut Land Records," examines both native and colonial mapping strategies during the colonial period. She teaches cartography.

Morris Sheppard Arnold was educated at Exeter, Yale, the University of Arkansas, Harvard, and the University of London. A former president of the American Society for Legal History, he taught at numerous American law schools, including the University of Pennsylvania, where he was professor of law and history and a vice president of the university. The French government recently named him Chevalier de l'Ordre des Palmes Académiques for his work on eighteenth-century Louisiana. A United States District Judge for seven years, he has been a judge of the United States Court of Appeals for the Eighth Circuit since 1992.

Gregory A. Waselkov, an associate professor of anthropology at the University of South Alabama, attended the University of Missouri at Columbia and the University of North Carolina at Chapel Hill. He has directed many archaeological excavations of historic sites in Alabama, including the early-eighteenth-century French colonial capital, Old Mobile. Other research, more ethnohistorical in nature, has focused on the colonial-era Creek Indians, recently manifest in a book coedited with historian Kathryn Holland Braund, *William Bartram on the Southeastern Indians*.

Patricia Galloway, who has published in southeastern North American ethnohistory and French colonial history, is special projects officer with the Mississippi Department of Archives and History. There she has supervised archaeological site surveys, permanent museum exhibits on Mississippi's colonial history, and numerous editing projects in history and archaeology. Her most recent book, *Choctaw Genesis 1500–1700*, was the 1996 recipient of the Erminie Wheeler-Voegelin prize for ethnohistory and the James Mooney award for southern anthropology. She is presently working on a second doctorate, in anthropology, at the University of North Carolina in Chapel Hill.

Peter Nabokov is an associate professor in the department of world arts and cultures and the American Indian studies program at the University of California at Los Angeles. Among his books are *Native American Architecture* (with Robert Easton), *Native American Testimony, Architecture of Acoma Pueblo, Indian Running, Tijerina and the Courthouse Raid,* and *Two Leggings: The Making of a Crow Warrior*. His current projects include an ethnographic study of the relationship between various Native American groups and Yellowstone National Park and an investigation of a century and a half of Pueblo Indian history through the experiences of a single Acoma Indian family.

Index

Aberley, Doug: *Boundaries of Home: Mapping for Local Empowerment,* 99

Aboriginal California and Great Basin Cartography (Heizer), 77

Ackia (Chickasaw village), 214

Ackomokki (Blackfoot leader), 42, 144, 148

Ackoweeak (Blackfoot leader), 144

Acoma people, 250

action maps: traditional, 281

Acts of the Commissioners of the United Colonies, 178

Adams, Dwight R., 242

Adler, Bruno F., 35, 39, 46, 48

adoption rituals: *(bassucuaa)* of Crow Tobacco Society, 265–66; calumet used in, 191

aerial photographs: and bioregional mapping, 99

agriculture: Chickasaw, 220n. 31; in Weantinock, 175

Aguilar (interpreter), 111

Aivilik Inuit: mapmaking styles of, 73; of Southampton Island, 246

Aivilingmiut: manuscript map by, 109n. 127

Alabama River: Indian nations along, 235

Alarcón, Hernando de, 18, 43

Alaska Commercial Company, 94

Algonkin Indians: maps by, 24

Allen, John: *Passage through the Garden: Lewis and Clark and the Image of the American Northwest,* 72

alliance belts, 86, 87; cartographic principles in, 84

alliance rituals: calumet used in, 191

alliances: between Illinois Indians and Quapaws, 191–92

allotment cartography: and native land loss, 247

All Roads Are Good (exhibition catalog), 265

alphabetic writing: introduced by Spaniards into Mexico, 113

Alvarado, Francisco de: Mixtec/Spanish dictionary by, 113

American Geographical Society, 41

American Museum of Natural History, New York, 24

Amerindian maps and mapping: Graham's debt to, 137; role in scientific mapping, 140. *See also* Indian maps; native maps

"Amerindian Maps: The Explorer as Translator" (Belyea), 72

Ammons, Mr. (Narragansett Indian), 242, 243, 244

Amoroleck (captive of John Smith), 211

Amos Bad Heart Bull (Oglala Lakota artist): drawing of Black Hills by, 87–88, 107n. 80; maps by, 106n. 65, **245**

Anasazi: astronomical motifs in rock art of, 252

Andree, Richard, 35, 39, 40; "Die Anfänge der Kartographie," 41; *Ethnographische Parallelen und Vergleiche,* 38

animal drives: depicted in rock art, 77–78

"animal homes": of Pawnee, 253

animals: portrayal of spirit guardian, 254

annals history, 117

Anthropogeographie (Ratzel), 35

anthropological contexts: recent scholarly encounters in, 3, 83–94

anthropological "upstreaming," 149, 150, 155n. 45

anthropologists, 83, 84, 85, 87, 88, 89, 94

Apalachicola River system: on Guillaume Delisle's map, 234

Apeony (Chickasaw village), 214

archaeoastronomers, 88

archaeoastronomy, 76, 252

archaeological encounters, 3

archaeological/ethnohistorical evidence: native maps as source of, 90–92

archaeological sites: depicted on John Smith's map of Virginia, 207–14; Sitting Rabbit's map and identification of, 79–80

archaeologist/ethnohistorians, 78

archaeologists, 74, 76, 80, 81, 193, 277; courses of Southeastern prehistoric and early historic paths traced by, 218; matching of sites from Smith's map of Virginia by, 213; native cartographic information on Smith's map of Virginia as interpretive tool for, 211

archaeology: as comparative discipline, 216

Archer, Capt. Gabriel, 43, 207, 213

architecture, 94, 249; *bousillage* French colonial construction, 190; of center shrines, 250; description of, on Quapaw/Arkansas painted buffalo hide, 188, 189; with vertical axis motifs, 253

Archives Nationales (Paris): Captain of Pacana's redrafted copy of map of Chickasaw towns in, 214

Archivo General de Indias, Seville: oldest extant North American map in, 43

Arctic coast: mapping of, 139

Arctic exploration: native maps used for, 23, 41

Arkansas: cartographic interpretation of the Quapaw/Arkansas buffalo hide, **196**; reconstruction of events via painted buffalo hide from, 187–98

Arkansas Post, 191; building of, 190; French houses within, 189, 200n. 13; on cartographic interpretation of the Quapaw/Arkansas buffalo hide, **196**; *métis* of, 198

Arnold, Morris S., 63, 79, 86, 228

Arrowsmith, Aaron, 141, 144; *Map Exhibiting all the New Discoveries in the Interior Parts of North America . . . ,* 135, 138, 143, 144, **145**

art: diachronic character of Indian, 196, 203n. 84; Eskimo, 94–95; prehistoric shell, 218n. 1; representational role of, 279; road motifs in, 265; sky dome motifs in, 252

artifacts: maps as, 10; repatriation of, 84

artistic ability: of Quapaw Indians, 191

artistic contexts: and future encounters, 279–82

ash: pictographs on ground in, 58

assimilation, cultural: between Quapaws and French, 197

Association of American Geographers, 41–42

astronomical motifs, 76

astronomy: Lakota knowledge of, 62; in native cosmography, 252; and star charts, 42

Athapascan people, 246; maps by, 102n. 15; toponymy of, 160

Atherton Company, 178

Atkinson, James R., 214

Audet, Michel, 246

auguries: casting of, 259

Avond, Father, 195

Awashonks (sachem of Sakonet), 174

Awawas (Paugussett), 165, 167

axes: directional, on Quapaw/Arkansas painted buffalo hide, 195–96; within Native American realm of cosmology, 252–53

axis mundi motif, 253

Aztec maps: differences from native North American maps, 94, 110n. 137

Aztec Mexico: cartographic histories in, 117–23, 131; community charters in, 123–31; maps in, 112–13; painters in, 111–12; paths for movement in maps of, 113–17, 131

Aztec painted books, 112
Aztecs: departure from Aztlan depicted on Mapa Sigüenza, 118–19, **119**, **120**; functions of cartographic paintings for, 113
Aztlan: Aztec mythical homeland of, 118; Aztecs' departure from depicted on Mapa Sigüenza, **119**, **120**

Babiquiviri Mountain: I'itoi's (Papago culture hero) home on, 257
back-sighting, 161
"bad conduct": avoidance of, 255
Bagrow, Leo: *Geschichte der Kartographie,* 38
Baker Lake Inuit: study of drawings by, 280
balance of power: Amerindian maps and European understanding of Indian, 225
Bally Husky (Cree), 258, 259
Bancroft Library, University of California (Berkeley): maps by Kohklux and wives in collection of, 26, 32n. 56, 53n. 52, 153n. 23
Barber, J. W.: *Historical Collections . . . of Every Town in Massachusetts:* redrawn version of Sepecan deed in, **171**
bark: pictographs on, 58
bark-covered houses: of Quapaws, 188
bark (or wood) slabs: pictographic records on, 83
Barsh, Russel Lawrence: *, The Road: Indian Tribes and Political Liberty,* 264

Bartram, William, 218n. 5
basket designs: Papago "road of life" woven into, 257
Bassett, William, 173
battle plans: by *tlacuiloque,* 112
battle records: and Aztec cartographic histories, 118
battle scene(s), 76; portrayed on Quapaw/Arkansas painted buffalo hide, 187
Baumhoff, Martin A., 77
beads (Gulf Coast whelk shell): long-distance trade in, 216
Bear Butte, South Dakota: importance to Sioux, 245
bear hunting, 98
Beaver Indians (Dunne-za): dream maps by, 13, 62, 108n. 108, 281; drum head drawn by shaman, **262**; incorporation of Christian cosmology by, 262
beaver preserves: maps solicited for monitoring of, 23–24
Belcher Islands: Wetalltok's map of, 16, **17**
belief systems: map-based, 88–89; moral choices exemplified in interpenetrations of, 263, 264
Bell, Robert, 45
Bellin, Jacques-Nicolas, 136, 137, 139, 141; Amerindian map derivations by, 142; *Carte de l'Amérique septentrionale,* 135, **136**, **137**
bells: use by Quapaws, 190
belts: wampum, 84
Belyea, Barbara, 37, 74, 101n. 6, 110n. 130; "Amerindian Maps:

The Explorer as Translator," 72; on "Eurocentric bias," 108n. 109
Berens River Chippewa, 255
Berghaus, Heinrich, 35
Berghaus, Hermann, 35
Best, George, 246
Beyond Geography: The Western Spirit against the Wilderness (Turner), 73
bias: in awareness of range of map types, 58; and errors, 82; Eurocentric, 108n. 109, 139; in history of cartography, 152n. 11
Bienville, Jean Baptiste le Moyne, Sieur de, 235, 239n. 40
Big Black Meteoric Star Bundle (Skidi Pawnee), 252
Big Horn Mountains, Montana: and Crow Indians, 265, 266
Big House ceremony or *Nkamwin:* of Delaware Indians, 260–61
bioregional mapping, 94, 99, 110n. 131; earliest example of, 100; worldwide projects by native peoples with, 99, 110n. 132
birchbark: enhanced map on, 65n. 18; incised, 47, 57; maps on, 15; message maps on, 12, 58, 87; Ojibwa records on, 83; in Shaking Tent rite, 259; stored maps on, 83
birchbark biting: accounts of, 46, 53n. 55
birchbark scrolls: and Great Medicine Lodge ceremony, 260; of Ojibwa peoples, 63, 84, 86; vertical-perspective pictographs from, 47
Black Elk (Oglala Lakota seer), 252;

fusion of Christian cosmologies and Lakota lore by, 263–64

Black Elk Speaks, 263, 264–65

Blackfeet people: visionary portrayals on tipis of, 254

Black Hills, South Dakota: Amos Bad Heart Bull's drawing of, 87, 107n. 80; depicted in Amos Bad Heart Bull's map, **245**; sacred significance to Sioux, 245

Black Warrior River: Choctaw villages along and connection to, 233, 235; Moundville polity on, 235

Blake, Joseph, 22

blankets, trade: cosmic tree portrayed with mother-of-pearl buttons on, 253

Blowsnake, Jasper (Winnebago), 260

Blue Lake (Wheeler Peak area): center shrine in, 250

Boas, Franz: maps in papers of, 24

Bockstoce, John, 73

Boische, Charles, Marquis de Beauharnois de la, 89

Boone, Elizabeth Hill, 86, 93, 94, 110n. 137, 234

borders: *lienzos* and marking of, 124

Bosh, Lawrence van den: manuscript map compiled by, **20**, 20–21, 22

Bossu, Jean-Bernard, 189, 190, 194; *Nouveaux voyages dans l'Amérique septentrionale,* 194

Bottle Creek mound site: 239n. 40

boundaries: creation of cadastral, upon land areas, 246–47; Indian/European differences in concepts of, 243; *lienzos* and disputes over, 127; political significance for tribal and colonial, 170; in Ridgefield deed, 168

Boundaries of Home: Mapping for Local Empowerment (Aberley), 99

Bourne, Richard, 173

Bragdon, Kathleen J., 160

Bray, William, 40, 78, 104n. 37; map redrawn by, **12**

breechclouts: for Quapaw men, 194, 195

Brinton, Daniel G., 85

Britain: new geography in, 34

British: possession of Louisiana colony east of Mississippi by, 237

British Columbia: Beaver Indians (Dunne-za) of, 262, 281

British explorers: advance into North America by, 135

British Museum, London: Indian maps in, 38, 42

Brody, Hugh, 246, 266, 281; *Maps and Dreams,* 13, 147

Broutin, Ignace: map by, 237, 239–40n. 49

Bry, Theodore De, 246

Buache, Philippe: *Carte physique des terreins les plus élevés de la partie occidentale du Canada,* 142, **143**

Buagrès (Quapaw chief), 197

buckskin: Pawnee sky chart on, 13, 84

buffalo hide: narrative scenes on robes of, 103n. 23; Quapaw painting on, 191, 228

built structures: space symbolized as, 63. *See also* architecture

Bureau of American Ethnology, 24, 46, 47

Bushnell, David, 211

Butler, Thomas, 173

cabanne de valer (house of valor): of Quapaws, 200n. 12

Cacique: deerskin map by, 42

cadastral boundaries: land areas and creation of, 246–47

Cadzow, Donald: tipi painting by Rock Thunder documented by, **263**

calendars: within Codex-Féjérváry, 130; within Codex Mendoza, 128; Mesoamerican, 133n. 21; and sky dome, 251

calmecac school: pictographic writing taught in, 112

calumets: represented on Quapaw painted buffalo hide, 187, 188; use by Quapaws, 190–91, 193, 195, 200–201n. 27

Campbell, Will: *Providence,* 247

Canadian Arctic Expedition (1913–1918), 57

Canadian Inuit: commercial art of, 95

Canadian scholarship, 35

Cantino planisphere, 44

Captain of Pacana (Alabama Indian): copies of maps created by, 226, 227; map of Chickasaw towns, redrawn by de Batz, 214, **215**

cardinal directions/orientations: within Codex Féjérváry, 130; within Codex Mendoza, 129; and deconstruction of maps, 91; within native realm of cosmology, 250, 253; question of, on Quapaw/Arkansas painted buffalo hide,

196, 197; simplified in Amerindian maps, 141
caribou hunting: and map biographies, 97, 98; and scapulimancy, 90
Caron, Bruce R., 243, 244, 256
Carpenter, Edmund, 246; *Eskimo Realities,* 73
Carte de l'Amérique septentrionale (Bellin), 135, **136**, 137
Carte des environs du Mississipi par G. De l'Isle Geogr. (1701), 233
Carte des environs du Mississipi par G. De l'Isle Geographe (1701): by Guillaume Delisle, **234**
Carte du Mexique et de la Floride (1703): by Guillaume Delisle, **236**
Carte physique des terrains les plus élevés de la partie occidentale du Canada (Buache), 142, **143**
Cartier, Jacques, 14, 18
cartobibliographies, 42
cartographic discourse, Native American, 241, 265–66; and conduct of life, 255–64; and domain of power, 242–48; and realm of cosmology, 248–54
cartographic encounters: native-white in the field, 14–26. *See also* creative cartographic encounters; mapmaking; maps
cartographic history(ies), 235; in Aztec Mexico, 117–23, 131
cartographic information: from Indians in European maps of Southeast, 207, 209, 213; in Lewis and Clark's route planning, 72; on wampum belts, 84

cartographic memory: as experienced environment by Native Americans, 246
cartography: German scholarship in, 34–35; historians of, 81; history of, 38; and language, 74; moral, 255; native land loss and allotment, 247; and natural language, 278; prehistory of, 75
cartouche: on *Carte de l'Amérique septentrionale* (Bellin), **137**
carvings: commercial demand for Eskimo, 94
Casasinamon (also known as Robin; Pequot sachem), 178, 179, 180
Cashawasset (also known as Harmon Garrett), 178, 181, 182
Cassini, Jean Dominique, 233
Castillo, Bernal Díaz del, 111, 112
Catawba Indians: Deerskin Map by, 29n. 36; Nicholson collection of maps by, 224, **225**, **226**, 228; maps by, 22, 216, 217
categories/categorization, 100; and map content, 277, 278; native mapmakers vs. European mapmakers, 62
Catholic missionaries: ladder charts used by, 261–62
cation-ratio varnish-dating technique, 77
Catoonah (Ramapo sachem), 167, 169
Cauc-chi-chenis (East Cree Indian): beaver preserve mapping by, 23
Cazones River: depicted on Mapa Local of Tochpan, 114, 115
celestial charts/maps, 93; incorporated into Navajo textiles, 95; by

Pawnee, 51n. 33; of Skidi Pawnee, 252. *See also* sky charts; star charts
Cenis Indians, 19; maps by, 20, 21
center shrines, 250
centers: within Native American realm of cosmology, 249–50
ceramics: ethnic identity and decoration/form of, 218; long-distance trade in fine, 216
ceremonial ground: symbolism of Southeastern Indian, **251**
ceremonial performance: representational role of, 279
Chalco: in Mapa Sigüenza, 121
Chamberlain, Von Del, 88
Champlain, Samuel de: account of Indian mapmaking by, 43
chantway dry-paintings: Navajo, 250, 255
Chapultepec: depicted on Mapa Sigüenza, 118, 119, 120
charcoal on spruce wood chip: vertical-perspective pictographs, 47
charities: Indian art with cartographic themes marketed by, 109n. 117
Charles (Sepecan Indian), 174
Charles II (king of England), 177; Connecticut charter issued by, 179
Charles V (king of Spain), 114, 128
"'Chart in His Way, A': Indian Cartography and the Lewis and Clark Expedition" (Ronda), 72
Cheesebrough, Samuel, 179
Chegeree (Miami or Twightwee): map by, 217
Cherokees: cosmology of, 203n. 81; four-directions motifs of, 250

Chickasaw/Alabama map of the Southeast (1737): redrawn by de Batz, 78, **208**
Chickasaw Indians, 187; captured by La Salle, 231; French and Indian alliance against, 192, 197; French war on (1736), 226; hostilities between Quapaws and, 192, 193, 197; maps by, 22, 78, 216, 217; Tonti's journey among, 235
Chickasaw towns: Captain of Pacana's map of (redrawn by de Batz), 214, **215**
Chicomoztoc: in Cuauhtinchan Map 1, 122
Chilkaht chief: map made by, 44
Chippewa: contrast in spatial conceptualization and land loss by, 246; Hallowell's work among, 258; maps by, 24. *See also* Ojibwa
Choctaws: civil war, 237, 240n. 49; ethnogenesis accounts by, 232, 249; paintings by, **227**, 228; territorial diminishment portrayed in literature, 247; Tonti's journey among, 235
Cholula: in Cuauhtinchan Map 1, 122
Christian cosmologies: Indian prophets and appropriations of, 262
Chumash (south-central California): astronomical motifs in rock art of, 252
Church, Benjamin, 174
Churchill, Winston, 11
circles: on Catawba map collected by Francis Nicholson, **225, 226;** network of, in Chickasaw-drafted map of 1723, 224, **226**; portraying "fire" of native polity, 224; representations on Southeast Indian maps, 207, 209, 212, 216, 217
circumferentor and chain, 161
Clark, William, 23, 34, 101n. 4, 145; map transcripts by, 82, 105–6n. 57
classification: and map content, 277; in native mapmaking, 62
Clastres, Pierre, 140, 147
Claus, William, 274
clerks' copies: of graphic maps, 171; of Indian deeds, 163
cloth: *lienzos* painted on, 124; maps of, 114
clothing: of Quapaws, 194–95
coastal charts, 158
Cockapatana (Paugussett), 165, 166, 167
Codex Féjérváry-Mayer (p. 1), **130**
Codex Mendoza: borderless depiction of Tenochtitlan within, 130–31; foundation of Tenochtitlan within, 128, **129**
cognitive science contexts: and future encounters, 278–79
Colden, Cadwallader, 60, 276, 277
colonial borders: and dependence on exact delineations of native territorial borders, 177, 178
colonial New England: property mapping in, 160
colonial settlement: role of maps in, 158; Southeastern maps used by Europeans for, 207, 209
colonial Southeast: Indian maps of, 205–18
colony maps: for European investors, 158
color: in face-painting by Quapaws, 190; on Quapaw painted buffalo hides, 188, 191; on wampum belts, 84; of war calumet of Quapaw, 191
Colorado: rock art panels in, 77
Comer, Capt. George, 25
commercial art: mapmaking as component of, 94; native North American map incorporation into, 279–80
Commissioners of the United Colonies: land disputes mediated by, 177
communication: cartographic, 13, 18, 135; colonial map, 75; future encounters and wide, 282; gestural, 146–49; pictographic, 42, 47, 58, 83, 111, 118–20, 126
community: maps of, in Aztec Mexico, 123–31
community charters: in Aztec Mexico, 123–31
compass rose: appended to Captain of Pacana's map of Chickasaw towns, 214, **215**
concentric cross-in-circle motifs: on conch shell, 27
Conceptions of Space in Social Thought: A Geographic Perspective (Sack), 279
conduct of life: within Native American realm of cosmology, 255–65
Connecticut: charter issued by Charles II to, 179; claims to Wecabaug by, 178
Connet (Sepecan Indian), 174

consecrated enclosure: Black Hills depicted as, 87
conservation initiatives: and bioregional mapping activities, 99, 110n. 132
consistency: in native maps, 141
constellations: in Amos Bad Heart Bull's Black Hills drawing, 87; depicted in rock art, 252; on Pawnee buckskin map, 252; woven designs of, 95. *See also* star charts
contexts: for indigenous mapmaking, 11
contextualization, 100
conventions: Aztec war, 117; cartographic, 140; and culturally specific perception, 140; English mapping, 166, 168; European mapping, 223, 227, 229; graphical, 224; intercultural opinions on mapping, 148–49; multiple-symbol, 230, 231, 233, 235; Southeastern Indian social circle, 216; understanding of native cartographic, 139, 149–50
conveyances: and Indian deeds, 157
Cook, Capt. James: second voyage to South Pacific, 36
Coon Come, Daniel, 148
Coon Come, Matthew (Quebec Cree), 148
cooperative networks: denoted on Southeast Indian maps, 207
Copp, John, 167
copper (Appalachians and Great Lakes): long-distance trade in, 216; Ojibwa records on, 83

Copper Indians, 57
Copway, George (Ojibwa chief), 86
copying, map: by de Batz, 226; clerks', 163, 171; errors in, 206; of maps produced by Delisle cartographic house, 233; photographic, 87
corn: used in Powhatan ceremony, 209–10, 219n. 11
corners: in Ridgefield deed, 168, 169; in Sepecan deed, 172
cornmeal: used in "road-making" and "road-closing" rituals, 256, 257
Coronado, Francisco de, 15
Coronelli, Vicenzo, 231
Corte-Réal, Miguel, 1
Cortés, Hernan, 111, 112, 114, 120
cosmic axis motif, 253
cosmic centrality: claim to, 249
cosmic ladders, 252
cosmograms, 249, 253–55; Chippewa, 260; for curing ceremonies, 255; Hopi altarpiece, 250; of Plains Cree, 263; in rite-of-passage rituals, 255; and road/trail/path/journey, 256
cosmographical maps, 93; accounts of, 40
cosmography, 2, 249; astronomy in native, 252; Eskimo concepts of, 73; in Powhatan ceremonies, 209–10
cosmological powers: bestowed by long-distance travel, 217
cosmology(ies), 75, 93, 249; Lakota, 81; meaning and direction from, 255; Native American cartographic discourse and realm of, 248–55; of

Navajo, 255; opposed vested interests of cartography and, 248; of Osages and Cherokee, 203n. 81; and territories, 241; and topography depicted on floors of ceremonial enclosures, **256**
cottonwood trees: for Plains Indian sun dances, 253
Coulange, Pierre Petit de, 190, 192
Council Fire, 174
Council of the Nation, 189
coureurs de bois: with Champlain, 151n. 8
court rolls (or terriers), 161
"Courtship, The" (Faulkner), 247
Cove Fishing Company, 177
Creates, Marlene, 96
creation myths and stories: in art, 252; and Great Medicine Lodge ceremony, 260; and Wallam Olum, 85
Creations Journey (exhibition catalog), 265
creative cartographic encounters: recent, by native peoples, 94–100
Cree Indians, 89; maximization of social stability among, 258–59
Creek Indians: four-directions motifs of, 250; paintings by, **227,** 228
Crenay, Baron de: map of 1733 by, 237
cribbage boards: carved by Eskimos, 95
Crisp, Edward: Thomas Nairne's 1711 inset to map by, 235
Cronon, William, 160
Crosby, Constance A., 160
cross-like finials: on eaves of French buildings in Quapaw/Arkansas painted buffalo hide, 202–3n. 63

cross-staff, 161
Crow Indians: sequence of Tobacco Society adoption ritual symbolizing migration story of, 265–66
Crow Tobacco Society, 265
Cruikshank, Julie, 160
Cuauhtinchan: depiction of founding of, 123
Cuauhtinchan Map 1, 126; migration history detailed on, **122,** 123
Cuauhtinchan Map 2, **121,** 122, 123, 126, 127
Culhuacan: depicted on Mapa Sigüenza, 120
cultural anthropologists, 100; field studies by, 56
cultural assimilation: between Quapaws and French, 197
cultural hegemony: and mapping, 244–45
cultural practices: cosmologies and models for, 255
curing ceremonies: cosmograms used for, 255
Custer, George Armstrong, 245

Dakota Territory: reaction to boundary claims process by natives of, 244
Daly, Charles P., 40, 41, 42
dance houses: center post within, 253
dancers: detail of in Quapaw/Arkansas painted buffalo hide, 188, **189**
dance(s) and dancing: during Green Corn "square ground" ceremonies, 250; and possibility of action maps, 281; of Quapaws, 190; space symbolized as, 63

d'Arce, Louis Armand de Lom, Baron de La Hontan, 42, 60, 61
d'Artaguette, Diron, 191, 195
dates and dating: within histories, 117, 118; prehistoric Southeastern archaeological sites and problems in, 211
dating techniques, 78; cation-ratio varnish, 77; rock art, 76
Davidson, George, 26, 44, 45
d'Azevedo, Warren L., 264
death: Quapaw use of black paint to signify, 200n. 18
de Batz, Alexandre, 198; Chickasaw/Alabama map of the Southeast redrawn by, 78, **208;** maps copied by, 226; redrawing of Captain of Pacana's map of Chickasaw towns by, 214, **215**
decoding, cartographic: and divination, 89
deconstructions, map: with topographic orientations and locations, 90–91
deerskin(s): Chickasaw map (1723) on, 30n. 37, 216, 217; circles painted on, 217; Quapaw *matachés* made of, 191
Delanglez, Jean, 233
Delaware Indians, 89; map and record keepers among, 83; maps by, 40, 86; *Nkamwin* or Big House ceremony of, 260–61; Wallam Olum of, 85
Delf, Norona, 63
Delisle, Claude: cartographic house founded by, 233
Delisle, Guillaume: *Carte des environs du Mississipi par G. de l'Isle Geographe* (1701), **234;** *Carte du Mexique et de la Floride,* **236;** detail of lower Mississippi valley by (ca. 1700), **231,** 231–32; recognized as *Premier Géographe du Roi,* 233
Delisle, Joseph-Nicolas, 233
Delisle cartographic house (Paris): Indian mapping information used by, 83; mapping of North America by, 233; mapping of the Southeast by, 223, 234–37
Delisle Papers, 238–39n. 31; Charles Levasseur expedition journal in, 235
deluge myth: and Wallam Olum, 85
Demallie, Raymond J., 248
Dene worldview, 142
Denison, George, 178, 179
Derrida, Jacques, 147
descent: lack of concern over, in native maps, 59
designs: in Navajo textiles, 95; in wampum belts, 84
De Soto, Hernando: expeditionary accounts of, 229
Devil's Tower: explanation of placement in Amos Bad Heart Bull's Black Hills drawing, 87; mapping of, 81
Dewdney, Selwyn, 13, 63, 86, 260
diachronic character: of Indian art, 196, 203n. 84
dialogue: bases for intercultural, 150, 155n. 42; encounter replaced with, 73, 74; of "ephemeral" maps, 155n. 40

INDEX

"Die Anfänge der Kartographie" (Andree), 41
directional mountains: within Navajo cosmology, 255
directional posts: in Navajo hogans, 255
direction(s): for expeditions, 23; in Native American cosmologies, 253; within Native American realm of cosmology, 250; represented on Southeast Indian maps, 207; topological manipulation of, 216
dispossession, land: and Indian deeds, 163–64. *See also* boundaries; surveying
distance(s): between cultures, 149; European measurements of, in *Carte de l'Amérique septentrionale*, 137; on maps made at Fort Churchill, 24; in maps made from memory, 26; measurement of in England, 161; native knowledge of, 92; topological manipulation of, in Southeastern Indian maps, 216
distance scale: appended to Captain of Pacana's map of Chickasaw towns, 214
distortion: mapping, 82, 105–6n. 57
diversification: in awareness of forms and contexts, 94
divination, 14, 46, 89, 107nn. 90, 91; cosmological orientation during rites of, 254; among Naskapi of Labrador, 259; and hunting, 89–90; Shaking Tent rite, 259. *See also* scapulimancy

divinatory almanacs: of Aztecs, 112
documentation: of Aztec lands, 126, 131; of official meeting between Spaniards and Moctezuma's emissaries, 111
Dodge, Col. Richard, 12
domain of power: and Native American cartographic discourse, 242–48
Domenique (Montagnais), 15
Doña Marina (interpreter), 111
doorways: in building on Quapaw/Arkansas painted buffalo hide, 189
Dorsey, George A., 42
Dorsey, James Owen: maps in papers of, 24
dot clusters: on Zuñiga Map of 1608, 212
dotted lines, 144; on John Smith's map of Virginia, **210**
Douay, Anastasius, 21
drainage systems: Great Lakes–St. Lawrence, 86, 87
"Draught of the Northern Parts of Hudson's Bay" (Norton), 139, 142, 148
drawings: of Black Hills by Amos Bad Heart Bull, 87–88; cosmological, **256**; of rapids and portages, 89; Southeastern Indian ceremonial ground, **251**
dreaming: use by Dunne-za (or Beaver Indians), 92–93
dream maps, 13, 108n. 108, 281; of Dunne-za (or Beaver Indians), 62
dreams: Aztec books of, 112
Dröber, Wolfgang, 35, 39

drums: of Quapaws, 190
Dunne-za (Beaver Indians): bioregional mapping among, 99; dreaming and mapmaking among, 62, 92–93, 281; incorporation of Christian cosmology by, 262
duplications: and map interpretations, 91
duration: dimension within Mapa Sigüenza, 120
Dutch colonies, New Netherlands: Indian deeds negotiated throughout, 182
dwellings. *See* houses
Dyirbal (Australian aboriginal language), 278

eagle feathers: on Quapaw calumets, 191
early-encounter landscapes: reconstructions of, 78
"Earring House": in Sitting Rabbit's map, 78
earth maps, 88; of Lakota, 62
earth mounds, 94
Earthnaming ceremonies: and sacred sites, 243
"earth navels," 249
earth tracings: vertical-perspective pictographs, 47
East Cree Indians: maps by, 23
Easter (or Pawcatuck) Pequots, 181
Eastern Algonquians: place names of, 160
Eastern California: rock art of, 77
Eastern Choctaw people: native name given to, 235

Easton, Robert: symbolism of Southeastern Indian ceremonial ground drawn by, **251**
ecological functions: of scapulimancy, 90; of toponyms, 160
Écores Rouges: Quapaw villages in, 193
Eliade, Mircea, 249, 253
elites, Southeastern Indian: description of maps by, 216–17
empire: ideology of, and place of Tenochtitlan within, 128–29
enclosure: and use of graphics, 161
encounter(s): artifacts and second, 87; captivity, 18–19; cartographic, 14–26; and criticism, 85; dialogue as replacement for, 73, 74; economic, 94–95; and effects of hoaxes, 82; face-to-face, 16–18; impact on spatial representations, 94; lengthy cartographic, 235; maps and mapmaking in process of, 3; personal, 40, 45, 73, 80, 101n. 4; posthiatus, 71–72; recent creative cartographic, by native peoples, 94–100; recent scholarly, in anthropological contexts, 83–94; recent scholarly, in historical contexts, 72–83; and repatriation of artifacts, 84
enemies: portrayed on Choctaw painting, 228
England: geographical information sought by, 19
engraved shell cup: from Spiro, Oklahoma, 78
engravings: by Alaska coast Inuit, 280; commercial demand for Eskimo, 94; on letter openers, 109n. 113; map, 233, 234
environmental information: about nodes in Wallam Olum, 86
ephemeral maps, 60, 205
erosion: effect on Southeastern Indian sites by, 211
errors: convention and culture-specific notion of, 142; copyists', 206; in distance on Captain of Pacana's map of Chickasaw towns, 214; geographical, 81–82; mapmaking, 23, 43; in patterns on bitten bark, 46
Eskimo: spatial concepts of, 73; walrus ivory carvings by, 94
Eskimo maps: in sand and stones, 50n. 21
"Eskimo Maps from the Canadian Eastern Arctic" (Spink and Moodie), 73
Eskimo Realities (Carpenter), 73
Essai sur l'origine des langues (Rousseau), 146
ethnic identity: and ceramic decoration and form, 218
ethnic relatedness: denoted on Southeast Indian maps, 207
"ethnocentric ideology," 140
ethnogenesis: of Crow people, 265
ethnographers, 217
Ethnographische Parallelen und Vergleiche (Andree), 38
ethnography: applied to Quapaw/Arkansas painted buffalo hide, 194–95
ethnohistorians, 81, 84, 217, 277
Euclidean geometry, 61, 64; and distance treatment on Capt. Pacana's map of Chickasaw towns, 214
European influence: in making of Quapaw/Arkansas painted buffalo hide, 195, 196
European mapping conventions: signal contribution by event transcriptions to, 227
European maps: influence of native maps on, 81; with native supplementations, 33
European-style maps: Eskimo engravings on ivory of, 95
event-oriented *(res gestae)* history, 117
"event transcriptions," 227–29, 234, 237
evidence: anthropological, 86; maps as, 274
expeditions: Fifth Thule, 26; Lewis and Clark, 72, 82; Oñate, 18–19
exploration: significance of Indian and Inuit maps recognized due to, 72–73; by Smith and Virginia colonists, 212
explorers: debriefed by Delisles for mapping of Southeast, 223, 235; influence of Southeastern Amerindian maps on, 229–32, 235; native line maps made for white, 79; native message maps made for white, 87
extrasensory experiences: of terrain, by native peoples, 92

face-painting: colors chosen for, 190; represented on Quapaw/Arkansas painted buffalo hide, 188

farming: Chickasaw preferences in, 220n. 31
Father Peyote, 264
Faulkner, William: "The Courtship," 247
fauna and flora: four directions associated with, 250
feathers: on Quapaw calumets, 191
fiction: maps appearing in, 280
Fidler, Peter, 42, 138, 139, 143, 144, 145, 148, 149
field ice, 92
field locations, 80
Field Museum of Natural History, Chicago, 42
fields, agricultural: shown on Captain of Pacana's map of Chickasaw's towns, 214
Fifth Sun (Aztec), 127
Fifth Thule Expedition: maps collected on, 26
First Chickasaw War (1736), 194
first-generation Baker Lake Inuit artists: drawings of, 280
fishing rights: and Indian deeds, 177
Five Nations Confederacy, 84
Flaherty, Robert, 16; Belcher Islands map discovered by, 44–45; cartographic encounter with Inuit, 57
Flaherty Island, 16
Flatheads, 249; land loss through allotment cartography, 247
Fletcher, Alice, 88, 247
floods: effect on Quapaws, 193
folk narratives: representational role of, 279
Fools Crow (Welch), 280–81

foot soldiers, 11
forgeries: and Indian deeds, 163
"forked path": of Iroquois, 262–63
formée cross: within Codex-Féjérváry, 130
Forster, Johann Reinhold, 37; *History of Voyages and Discoveries made in the North,* 36
Fort Berthold: Hidatsa of, 243
Fort Churchill: maps made at, 24
Fort De Quesne, 104n. 37
Fort Desha, 194
Fort Detroit, 78; de Lèry's plan of, 104n. 37
Fort Laramie: all-Plains Indian treaty conference (1851), 243, 244
Fort Loudoun, 78
Fort Michilimackinac: de Lotbinière's plan of, 104n. 37
Fort Pitt, 78, 104n. 37
Fort Selkirk: map of journey to, 26, 32n. 56
Foucault, Michel, 147
Four Bears (Hidatsa chief), 243
four-directions motif, 249, 250; on drum head drawn by Beaver Indian shaman, **262**
four-tipi-pole metaphor: evoked in Crow tobacco adoption ceremony, 266
framing elements: and native conceptions of maps, 142, 153n. 23
France: Delisle's mapping and interests of, 235; geographical information sought by, 19; Louisiana abandoned by, 192; mapping of by Cassini, 233; new geography in, 34

Franklin, Sir John, 23, 96; lost expedition of, 97
Frauenstein, Georg M., 39
French: casualties of, during First Chickasaw War, 194; Illinois Indians and Quapaws allied with, 192, 201n. 35; mapping of Southeast by, 223
French and Indian Wars, 78
French-Canadian *voyageurs,* 11, 276
French explorers: advance into North America by, 135
French league: Ojibwa experiential understanding of, 276
fur trade explorers, 135
fur trade surveys, 135
future encounters, 4; artistic, literary, and performance contexts, 279–82; cognitive science contexts, 278–79; language, linguistics, and semantics in translational contexts, 276–78; and legal context, 274–76; necessary conditions for achieving of, 282–83; social science contexts, 279

gable-end rectangular buildings: used to symbolize villages, 231–32
galena (central Mississippi River valley): long-distance trade in, 216
Galloway, Patricia, 83
game trails, 76
Ganong, William F., 44
garments: narrative scenes in, 103n. 23
Garrett, Harmon (or Cashawasset), 178, 179
Gartner, William, 13

Geertz, Armin W., 264
Geertz, Clifford, 255
gender divisions: within Navajo cosmology, 255
genealogical information: in *lienzos*, 124, 126
General Allotment Act of 1887, 247
General Assembly: appointments by, 169; land purchase petitions to, 167
General Court (New London): and Indian deeds, 165; Paugussett complaints of land agreement abuses filed in, 167; testimony concerning Wecabaug and Squamicut in, 179, **180**
General History of the Things of New Spain (Sahagún): reconnaissance map from, 113, **116**
geographical consciousness: incorporated into American Indian liturgies and rituals, 259; myth and religion and native, 248; within sub-Arctic culture, 257–59
"geographical distance": corresponded with "supernatural distance," 217
geographical exploration: significance of Indian and Inuit maps recognized owing to, 72
geographic information: by Indians for French mapping of Southeast, 223, 235
Geographic Information System (GIS) computer software: use by Nisga'a to defend tribal sovereignty, 99
geographic space: in Mapa Sigüenza, 120, 121, 122, 123

geography: German scholarship in, 34–35; sacred, 249
Geological Survey of Canada, 24, 45
geologists: map solicitation by, 24
geometrical tables, 161
geometric designs: on skins by Plains Indian women, 201n. 31
geometry and geometrical properties, 59; Euclidean, 61, 64, 214; of Indian maps, 60–61; projective, 64; "rubber-sheet," 67n. 28; topological, 13
George III (king of England), 274
Germans: awareness of native maps by, 34–35; scholarly encounters by, 35–40
Geschichte der Kartographie (Bagrow), 38
gesture: maps, 11–12; within native North American cultures, 146–48; space symbolized as, 63
Ghost Hill *(Inyan Kara)*: on Amos Bad Heart Bull's map, 245
Gillespie, Beryl, 90
GIS. *See* Geographic Information System
Gitksan: mapping for empowerment among, 99
globe gores: by Coronelli, 231
glyphs: question of maps in, 76
gold: expeditions in search of, 245; map solicitation and seeking of, 21
gold rushes: effect on market for Eskimo ivory carvings, 94
Gómara, Francisco López de, 114
Good, J. Paul, 42
Good Journey, A (Ortiz), 265

Good Red Road, The: Passages into North America (Lincoln and Slagle), 264–65
Goodyear, Stephen, 175
gorgets (Spanish brass): long-distance trade in, 216
Gough, Richard, 92
gourds: worn by Quapaws, 203n. 74
Graham, Andrew, 36; *Plan of Part of Hudson's Bay and Rivers, Communicating with York Fort & Servern*, 135, 137, **138**, 139, 143
grants: native land purchase following receipt of, 162; negotiated from Weantinock Indians, 175; remapping of land in, 169
graphical maps: vs. word maps, 75
graphical representation: intercultural differences in, 275–76
graphic designs: historical analysis of, within Amerindian maps, 140
graphics: mapping through, 169–82
graphic surveys, 161
grass houses: origin myth for, by Wichita, 250
Gravier, Father, 190, 191, 195
Great Falls at Metichawon, 176
Great Medicine Lodge ceremony *(Midewewin)*, 260
Great Plains: rock art in southwestern, 77
Great Slave Lake, 92, 108n. 105; depicted in Matonabbee's map, 91; mapping of, 139; shaping of, 93
Green Corn "square ground" ceremonies: in Southeastern native traditions, 250

INDEX

Greenland: vertical-perspective pictographs from, 47
grid maps, 141
"ground map" (or dry painting): as cosmograms in rite-of-passage rituals, 255
ground models, three-dimensional: vertical-perspective pictographs, 47
ground plans, 76
ground tracings, 65n. 17
Gulf Coast: Iberville's meetings with Indian leaders on, 233
Gulf of Mexico: depicted in Mapa Local of Tochpan, 114–15

"Hailway: The Night Sky" (Klah), 95
hairstyles: of Quapaw women dancers, 190
Halchidhoma: maps by, 44
Hallowell, A. Irving, 246, 257, 258, 260
Hammond, George P., 43
Handbook of North American series: *Subarctic* volume of, 90
Handsome Lake (Iroquois prophet), 262
Harley, J. B., 140; *History of Cartography* series, 158; redefinition of concept of map by, 64
Harney Peak, South Dakota: importance to Sioux, 245
Harvard Indian College, 173–74
Hawwasues: Weantinock deed by, 176
healing ceremonies: and patterns in sandpaintings, 95; Plains Indian sun dances, 253
healing powers: and "animal homes," 253

Hearne, Samuel, 23, 24, 137, 138, 139, 148
heaven: dream visits to, 93
Heckewelder, John Gottlieb Ernestus, 40
Heizer, Robert F.: *Aboriginal California and Great Basin Cartography*, 77
Helm, June, 90, 91, 92, 101n. 6, 139
Helms, Mary, 217
Henday, Anthony, 137, 139
Henderson, James Youngblood: *The Road: Indian Tribes and Political Liberty*, 264
Hennepin, Louis, 21; *A New Discovery of a Vast Country in America*, 22
Henriksen, Georg, 90
Hester, Thomas R., 77
Heye, George, 84
Heye Foundation: repatriated items from, 84
Hidatsa Indians: Crow Indians split from, 265; government encounter over boundary claims of, 243
hidden agendas: with mapping and cultural hegemony, 244–45
hides: maps on, 62, 88, 118; pictographs on, 58; Quapaw paintings on, 191
hierarchies: and map content, 277
Hind, Henry Youle, 15
Hind, William, 15
Historical Collections . . . of Every Town in Massachusetts (Barber): redrawn version of Sepecan deed in, **171**
historical contexts: recent scholarly encounters in, 72–83
historical encounters, 3
historical geographers, 78

history: Aztec maps of cartographic, 117–23; Mapa Sigüenza as, 119–20
History of Cartography series (Harley and Woodward), 158
History of Voyages and Discoveries made in the North (Forster), 36
hoaxes: exploration, 82
Hoffman, Walter J., 260
hogans: directional posts within Navajo, 255
Holmes, John, Jr., 167
Holy People, 95
"Holy Road, The," 244
Honduras: map that led Cortés to, 114, 116
honesty: among Inuit hunters, 98, 99
Hopi Indians (Arizona): altarpiece cosmograms of, 250; center shrines of, 250; Prophecy Rock of, 264; Snake Dance of, 256
horses: stealing forays, 76; trade networks in, 80
Housatonic Water Company, 177
houses: blessing of, in Shalako ceremony, 256; of Chickasaws, 194; distribution pattern of Powhatan, 211, 212, 213; ground plans of lodges and, 76; Kwakiutl, 253; Quapaw, 228; represented on Quapaw/Arkansas painted buffalo hide, 188, 189. *See also* villages
Howard, Gen. Oliver O., 243
"How the First Letter Was Written" (Kipling), 280
Hudson Bay: Best's 1578 map of entrance to, 246; Powon's map of western shore of, **25**

Hudson's Bay Company: archives of, 36, 37, 42; winterers from vs. other North American explorers, 139

Huitzilopochtli (Aztec patron god), 127; depicted on Mapa Sigüenza, 118

hunters: map biographies of, 97–98

hunting, 76; and divination, 89–90; map of musk-ox, 25; maps, 25, 46, 53n. 54; memory maps and ground demarcations for, 100; motifs in rock art panels, 77–78; and scapulimancy, 90; trails seen in dreams, 93

Huron: importance of dreams to, 281

hydrological representation: in native maps, 62. *See also* drainage systems

Iberville, Pierre Le Moyne, sieur de, 231, 233, 235

"ice blink," 92

ice patterns: detected by Inuit, 92

identity, community: and *lienzos,* 127

Idotlyazee, 93

Iglulik people: mapmaking styles of, 73; maps by, 25, 50n. 31

I'itoi (Papago culture hero), 257

Illinois Indians, 188; alliance between Quapaws and, 191–92; clothing of, 194–95; face-painting by, 190; French-allied, 217

imagery: Mapa Local of Tochpan and representation with, 114–15; on Mapa Sigüenza, 118–19. *See also* symbols and symbolism

incorporations, map. *See* map incorporations

indexes: Phillips's, 42

Indian art: diachronic character of, 196, 203n. 84

Indian deeds, southern New England, 157, 182; and forms of native and non-native mapping, 159–61; mapping through graphics, 169–81; mapping through words, 164–69; process of making, 161–64

Indian mapmaking: and ethical imperative to share, 109n. 129

Indian maps and mapping: and cognitive science contexts, 278–79; of colonial Southeast, 205–18; future studies of, in legal contexts, 274–76; growth of interest in, 100; linguistics, language, and deeper understanding of, 278; post-1970 revival of interest in, 48; recent developments in understanding of, 282; scholarly interest in, 35; sixty-year hiatus in scholarly interest in, 55–56; used in land negotiations/agreements, 274–76

"Indian Maps of the Colonial Southeast" (Waselkov), 78

Indian Office archives (Washington): and question of Chief Joseph's missing map, 243

Indian Territory: Skidi band (Pawnee) relocated to, 83

indigenous cartography: alien perspectives on, 10

indigenous maps: finding of, 5–6n. 4

informants: French explorers and importance of Indian, 231

initiation rituals: and cosmograms, 255

Institute of Indian Arts, Santa Fe, 265

instructional purposes: mapmaking for, 12

intelligence gathering: and Captain of Pacana's map, 227, 237. *See also* military maps and mapping

intercolonial competition, 19

internodal linkages: in Wallam Olum, 86

interpretations, 42–43; ethnographic, 90–91; of native maps, 72–73; and perspectives, 82; and reinterpretation of native maps, 142; and scaleless maps, 89; silences and problems in, 99, 110n. 130; of Smith's map of Virginia, 212–14; of Wallam Olum, 85; by Waselkov, 78–79

interpreters: and Indian deeds, 163

intersection: vs. spacing in Amerindian map design, 141

Inuit Land Use and Occupancy Project, 96, 97, 98, 99

Inuit maps and mapmaking, 15, 16–18, 24, 26, 40, 41, 110n. 137; and cognitive science contexts, 278–79; future studies of, in legal contexts, 274–76; growth of interest in, 100; linguistics, language, and deeper understanding of, 278; post-1970 revival of interest in, 48; precision in, 146; recent developments in understanding of, 282; scholarly interest in, 35; sixty-year hiatus in scholarly interest in, 55–56; sky maps, 92; used in land negotiations, 275

Inuit people: cartographic elements in

commercial art of, 96; of Eastern Hudson Bay, 246; memorial culture of, 140–41
Iron Bull (Crow Indian chief), 249, 265
Iroquois (upper New York state): cosmic axis motif of, 253; wampum belts stored by, 83
Isham, James: *Observations on Hudson's Bay,* 141
Itiwana or "Middle Place": of Zuni Pueblo, 250
Itsikamáhidic (Hidatsa culture hero), 80
ivory: carvings by Eskimo on, 94; pictographs on, 58
Ixtlilxochitl, Fernando de Alva, 126

Jackson, Marion, 280
Jacob, Christian, 140, 147; *L'Empire des cartes,* 139
James II (king of England), 167
James River: Indian villages along, 213
Jefferson, Thomas, 198; *Notes on the State of Virginia,* 213
Jemez Pueblo: depiction of tribal universe by, **252**; sky domes of, 252
Jenness, Diamond, 57
Jesuit Relations and Allied Documents, The (Thwaites), 46
Jesuits: among Quapaw, 195
Jesus Christ: Dunne-za prophet's dreams about, 262; peyotists' view of, 264
Johann Georg Kohl Collection of Maps, Geography and Map Division, Library of Congress, 37
Johnny (interpreter), 16

Johnson, Albert, 252
Jolliet, Louis: detail of lower Mississippi valley by, **230**
Joseph, Chief (Nez Percé), 248; "missing map" of, 243
journey: and road metaphor in Indian thought, 264; as trope in American Indian consciousness, 256
Joutel, Henri, 190, 191, 203n. 74

Kalli-herua (alias Erasmus York): map by, 39, 50n. 22
Kansa Indians: maps by, 24
Kappa village (Quapaw), 187, 189, 193
Karpinski Collection of Maps, Newberry Library: photostatic copies of Delisle's Choctaw maps in, 233
Kaskaskias (Illinois Indians), 192
Key into the Language of America, A (Williams), 159
King Philip's War (1675): effects of, 165; events leading to, 174
Kioocus (Blackfoot [Siksika] leader), 144
Kiowas: sand map by, 50n. 21
Kipling, Rudyard: "How the First Letter Was Written" *(Just So Story),* 280
Kitchin, Thomas: "A New Map of the Cherokee Nation with the Names of the Towns and Rivers . . . ," **206**
kivas: of Jemez Indians, 252; Southern Pueblo subterranean, 253
Klah, Hosteen, 95
Klondike gold rush: effect on market for Eskimo ivory carvings, 94

Knife-Heart region, upper: early maps of, 80
Knight, James: maps made by Indians for, 36–37
Kohklux, Chief (Tlingit): maps by, 32nn. 55, 56, 52n. 48, 153n. 23
Kohl, Johann Georg, 35, 38, 41, 42, 43; maps collected by, 37
Kohl Collection, The: Phillips's index to, 44
Kootenay people: Fidler's encounter with, 149
Koroa: Quapaw raids against, 194

Labrador: bioregional mapping among Inuit of, 99; Speck's family hunting districts of, 100
labyrinth designs, 94
ladder charts: used by Protestant and Catholic missionaries, **261**, 261–62
ladders: cosmic, 252
Lafferty, Robert H., III, 78
La Hontan, Baron, 43
Lake Athabaska: mapping of, 139
Lake Nipigon, Ontario, 108n. 105; Windigo's map of, 45, 52–53n. 51
Lake Superior region: early French uses of Indian information about, 81
Lake Texcoco: on Mapa Sigüenza, 118
Lake Winnipeg: Berens River Chippewa of, 258
Lakoff, George: *Women, Fire, and Dangerous Things: What Categories Reveal about the Mind,* 277–78
Lakota, 89; cartographic research on stellar theology by, 81; and government encounter over boundary

claims, 244; map interpretations by, 88; painting of cosmos on tipi cover by, 254; star and earth maps of, 62

Lakota theology: and depiction of Black Hills, 87

La Marteblanche (Cree), 135

Lamb, William Kaye: *Report of the Public Archives for the Year 1951,* 56

land acquisition: Indian deeds integral to, 162

land agreement/negotiations maps: role of, 274–76

land deeds. *See* Indian deeds, southern New England

land holdings: documentation of, in pre-Columbian Mexico, 112

Land of the Dead (Chippewa), 260

land purchases, state: and native tribal unity, 242, 244

land reclamations efforts: and mapmaking, 96

land records: in Aztec Mexico, 113

land sales, native: impact of European boundary delimitations on, 60. *See also* boundaries

landscape: maps of, overlapped with maps of proper action, 257; symbolic organization of, 102n. 15; topological view of, in Southeast Indian maps, 207

land speculators: claims to Wecabaug by, 178; Indian deeds negotiated with, 162

land tenure, 162

land titles: *lienzos* serving as, 124, 126, 127, 131

land transfers, 157

land-use categories: on map biographies, 98

language: and cartography, 74; and impact on cartographic information, 91–92, 102n. 15; linguistics, and semantics in translational contexts of future encounters, 276–78; and origins of mapmaking, 103n. 21; vs. gesture, 146; trade, 2

La Salle, Nicolas de, 192, 194, 231

La Salle, René Robert Cavelier, 82

lateral (or mirror) inversion: and star charts, 88

La Vérendrye, Pierre Gaultier de Varennes et de, 89, 142; composite of Cree maps of Lake Superior region by, 135, 136–37

lead: Ojibwa records on, 83

leaders: depicted on Mapa Sigüenza, **120**; identified by name-glyphs, 118, 125, 128

leases: and Indian deeds, 157

Leavenworth, Col. J. H., 248

Le Bouteux, 198

Le Clercq, Chrétien, 61

"Lecture on Exploration in the Nipigon Country" (Bell), 45

ledger art: by Plains Indians, 106n. 65

Leech Lake, Minnesota: maps of, 86; route of midé religion to, 63

legal challenges/suits: and role of bioregional mapping, 94; and role of *lienzos,* 127

legal context: and future encounters, 274–76

L'Empire des cartes (Jacob), 139

Lèry, Gaspard Chaussegros: Fort Detroit plan by, 104n. 37

letter openers: Eskimo engravings on, 95, 109n. 113

Levasseur, Charles: expedition of 1700, 235

Lévi-Strauss, Claude, 147

Lewis, G. Malcolm, 101n. 6, 219n. 11; and decoding of culturally hybrid map depictions, 246; and linear-river-plus-"tadpole"-lake configuration, 235; on lines within native graphic style, 172; on "precise boundary delimitations," 243; and *terra semicognita,* 137; and "unacknowledged assimilations," 142

Lewis, Meriwether, 23, 34, 101n. 4

Lewis and Clark expedition: cartographic information used in, 72; and geographical errors, 82

Libby, Orin G., 79, 104–5n. 39

Lienzo of Tlapiltepec: spatial presentations in, 123

Lienzo of Zacatepec, 123, **125**, 126–27

lienzos, 112, 123, 124, 125, 126–27, 131; spatial presentations juxtaposed on, 123

Lilly, Eli, 85

Lincoln, Kenneth: *The Good Red Road: Passages into North America,* 264–65

linear-river-plus-"tadpole"-lake configuration: as Amerindian source indicator, 235

line maps, 79, 91

lines: communication routes portrayed by, on Southeast Indian maps, 217;

dotted, 144; network of, in native maps, 141–42; on Quapaw/Arkansas painted buffalo hide, 187; on Weantinock deed, 176; on Wecabaug map, 180
linguistics: language, and semantics in translational contexts of future encounters, 276–78
literary contexts: and future encounters, 279–82
literature: representational role of, 279
locations: within histories, 117, 118; mapping of by Southeastern Indians, 206
Logan, William E., 24
Lone Wolf (Kiowa head chief), 248
long-distance trade/exchange: in late prehistoric and early historic Southeast, 216
longhouses: of Quapaw, 228; symbols on Smith's map of Virginia, 212
long-life maps, **59**, 60
long-poled ladders: access to nether worlds by, 253
Lophophora Williamsii (hallucinogenic cactus), 264
Lords of the Council for Trade and Plantation, London, 22
"lost map," 242, 243
Lotbinière, Michael Chartier de: Fort Michilimackinac plan by, 104n. 37
Louis (Montagnais guide), 15
Louisiana: abandoned by France, 192; cultural symbiosis between Indians and Europeans in, 198; French policy and importance of Indian informants to, 231; Quapaw hide paintings in, 191
Lowery, Woodbury, 43, 44; *The Spanish Settlements within the Present Limits of the United States: 1531–1561, with Maps,* 43
Ludlow, William, 245
"luminous track": within pictographic design, 147, 148

Maclagan, David, 108n. 108
MacMurray, Maj. J. W., 248
magnitudinal ordering: and purpose in native maps, 62
Makenunatane (Dunne-za prophet), 262, 281
Malecite Indians: maps, 44
Mallery, 48, 146; "Picture Writing of the American Indian," 46, 58
Manahoac villages: on Smith's map of Virginia, 211
Mandan people: mapmaking by, 79, 80
Manitoba Historical Atlas, 61
manuscript maps, 42
Mapa Local of Tochpan, 117
Mapa Sigüenza, **119, 120**, 122, 123, 127, 128; described, 118–19; detail of founding of Tenochtitlan on, **120**
map assimilations, 33–34; challenges to, 72–73, 74; and dissimulation of native forms, 142–43; and geographical errors, 82; processes of, 81
map biography(ies), 96–97
map collecting: active periods of, 45; by archives, 24, 46; of celestial charts in Navajo textiles, 95; by Francis Nicholson, 224, **225, 226;** and importance of Indian maps, 38; by libraries, 37; by museums, 42, 46, 87; systematized set, by Ruggles, 56
map comparisons: cross-cultural, 150, 155n. 42
map content: importance of categories, hierarchies, and classifications on, 277; and purpose in native maps, 62
Map Exhibiting all the New Discoveries in the Interior Parts of North America . . . (Arrowsmith), 135, 138, 143, 144, **145**
map-as-experience, 242
map incorporation, 18, 33–34; of Amerindian information onto European maps, 223, 234–37; analysis of, 81; of "event transcriptions" in European maps, 227–29; for expeditions, 101n. 4; with line maps, 79; native North American, into commercial art, 94, 279–80; of "sociograms" into European maps, 224–27, 237; from Southeast Indians into European maps of Southeast, 207, 209; into two-dimensional art, 96
map key: on John Smith's map of Virginia, 211
map legends: of Coronelli, 231
mapmaking, 9, 10; by birchbark biting, 46; by Chief Kohklux and wives, 26, 32n. 56, 52n. 48, 56, 153n. 23; via dreams, 13, 281; errors in, 81–83, 105–6n. 57; Eskimo styles

of, 73; Eurocentric perspectives on, 93; within evolutionary context, 279; on ground with sticks, 205; growth of interest in Indian and Inuit, 100; for instructional purposes, 12; interactive nature of, 13, 45; by Inuit, 15, 16–18, 99; and land reclamation efforts, 96; by Mandan, 79, 80; and map biographies, 97; vs. mapping, 147; metaphysical dimensions within native geographical awareness in, 266; by Nootkan Indians, 44; origins of, 103n. 21; pre-encounter and indigenous post-encounter, 11–14; Quapaw potential for, 197; relationships among spatial cognition, production and, 279; and scale, 61; and vertical-perspective pictographs, 47–48

map media: bark, 38, 61, 118, 205, 274; birchbark, 12, 15, 44, 46, 50n. 21, 53n. 54, 87, 275; blazed tree trunks, 40; buckskin, 13, 42, 84, 252; charcoal on bark, 20, 21; charcoal on parchment, 36; cloth, 114, 118; with crayon, 31n. 50; deerskin, 40, 42, 216, 224, 228, 261; hides, 62, 88, 118; ink, 31n. 50, 51, 52; ink and five-color crayon, 87; ivory, 94–95; moose-hide, 281; painted/carved on rock, 205; painted on skin, 205; paper, 95, 275; with pencil, 31n. 50, 32n. 56, 51, 52; pencil on paper, 24, 52n. 48, 96, 243; sand and stones, 50n. 21; skins, 15, 22, 38; textiles, 95

mapping: Amerindian information and European conventions of, 223, 227; celestial, 252; Christian, 262–63; consistency in native, 141; and cultural hegemony, 244–45; by Delisle cartographic house, 233–37; development of, 86; English conventions of, 166; European conventions of, 223, 227, 229; forms of native and non-native, 159–61; by gesture, 147; from infrared photographs, 224; of lakes and portages from Lake Superior to Lake Winnipeg, 135; language and origins of, 75; from memory, 140–41, 148; Powhatan villages and difficulties of, 212; property, 160; by Quapaws, 197; of routes from Xicalanco to Nicaragua, 114; scientific, 135; as spatial representation, 159; of spiritual teachings, 261; through graphics, 169–82; through naming, 159, 160, 166, 169; through words, 164–69, 172, 181

"mapping the sacred": in preindustrial societies, 243

maps, 9, 10; art used in understanding of, 280; in Aztec Mexico, 112–13; on birchbark, 12, 15; birchbark message, 87; on blazed trees, 12; chance and production of, 14; communications by means of, 2; by Coronelli, 231; to define and situate polity and people, 125, 126, 131; delimiting tribal locations with, 246; of distant circumstances, 259; dream, 13, 108n. 108, 281; earliest accounts of, in indigenous context, 40; from early colonial period in Mexico, 112–13; engraved, 233, 234; engraved by Eskimos on ivory, 94–95; ephemeral, 60; as evidence, 274; first systematized assembled set of native North American, 56; future encounters and operational definition of, 282; gesture, 11–12, 147; grid, 141; "ground," 255; to heaven, 93; historically important, 42; of Indians, redrawn for publication, 42; interpretations and understanding of, 78; in land negotiations/agreements, 274–76; line, 79, 91; "lived" by Indian peoples, 255, 256; and location, 117; looser definition of native, 63–64; manuscript, 42; from memory, 96, 100, 148; metaphysical dimensions of native geographical awareness in, 266; narrative, 86; oldest extant North American, 43; outline, 95; pictographs categorized as, 58; political, 236–37; for proper action overlapped with maps of landscape, 257; property, 75; and Quapaws, 197; question of linguistic concept of, for native North Americans, 54n. 62; reconnaissance, 113, **116;** redefined by Harley and Woodward, 158–59; reference (Mississippi-Alabama region), **232;** as rock art design, 76; scaleless, 89; scientific, 135, 140,

144; short-life, **59**, 60; stick, 15; on stones, 274; storage of, 83; survival of originals of, 18; of *terrae incognitae,* 19, 33; of *terrae semicognitae,* 22, 23, 99; thematic, 224; topographic, 13; traditional action, 281; transcripts of, 18, 23, 142; transport, 141; two-roads, 265; types of Southeastern Indian, 205, 206–7; and wampum belts, 84; "way-finding," 114, 116, 117, 120

Maps and Dreams (Brody), 13, 147

map searches: and acquisitions, 26, 32n. 55, 56

map soliciting: in captivity, 18–19, 21; by geologists, 24; by Indians from Indians, 15; for strategic reasons, 22; for surveys of the West, 23

Maguire, Rochfort, 57

Marchand, Virgil, 265

marine shells (Gulf and Atlantic coasts): long-distance trade in, 216

maritime charts, 158

Marquette, Father Jacques, 21, 188, 230

Marshall, Joseph, III, 244

Mashantucket Pequots, 181

Massachusetts: claims to Wecabaug by, 177–78; Indian deeds in Western, 163; rights to Pequot country to Connecticut from, 179

Massasoit. *See* Ousamequin

matachés, 191

Matonabbee, 93; map by, 90, 91

Mattatuck (1685): native mapping through words, 164–67

Mattatuck deed: related to nineteenth-century topographic map, **164**

McCutchen, David, 85

McGrath, Robin, 95, 96

McNall-Findley crime, 243

McNickle, D'Arcy: *Wind from an Enemy Sky,* 247

measuring chains, 246

medicine bundles: Pawnee, 252

medicine men and women: sandpaintings and healings by, 95

memorial culture: renewal of, 153n. 17

memory: experienced environment upon Native American cartographic, 246; maps from, 96, 100, 148

Mercator, Gerardus, 41

Mesoamerica: power of Tenochtitlan within, 127

message maps, 12, 46, 79, 87; birchbark, 58

Metacom or "King Philip" (Pokanoket sachem): betrayal of, 174; land conveyed in Sepecan deed by, 171; map of Sepecan by, 172. *See also* King Philip's War

metaphysical dimensions: within native geographical awareness in mapmaking, 266

Metichawon: Indian deed drawn at, 175

métis citizens: of Arkansas, 197, 198

metrical systems: intercultural differences in, 276

Mexico: invasion of, by Cortés, 111

Miantonomi: murder of, 177

Micmac Indians: maps by, 44, 61

Micronesian stick charts, 73

middlemen: as long-distance traders, 216

midé migration charts: of Ojibwa, 86, 87

midé religion: diffusion route of, from Atlantic Ocean to Leech Lake, Minnesota, 63

Midewewin (Great Medicine Lodge ceremony), 260

migration charts: of Ojibwa, 86

migration legends/stories, 234–35; Aztec, 118–20, 131; Cuauhinchan maps as, 123; and Great Medicine Lodge ceremony, 260; and Wallam Olum, 85, 107n. 75. *See also* origin myths/stories

migration scrolls: of Southern Ojibwa midé, 13

Miguel (Plains Indian): map made in captivity by, 19, 21, 43, 51n. 39

military maps and mapping, 12, 40. *See also* reconnaissance maps

military strategists, European: map portrayals of Amerindian population distributions for, 237, 239–40n. 49

military victories: portrayed on Codex Mendoza, 128

Milky Way: woven designs of, 95

mimicry: Inuit kinesic or gestural, 146

Miner, Thomas, 179

mineral resources: extraction from Black Hills, 245–46

mineral seekers: native line maps made for white, 79

Mingo Ouma (Chickasaw war chief): copies of maps created by, 216, 226, 227

Minor, John, 175, 176, 178

mirror (or lateral) inversion: and star charts, 88

Misquamicut Company, 178

missionaries: maps to promote Christianity by, 261

Mississauga Indians: and Toronto Purchase of 1787, 274

Mississippi-Alabama region: reference map of, **232**

Mississippian period (A.D. 700–1540): and hypothesis of pre-encounter political system, 78; long-distance trade and exchange during, 216

Mississippi River: errors in maps of lower, 82; map of lower, **20**

Mississippi valley, lower: detail of, by Guillaume Delisle (ca. 1700), **231**, 231–32; detail of, by Louis Jolliet, **230**; explorers debriefed by Delisles and mapping of, 223, 235; French settlements of, 22

Missouri River: Sitting Rabbit's untitled painting on canvas of, 104–5n. 39

Missouri River, upper: Blackfoot sketches of vs. Arrowsmith version of, 144–45

Mistassini Cree: maximization of social stability among, 258–59; on role of divination, 90

Mixtecas: *lienzos* of, 123; surviving corpus of histories from, 112

mnemonic devices: on birchbark scrolls, 260; cosmological, 252; and Indian mapmaking, 83; Indian paintings used as, 192; toponyms as, 159, 164, 172, 181; and wampum records, 84

Mobile River delta: Levasseur's expedition through, 235

mobiles: cosmological, 254

Moctezuma, 111, 112, 114

Mohawk Indians: boundary disputes by, 276, 277

Mohegan-Narragansett war, 177; Mohegan land claims following, 178

Mohegans: land battle between Narragansetts, southern New England colonies and, 177–81

Molina, Alonso de: Nahuatl/Spanish dictionary by, 113

Monacan villages: on Smith's map of Virginia, 211

Montagnais band: bitten bark maps by women within, 46; hunting district in Labrador, 100; scapulimancy practiced by, 89

Montigny, Dumont de, 198

monuments: in Tenochtitlan, 127

Moodie, D. Wayne, 101n. 6; "Eskimo Maps from the Canadian Eastern Arctic," 73

moon: represented on Quapaw/Arkansas painted buffalo hide, 187, 195

Moore, Omar, 90

moral cartography, 255

moral choices: and influence of Christian mapping on Indians, 262; and interpenetrations of belief systems, 263, 264

mortgages: and Indian deeds, 157

mortuary ritual, 83; and eastern Choctaws, 235

Mosco (Algonquian interpreter), 211, 212

mother-of-pearl buttons: portrayal of cosmic tree on trade cloth blankets with, 253

motifs: within American Indian cosmology, 249–55; Christian/Lakota within "The Two Roads Map," 263–64. *See also* symbols and symbolism

Motolinía (Franciscan friar), 112, 116

Moundville polity (Black Warrior River): history and archaeology of, 239n. 38; and question of Choctaw heritage, 235

mourning: face-painting by Quapaw for, 190

movement, paths for: in Aztec Mexico, 113–17; and road-making rituals, 257

Multnomah (now Willamette) River: Indian map of, 101n. 4; and mapping error, 82, 105–6n. 57

Musée d'Homme (Paris), 228; Quapaw/Arkansas painted buffalo hide in, 187

Museum of the American Indian: bitten bark maps in, 46

museums, 3; collecting of maps by, 26, 42, 46; midé migration charts in, 87

musk-ox hunting: map of, 25

Mutch, James, 25

mythic racetrack: around Black Hills, South Dakota, 245

myths: native geographical consciousness grounded in, 248; in Wallam Olum, 85

Nabokov, Peter, 63, 93, 94
nadir: in American Indian cosmologies, 253
Nairne, Thomas: 1711 inset by, on Crisp's map of North America, 235
Nambikwara chief: gestures of, 147
name glyphs: for Aztec leaders/rulers, 118, 125, 128
Nanook of the North (Flaherty), 16
Nanticoke River: depicted on Smith's map of Virginia, 212
Narragansett Indian council (Rhode Island): defense of unified existence by, 242, 244
Narragansetts: land battle between Mohegans, southern New England colonies, and, 177–81
narrative maps, 86
Naskapi band (Quebec and northern Labrador): divination practices of, 14; hunting district of, 100; scapulimancy practiced by, 89, 90, 259
Natchez: Quapaw engagements against, 194
Natchez Revolt of 1729, 237
Natchez Trace Parkway, Mississippi, 214
National Anthropological Archives: native maps in, 24, 46
National Archives of Canada, 61
National Museum of the American Indian, Washington D.C., 265

National Museum of Denmark: Inuit maps in, 26
National Museum of Natural History, Washington, D.C. (Department of Anthropology), 46
Nations amies et ennemies des Tchikachas (Nations Friendly and Hostile to the Chickasaw): Alabama map of Southeast (de Batz), **208**
Native American Church: peyote ceremony of, 264
native cartography, circa 1970: dimly perceived aspects of, 58–59; overlooked aspects of, 62–64; perceptions of, 56–59; unperceived aspects of, 59–62
native maps: compared with European-style maps, 60–61; directional orientation and lived experience in deciphering of, 258; first systematized assembled set of, 56; specific purposes of, 62. *See also* Indian maps and mapping
native peoples: central role of, in future encounters, 282
native territorial borders: dependence of colonial borders on exact delineations of, 177, 178
native-white communications: challenges and dimensions of, 1–2
Natural History Society of Montreal, 45
Navajo: celestial charts depicted on textiles of, 95; conception of cosmos by, 255; creation myths in art of, 252; dry paintings of, 250
navigation, 93
nebulae: depicted in rock art, 252

Neolin (Delaware prophet), 261
Netsiliks: maps by, 51n. 31
Nevada: rock art of, 77
New Bedford Whaling Museum (Massachusetts), 73
New Discovery of a Vast Country in America, A (Hennepin), 22
New England: property mapping in, 75. *See also* southern New England
"New Map of the Cherokee Nation with the Names of the Towns and Rivers, A" (Kitchin), **206**
New Purchase Cession (colony of Georgia): colonial and Indian surveying of, 218–19n. 5
Nez Percé: and General Allotment Act of 1887, 247; land-body association of, 248; lost or missing map of, 242–43; maps by, 23
Nicholson, Francis, 21, 22, 42; Catawba Indian maps collected by, 224, **225**, 228; Chickasaw Indian map collected by, **226**; draft drawn upon deerskin presented to, 216, 217, 228
Ninigrads or Ninigret (Niantic sachem of Misquamicut), 179
Nipissing: importance of dreams to, 281
Nisbet, Jack, 146
Nisga'a: bioregional mapping among, 99
Nkamwin or Big House ceremony: of Delaware Indians, 260–61
Noank Pequots, 181
nodes: groups represented as, 224; in

Wallam Olum and Quapaw painted buffalo robe, 86
Nokopurrs: Weantinock deed by, 176
Nome gold rush: effect on market for Eskimo ivory carvings, 94
Norona, Delf, 75
North America: cartographic publications in, 38; Chickasaw map (1737) of Southeast, 78; limited European knowledge about interior of, 135, 136, 137; mapping of, by Delisle cartographic house, 233; scholarly interest in native maps in, 35; Thomas Nairne's inset to Edward Crisp's map of, 235
North Americans: scholarly encounters by, 40–48
Northern Pipeline Agency, British Columbia, 13
Northwest Coast: cosmic tree of, 253
Northwest Passage: Arctic peoples' maps for whites' search for, 96
Northwest Territories, 97
Norton, Moses, 91; "Draught of the Northern Parts of Hudson's Bay" drawn by, 139, 142, 148
Notes on the State of Virginia (Jefferson), 213
Nouveaux voyages dans l'Amérique septentrionale (Bossu), 194
Nuligak (Inuit), 56, 57
numbers: conventional symbols to represent, 228, 229
Nuttall, Thomas, 190

Observations on Hudson's Bay (Isham), 141

Ochagach (Cree), 135
Ogilby, John, 79
Oglala Sioux: storage of records by, 84
Ohio valley: analysis of pictograph painted on tree in, 78
Ojibwa (Mississagi River band): boundary dispute and maps used by, 274–75, 276
Ojibwa Indians, 89; birchbark maps by, 12; maps by, 45; and principles of native mapping, 81; record storage by, 83; sacred scrolls of Southern, 86
Ollokot ("the Frog") (Nez Percé), 243
Omaha Indians: maps by, 24
onanne ("road of life"), 257
Oñate, Don Juan: expedition of 1601, 43
oral narrative/tradition, 86, 95
Oregon Trail: impact on Lakota, 244
orientation, 93
origin myths/stories, 83; on birchbark scroll (Chippewa), 260; of Crow Indians, 232–37; of eastern Choctaws, 234–35; for grass houses built by Wichita, 250; of Iroquois, 253
origin problems: Choctaw, 232–37
Ortiz, Simon J.: *Good Journey, A,* 265
Osages: cosmology of, 203n. 81; hostilities between Quapaws and, 194; maps as body tattoos, 63
Osotouy (Quapaw village), 187, 193, 228
Ottawa-Ojibwa: importance of dreams to, 281
Ousamequin (Massasoit), 173
outline maps: Inuit, 95
outline shapes: vs. surveyed shapes, 92

pack ice, 92
Paconaus: Weantinock deed by, 176
painted books: role and description of, in pre-Columbian Mexico, 112
painted hides/skins, 57; designs by Plains Indians on, 201n. 31
painted manuscripts: Tenochtitlan depicted in, 127
painters: in Aztec Mexico, 111–12
paintings: cartographic content within Indian, 187; Creek and Choctaw (reproduced by Romans), **227**; maplike, 15; Navajo dry, 250
Palmer, Nehemiam, 179
panels: of rock art, 76–78
pan-Indian church (Oklahoma): incorporation of (1918), 264
Papago (southern Arizona): "road of life" motif of, 257
paper: *lienzos* painted on, 124
Papetoppe: Weantinock deed by, 176
Parameshe: Weantinock deed by, 176
parcels: land conveyance in southern New England by, 166–67
Parke, Thomas, 179
Parks, Douglas R., 252
Parry, Sir William Edward, 41
Pascagoulas, 233
Passage through the Garden: Lewis and Clark and the Image of the American Northwest (Allen), 72
Passamaquoddy Indians: maps by, 44
patents: land, 247
path, the: in American Indian consciousness, 256

paths for movement: in maps of Aztec Mexico, 113–17, 131
Path We Travel, The (exhibition catalog), 265
Patsah, Joseph: gesture in mapping by, 147, 148
patterns: decorative, 14; in sandpaintings, 95. *See also* motifs
Paugussett-affiliated peoples, 164, 165
Paugussett confederacy: Weantinock and Schaghticoke peoples within, 174
Paugussett territory: mapping through words in ceding to whites, 165–66
Pawnee, 89; correlations between celestial world and earthly organization of, 88; star charts by, 42, 87, 88; underworld representations of, 253
peace: calumet used in rituals for, 191; sacred "white" tree of, 253
Pearce, Margaret, 63, 75, 76, 275
Pentland, David, 101n. 6
Pequot War of 1637, 177
perception: culturally specific, and convention, 140; and sense of cultural ascendancy, 149
perceptiveness: about Indian maps, 78; in deconstruction of Matonabbee's map, 91
Percy, George, 213
performance contexts: and future encounters, 279–82
Périer, Governor, 193
Peron, François du, 281
Pessacus (Narragansett sachem), 177
Petermann, August, 35

petroglyphs, 58, 65n. 13; of Nevada, 77
Petun: importance of dreams to, 281
Peutinger map, 79
peyotism, 264
Philbrook Museum, Oklahoma, 254
Philip, King (Pokanoket). *See* Metacom or "King Philip"
Phillips, Philip Lee, 42
physical geography: in Germany, 35
Physical Geography (Somerville), 45
physical sensing methods, 92
pictographic paraphernalia: cartographic awareness of, within Great Medicine Lodge, 260
pictographs: categorized as maps, 58; communicating and recording via, 42; composite, 12; diversity of media with, 58; on map collaborations between Indians and Europeans, 246; painted on trees, 78; to represent rapids, 89; in Wallam Olum, 86
pictography: categories based on whites' perceptions, 57; first naming of, 47; scholarly understanding of, 41
"Picture Writing of the American Indians" (Mallery), 46, 58
piedmont towns: depicted in Smith's map of Virginia, 211, 213
Pierre (Montagnais guide), 15
pigment identification: through polarized microscopy, 88. *See also* paintings
pilgrimages: to Southwestern center shrines, 250; through mythologized landscapes, 256
Pittman, Capt. Philip, 194

place-names: site-specific nature of native, 159
Plains Cree (Saskatchewan): cosmograms of, 263
Plains Indians: designs on painted skins of, 201n. 31; four-directions motifs of, 250; ledger art by, 103n. 23, 106n. 65; paintings of, 76; sun dances of, 253
plane table, 161
planets: depicted in rock art, 252
Plan et scituation des villages Tchikachas . . . (Capt. Pacana's map of Chickasaw towns): redrawn by de Batz, 214, **215**
Plan of Part of Hudson's Bay and Rivers, communicating with York Fort & Severn (Graham), 135, 137, **138**, 139, 143
Plymouth Colony, 173; Sepecan deed from, 171
Plymouth Colony Records, 171, 174
Plymouth Court Records, 173
pochteca (long-distance merchants): "way-finding" aids used by, 117
Poisson, Father du, 191
Pokanoket: Sepecan land holding of, 172
polarized microscopy: for pigment identification, 88
political boundaries: lines signifying, 172
political geography: Amerindian "sociogram" representation for portraying of, 224–27, 237
political leaders: arcane symbolism on

maps drawn by Southeastern Indian, 217
political relationships/systems: delineated by circles in Southeast Indian maps, 207; information about, 19; prehistoric map and hypothesizing about, 78
polities: political maps representing, 236–37
Ponca Indians: maps by, 24
Pontiac's rebellion, 192
Pookmoosh Band of Micmac: gesture map by chief of, 11–12
Pootatuck people: and land dispossession, 177
Pope, John, 190, 195
Popple, Henry: map of lower Mississippi region by, 235
population: Quapaw (1750), 228; use of multiple symbols for indication of, 229
portages: native drawings of, 89
Postclassic Mexico, Late: political and social pattern of, 123
Potomac River: Southeastern Indian villages along, 213
pottery decoration, 83; and eastern Choctaws, 235
Powell, John Wesley, 264
power: Indian deeds and establishment of, 163–64
Powhatan (Chief): John Smith's captivity by, 209–10
Powhatans: John Smith's encounter with, 207, 209, 210, 211, 212
"Powhatan's Mantle" (artifact), 209
Powhatan's Mantle (Wood et al.), 80, 81

Powon (Chipewyan guide): map by, **25**
Prando, Father Pierpaolo, 249
prayer deposits, 250
prayer rituals: calumet used in, 191
precious metals: information about, 19
precision: of Inuit maps drawn for explorers, 146
pre-Columbian cartographic tradition, 112–13
prehistoric cartography, 75
preliterate spatial orientation, 284n. 17
preservation: of maps by Delisle cartographic house, 233; of records by native people, 84
printmakers: Inuit, 96
projective geometries, 61, 64
property mapping. *See* Indian deeds, southern New England
Prophecy Rock *(Wutakyawvi)*: of Hopi, 264
prophesies, books of: Aztec, 112
Protestant missionaries: ladder charts used by, **261**, 261–62
provenance: lack of concern over, in native maps, 59; of Quapaw/Arkansas painted buffalo hide, 187, 188–92
Providence (Campbell), 247
psyche: definition of and mapping of, 108n. 108
psychology: historians of cartography and literature of, 279
Ptolemaic system: of spatial coordinates, 141
puberty rituals: and cosmograms, 255
Pudlat, Pudlo: "The Settlement from a Distance," 96

Puebla *lienzos,* 123
Pueblo Indians: and possibility of traditional action maps, 281; sense of centrality of, 250; synchronization of solar movement by, 251
Pukwat Skwes (Mistassini Cree mythic woman), 258
Purcell, Joseph: map of 1775 by, 237
Purchas, Samuel, 209
Pushruk, Tony: engraved letter opener by, 109n. 113
Pynchon, Thomas, 243

Quaker teachings: influence on Iroquois, 262
Quapaw/Arkansas painted buffalo hide, 79, 187–88, **188, 189**, 228; cartographic component within, 195–98; distinctive nodes in, 86; establishing provenance of, 188–92; ethnography and possible acculturation of, 194–95; incorporation of Indian cartographic tradition on, 63; places and events on, 192–94
Quapaw Indians: alliance between Illinois Indians and, 191–92; clothing of, 194–95; face-painting by, 190; hide paintings of, 191; hostilities between Chickasaws and, 187, 192, 194, 197; raids by, 194; use of calumet by, 200–201n. 27
Quapaw villages: cartographic interpretation of the Quapaw/Arkansas buffalo hide, **196**
Quenaponen (Mississauga spokesman), 274

Radin, Paul, 260
rainbow sky dome: depiction from Jemez Pueblo kiva wall: **252**
rapids and portages: native drawings of, 89
Rapiscotoo: Weantinock deed by, 176
Rasmussen, Knud J. V.: maps collected by, 32n. 54
Ratzel, Friedrich: *Anthropogeographie,* 35
Reagan, Alfred, 252
realm of cosmology, Native American, 248–49; centers, 249–50; directions, 250; sky dome, 250–52; vertical axis, 252–53; whole cosmologies, 253–55
reconnaissance maps: in Sahagún's *General History of the Things of New Spain,* 113, **116**
reconstructions: native maps used in, 78
record keeping: by winterers of Hudson's Bay Company, 146
Records of the Colony of New Plymouth in New England: Book of Indian Records for Their Land: redrawn version of Sepecan deed in, **171**
Red Bean Man, 250
Red Score. *See* Wallam Olum
Red Sky (Ojibwa keeper), 87
reference map: of Mississippi-Alabama region, **232**
registration: of Indian deeds, 158, 163, 169–70
Relación geográfica: map of, 123
religion: healing ceremonies and sandpaintings, 95; migratory routes of, 63; native geographical consciousness grounded in, 248; and sacred scrolls, 86
remapping: and native territorial diminishment, 47
remote sensing: question of use by sub-Arctic Indians, 92
Report of the Public Archives for the Year 1951 (Lamb), 56
representation: and symbolization, 64. *See also* motifs; pictographs; symbols and symbolism
representative figures: produced by Plains Indian men, 200n. 31
Reserve #8 (Lake Huron region): northern boundary dispute by Mississagi band of Ojibwa over, 274–75, 276
Revolutionary War: Passamaquoddys in, 46
Rey, Agapito, 43
Rhode Island: charter received by, 181; claims to Wecabaug by, 177; and Connecticut charter, 179
Richmond, Trudie Lamb, 175
Ridgefield (1708): native mapping superseded by English mapping in, 167–69
Ridgefield deed: related to nineteenth-century topographic map, **168**
Ridington, Robin, 262, 281
ritual-divinatory codices, 130
rituals, 94; calumet used in Quapaw, 190–91; Indian maps made for, 83; puberty, 255; "road-making" and "road-closing," 256
ritual sites: and centers, 249
road, the: in American Indian consciousness, 256
Road, The: Indian Tribes and Political Liberty (Barsh and Henderson), 264
"road-closing": Southwest ritual practice of, 256
"Road to Life and Death" (Winnebago), 260
"road-making": Southwest ritual practice of, 256
Road Man, 264
roads: metaphor of, appropriated by Indian prophets, 262; in pan-Indian church, 264
Robinson Treaty of 1850, 274, 276
Rochester: Sepecan remapped as, 174
rock: possibility of maps in, 76–77
rock art, 3, 57, 246; astronomical motifs in, 252; dating techniques for, 76, 77; posthiatus synthesis on North American, 76; question of map incorporation in, 9
rock paintings: of Southeastern Indians, 205
Rock Thunder (Plains Cree elder): tipi painting by, **263**
Rocky Mountains: native sketches of, 144
Romans, Bernard, 229; Creek and Choctaw "paintings" reproduced by, **227**
Ronda, James: "'A Chart in His Way': Indian Cartography and the Lewis and Clark Expedition," 72
Ross, Sir John, 41
Rousseau, Jean-Jacques: *Essai sur l'ori-*

gine des langues, 146
rubber-sheet geometry, 67n. 28
Ruggles, Richard, 56, 61, 101n. 6
rulers/leaders: credentials given within Aztec maps, 126, 128
Rundstrom, Robert A., 99, 100, 140, 146
Rupert's Land Research Centre Colloquium (Fort Churchill), 90
Russia: map monographs published in, 39

sachems, regional, 173; and Indian deeds, 162
Sack, Robert: *Conceptions of Space in Social Thought: A Geographic Perspective,* 279
sacred bundles, 51n. 33
sacred destiny, tribal: native cartographic case for, 249
sacred fire: four-directional, of Southeastern Indians, **251**
sacred geography, 249
sacred pole: connection to cosmic realms by, 253
sacred route: of Chippewa, 260
sacred scrolls: of southern Ojibwa, 86
sacred sites: and Earthnaming ceremonies, 243
Sahagún, Bernardino de: *General History of the Things of New Spain:* reconnaissance map from, 113, **116**
Salle, Sieur de la, 19
sand: designs in, 58; Kiowa map in, 50n. 21; paintings in, 57, 95; pictographs on ground in, 58
San Felipe Pueblo: prayer deposits in, 250

Santa María Zacatepec: Lienzo de Zacatepec in archive of, 126–27
Santo Domingans (New Mexico): cosmogram of, 253, **254**
Saskatchewan River: Blackfoot sketches of upper, 144; mapping of, 139
Sassamon, John, 172–73; murder of, 174
satellite imagery digitization: and territorial sovereignty defense, 99
Saya (Beaver culture hero), 262
scale, 61: convention and culture-specific notion of, 142; from native maps, 17, 18
scalp dance: portrayed on Quapaw/Arkansas painted buffalo hide, 187
scalping: and war parties, 193
scapulimancy, 53n. 56, 89, 107n. 91, 259; accounts of, 46; ecological and social functions of, 90
Schaghticoke people, Reservation, 177
scholarly encounters: by Germans, 35–40; by North Americans, 40–48; recent, in anthropological contexts, 83–94; recent, in historical contexts, 72–83; recent creative cartographic, by native peoples, 94–100; sixty-year hiatus in, 55–56
Schoolcraft, Henry R., 47
scientific cartography, 140
scientific exploration: and map commissions, 23
scientific mapping, 135; function of dotted lines in, 144; role of Amerindian maps in, 140
scientism, 6n. 7

scientists: map use by, 11; references to native maps by, 34
Scroggs, John: journal of, 36
scrying, 53n. 56; accounts of, 46. *See also* divination; scapulimancy
sea ice, 92
second-generation Baker Lake Inuit artists: drawings of, 280
seeing, ways of, 142; Amerindian, 152n. 16;
Seeley, Joseph, 167
Selkirk, Lord, 246
Selmo, Sapiel (Passamaquoddy chief): maps made by, 46
semantic categories: intercultural differences in, 275
semantics: language, and linguistics in translational contexts of future encounters, 276–78
Sepecan (1666): native graphic mapping in transaction of, 171–74
Sepecan deed, **170**; redrawn, two versions of, **171**
Sepecan Neck, 174
sequential spatial system: in Mapa Sigüenza, 120, 121, 122, 123
"Settlement from a Distance, The" (print by Pudlat), 96
settlement hierarchy: map symbolism and English notions of, 212
settlements: Chickasaw farming practices and patterns within, 220n. 31; Indian resistance to colonial, 174; symbols by European cartographers for, 229, 233, 235. *See also* houses; villages

INDEX

settlers: native line maps made for white, 79
Seven Cities of Cibola, 15
Seven Years' War: effects of, 192
Shaking Tent, 259
Shalako ceremony (Zuni Pueblo), 256
shamans: "animal homes" and Pawnee, 253; in Shaking Tent rite, 259
Shamenunckqus (alias Bapistoo), 176
shape: native mapping and complexity of, 92, 108n. 105
shell art, prehistoric: from Southeast, 218n. 1
shell beads: on "Powhatan's Mantle," 209
shells and disks, stylized: in Aztec treatment of water, 114
Sherzer, Joel, 277
Shoopack: Weantinock deed by, 176
short-life maps, **59**, 60
Shoshones, 249
shrines, 94; center, 250
sign language, 146, 149
silences: on maps, 110n. 130
Sinte Gleska College (South Dakota): Lakota stellar theology research at, 81
Sioux, 249
Sioux Treaty of 1868: Custer expedition in violation of, 245
site locations: correlation of locations on Smith's map of Virginia with archaeological, 211
Sitting Rabbit (Mandan), 80; map by, 79, 80
Six Nations Iroquois Confederacy: wampum belts repatriated to, 84

Skidi band of Pawnee (Nebraska and, later, Oklahoma): celestial mapping by, 252; Chamberlain's observations on star chart of, 88–89; storage of sky chart of, 83–84
skins: layout of image dictated by shape of, 228; maps on, 15
sky charts: uses of, 13–14; storage of, 84
sky dome: within Native American realm of cosmology, 250–52
sky maps: by Inuit, 92
Slagle, Al Logan: *The Good Red Road: Passages into North America*, 264–65
slate: Ojibwa records on, 83
slaves: trade networks in, 80; traffic in Indian, 217–18
sleeping positions: of Mistassini Cree, 259
Smith, Capt. John: archaeological analysis of map of Virginia by, 207–14; detail from map of Virginia by, **210;** Indian component of map of Virginia by, 80; Powhatan encounters by, 207, 209, 210, 211, 212
Smith, M. E., 124
Smithsonian Institution: *Handbook of North American Indian* series published by, 90
Smoholla (Wanapum prophet), 248
smokehole: of sky dome, 252
Snake Dance, Hopi: cosmogram used in, 256
Snakes (Shoshones), 249
snow: pictographs on ground in, 58
social landscapes: Native American

map information on, 81; Native American maps and information about, 218
social relationships: delineated by circles in Southeast Indian maps, 207, 209, 212, 216, 217
social science contexts: and future encounters, 279
social stability: maximizing of, among Chippewa and Cree, 258–59
"sociograms," 224–27, 237
soil replenishment: and late-prehistoric pattern of field placement, 220n. 31
soil types: correlated to agricultural fields, on Captain Pacana's map of Chickasaw towns, 214
solar movement: within Navajo cosmology, 255; synchronization of, by Pueblo communities, 251
Somerville, Mary: *Physical Geography*, 45
songs: cartographic awareness in, 260
Soto, Hernando De. *See* De Soto, Hernando
South Dakota: map depicting Black Hills by Amos Bad Heart Bull, **245**
Southeastern Amerindian maps, 205–18; and Choctaw origin problem, 232–37; "event transcriptions," 227–29, 234; influence on early explorers by, 229–32, 235; "sociograms," 224–27, 237; types of, 205, 206–7, 224
Southeastern Indians: symbolism of ceremonial ground of, **251**

Southern California Indians: cosmology and topography depicted on floors of ceremonial enclosures by, **256**

southern New England: property mapping in, 157, 181, 182

southern New England colonies: land battle between Narragansetts, Mohegans, and, 177–81

southern New England land deeds: maps as exceptions in negotiation of, 275. *See also* Indian deeds, southern New England

Southern Ojibwa, 89; birchbark scrolls of, 63; sacred scrolls of, 86

Southwestern Pueblo, 249; four-directions motif among, 250; long-poled ladders of, 253

souvenir trade: background to Alaskan, 94

sovereignty, tribal: use of Geographic Information System software for defense of, 99

space, 93; awareness of diverse representation of, 64; emergence of concepts of, 103n. 21; Eskimo concept of, 73; European concepts of, 60; Great Lakes and sub-Arctic techniques for transcending of, 259; human cognition of, 278; kinds of, on Mapa Sigüenza, 120, 121, 122, 123; material representations of, in three native spheres, 93; relationships between things and events in, 2; road symbol and moral and spiritual movement of Indian people through, 265; role of in native life and beliefs, 94; social structuring of, 279; uses of material representations of, 241, 246; visual organization of in Aztec maps, 131

Spain: geographical information sought by, 19

Spalding, Eliza: Protestant ladder chart (ca. 1842), **261**

Spanish Settlements within the Present Limits of the United States: 1531–1561, with Maps (Lowery), 43

spatial concepts: of Eskimo, 73

spatial coordinates: Ptolemaic system of, 141

spatial deixis, 75, 277

spatially arranged information: descent and survival states of, communicated by Indians and Inuit, **59**

spatial orientation: preliterate, 284n. 17

spatial relationships/patterns: pictographic representation of, 47; on Quapaw/Arkansas painted buffalo hide, 188

spatial representation: unexpected forms of, 56

spatial symbolism: modes of, 63

spatial understanding: facilitated by maps, 3, 159

Speck, Frank G., 46, 100

speech: vs. sign language, 146

Spengler, Oswald, 74

Spink, John: "Eskimo Maps from the Canadian Eastern Arctic," 73

spiritual sites: for Paugussett peoples, 177

spiritual teachings: mapping of, 261

Spokane people: use of gesturing by, 146–47

Sproat, Gilbert M., 44, 45

St. Cosme, Father, 190, 194

St. Lawrence Iroquois: maps by, 44

Stamp, Sir Dudley, 63

Stanton, Thomas, 178, 179, 181

star charts: by Pawnee Indians, 42, 87; questions about, 252

star maps, 88; of Lakota, 62

stars: depicted in rock art, 252

State Historical Society of North Dakota, 79

stellar theology: Lakota, 81

Steltenkamp, Michael F., 263

stewardship, natural resources: computer and satellite technology as aid in native, 99

stick charts: Micronesian, 73

sticks: maps of, 15; for telling time, 88; used in Powhatan ceremony, 209–10, 219n. 11

Stieler, Adolf, 35

Stone of the Sun (Aztec), 127, 133n. 36

Stone of Tizoc (Aztec), 127

storage: native guardians selected for maps, 83

straight lines: as alien representation to Indian understanding, 276

Stuart, John: map of 1775 by, 237

sub-Arctic culture: geographic consciousness of, 257–59

subterranean underworlds: representations of, 253

sun: represented on on Pawnee buckskin map, 252; on

Quapaw/Arkansas painted buffalo hide, 187, 195
sun dances: of Plains Indians, 253
sun/moon representation (Quapaw painted buffalo hide): religious significance of, 195
"supernatural distance": corresponded with "geographical distance," 217
surveyed shapes: vs. outline shapes, 92
surveying: instruments for, 161, 247; intercultural misunderstanding over measurements used in, 276; as profession, 161
surveying parties: Indian headmen and assurance of accuracy from, 207
surveyors: colonial, 158; of medieval English open-field system, 160–61; Quapaw assistance to, 197
surveyor's transit, 247
surveys: in New England map history, 158; of Wecabaug, 178–79
survey-textbook industry, 161
survival states, **59**; lack of concern over, for native maps, 59; of maps from early colonial period in Mexico, 112; of originals of maps, 18
symbols and symbolism, 64; within adoption ritual of Crow Tobacco Society, 265–66; changes in European representational conventions of Indian villages, 229–31, 235; color and representation of, 84; on drum head drawn by Beaver Indian shaman, **262**; and English notions of settlement hierarchy, 212; four-directional, 250; Indian-house, from early maps of Americas, **229**; on maps by Southeastern Indian elites, 217; mnemonic approach to, 172, 182; for number representation, 229; of place names on Mapa Local of Tochpan, 115; in Skidi star chart, 88; for "sky dome" or "celestial vault," 251; of Southeastern Indian ceremonial ground, **251**; in tipi painting by Rock Thunder (Plains Cree elder), **263**; on Weantinock deed, 176; on Wecebaug map, 180. *See also* pictographs
syntax differences: and spatial deictic systems, 277

Tacchigis (Cree), 135
Taensas: cartographic information provided to Iberville by, 231, 233
Tanner, Adrian, 90, 258
tattoos, body: Osage maps and, 63
tax and tribute records: in pre-Columbian Mexico, 112
Tchoukafala (Chickasaw village), 214
Teit, James, 257
Templo Mayor (Aztec), 127
Tenochtitlan, 111; founding depicted on Mapa Sigüenza, **120**; founding of, Codex Mendoza, 128, **129**; on Mapa Sigüenza, 119; power residing within, 127–131; significance of location of on Mapa Sigüenza, 121
Tequixtepec: importance of *lienzos* to, 127
terrae incognitae: maps of, 19, 33

terrae semicognitae: maps of, 22, 23, 137; during Age of Exploration, 99
terrain: extrasensory experiences by native people of, 92; in Mapa Sigüenza, 120
terrestrial maps, 76, 93
terriers (or court rolls), 161
territorial reinscription, 248
territory: and Aztec cartographic histories, 123, 131; graphic representations of, 241; Indian deeds and representation of, 157, 163, 164
testimonies: and Indian deeds, 157
Tewa: sky dome motif in art of, 252
Texcoco: in Mapa Sigüenza, 121
thematic cartography, 35
thematic maps, 224
theodolites, 161
Thompson River Indians (British Columbia): map of life's road by, **257**
Thrower, Norman, 76–77
Thwaites, Reuben Gold: *The Jesuit Relations and Allied Documents,* 46
ticked lines: interpretations of, 77
time: archaeological site loss through, 211–12; Aztec symbols for passage of, 118, 130; diagram of, within Codex-Féjérváry, 130; Great Lakes and sub-Arctic techniques for transcending of, 259; representation of in Codex Mendoza, 128; road symbol and moral and spiritual movement of Indian people through, 265; sticks for telling of, 88. *See also* calendars

tipi paintings, 94; by Rock Thunder (Plains Cree elder), **263**
tipis: cosmological portrayals on, 254; narrative scenes on, 103n. 23
Tizoc Stone (Aztec), 133n. 36
tlacuiloque, 111, 112
Tlatelolco: on Mapa Sigüenza, 119, 120
Tochpan: Mapa Local of, 114, **115**
tonalamatls, 112
Tongigua (Quapaw village), 228
Tonti, Henri de, 194, 235
Toolhulhulsote (Nez Percé spiritual adviser), 248
topographic features: in *lienzos,* 124
topographic maps, 13, 97
topographic orientations/locations: and map deconstruction with, 90–91
topography: and cosmology depicted on floors of ceremonial enclosures, **256;** of lake in Mapa Sigüenza, 121; in maps and profiles, 47; painted on hides, 88; social environment rendered in Amerindian maps with, 224, 237
topological geometry: of some wampum belt designs, 84
topology: and native maps, 61; within Southeast Indian maps, 207
toponyms: Athapascan, 160; within Aztec maps, 117; English, 160; native use of, 159, 164, 169, 181; places identified by pictorial, 125, 131
Toronto Purchase of 1787, 274
totemic representation: in Amos Bad Heart Bull's drawing of Black Hills, 87

Totosin. *See* Watachpoo
Tourima (Quapaw village), 187, 193, 228
town proprietors: and native land transactions, 162
trade: during late-prehistoric and early historic eras of Southeast, 216; at Metichawon, 175; networks of, 80; patterns of, on maps, 19
trade languages, 2; syntactical limitations of, 277
traders: native, 2; native line maps made for white, 79
traditional territories: bioregional mapping and legal challenges over, 94
traditions: maps and preservation of, 13
trail, the: in American Indian consciousness, 256
trail: maps, 46; patterns, 76
Transactions of the Historical and Literary Committee of the American Philosophical Society, 40
transcript map(s), 26, 18–19, 142; collections of, 43; of Delaware pictographic record on bark, 85
translational contexts: and future encounters, 276–78
translation(s): between cartographic conventions, 140; of native map conventions, 143–44
transport maps, 141
trappers: native line maps made for white, 79
travel lines: and caribou hunting, 98
treaty negotiations: Native American map use in, 275

trees: cosmic, 253; maps on, 12; paintings on, 228; pictographs on, 58, 78
triangulation: development of, 161
Tribale from Survival: catalog of, 109n. 117
tribalism: and the Road, 264
Tricentennial Run, Taos Pueblo (1980): blessing with cornmeal before, 257
Triple Alliance empire (Aztec), 127, 132n. 8
Tuan, Yi-Fu, 249, 255
Tunicas: Quapaw engagements against, 194
Tunxis people, 165
Turner, Frederick: *Beyond Geography: The Western Spirit against Wilderness,* 73
Tuscan crosses: on John Smith's map of Virginia, **210,** 211
Tuxpan (ancient Tochpan), 114
Tuxpan River: depicted on Mapa Local of Tochpan, 114, 115
two-dimensional art/renderings: in Eskimo engravings on ivory, 94; map incorporation into, 96
two-roads map, 265
"Two Roads Map, The," 263
Tyrrell, James B.: map of Hudson Bay by Powon for, **25**
Tyrrell Collection, Thomas Fisher Rare Book Library, University of Toronto: Algonkin Indian map in, 24

Ukakhpakhti? (Quapaw village), 228
Uloksak (Copper Indian), 57
Uncas, 177, 178, 179, 181

underworlds: representations of, 253
United Colonies, 181
United States: new geography in, 34
universe, models of. *See* cosmograms
"upstreaming." *See* anthropological "upstreaming"

Vaudet, Gilbert, 150
Vaudreuil, Governor, 193
Velasco manuscript map (c.1610), 66n. 20, 103n. 19, 219n. 13
Vennum, Thomas, Jr., 260
Vergès, Bernard de, 197
vermilion pigment (mercuric sulfide): trading of in French Louisiana, 200n. 18
vertical axis: within Native American realm of cosmology, 252–53
vertical perspective, native, 47, 92
victory monoliths: in Aztec Mexico, 127
villages: arrangement of on Quapaw/Arkansas painted buffalo hide, 193; details of, in Quapaw/Arkansas painted buffalo hide, **189,** 200n. 13; problems in distinguishing Powhatan chiefs' vs. commoners' villages, 213; symbols used in representing, 229–32, 233, 234–35
Virginia: detail from John Smith's map of, **210**
vision quest: of Black Elk, 252
vision questing locations: and lands converted into boundaries of political territory, 243
Vitry, Father de, 195
Vivier, Father, 228

Vorsey, Louis De, Jr., 82
voyageurs, French-Canadian, 11

Walker, Amelia Bell, 251
Wallace, Anthony F. C., 261
Wallam Olum (or Red Score), 87, 107n. 75: internodal linkages in, 86; interpretations of, 85
Wampanoag people, 171, 174; Sepecan land holding of, 172
Wampotoo: Weantinock deed by, 176
wampum/wampum belts, 57; pictographs stylized in, 58; repatriation of, 84; storage of Iroquois, 83
Wanapum: in Columbia Plateau, 248
war belts, 86, 87; cartographic principles in, 84
war plans: Aztec, 116
warriors: paintings memorializing achievements of, 192; portrayed on Choctaw painting, 228
Waselkov, Gregory A., 79, 80, 90, 224, 225, 277; "Indian Maps of the Colonial Southeast," 78
Watachpoo (or Totosin): purchase of Sepecan lands from, 173–74
water: Aztec conventional treatment of, 114; lines signifying, 172
Waterman, T. T., 254
"water sky," 92
Watlala Indians: maps by, 82
"way-finding" maps, 114, 116, 117, 120
Weantinock: graphic map drawn at (1703), 174–77
Weantinock deed: redrawn version of, **175;** related to nineteenth-century topographic map, **176**

weavings: of celestial charts, 95
Webber, Alika, 259
Wecabaug: clerk's copy of map, 171; graphic map signifying tribal and colonial boundaries, 177–81
Wecabaug testimony: original (1662), 179, **180**; related to nineteenth-century topographic map, **181**
Welch, James: *Fools Crow,* 280–81
Wellmann, Klaus, 76, 77
Wesawegun, 179
Western Massachusetts: Indian deeds in, 163
Wetalltok (Quebec Inuit): Belcher Islands map on verso of missionary lithograph, **17;** missionary lithograph with map on verso of, **16**
Wetalltok Bay, 16
Wethersfield: Indian deed description from, 163
Wetsuweten: mapping for empowerment among, 99
Wewinapouch: Weantinock deed by, 176
whalers: Inuit maps made for white, 95
whaling captains: map collections by, 24
White, Leslie, 253
White Bird (Kiowa medicine man), 248
White Eyes (Delaware chief), 40
whole cosmologies: within Native American realm of cosmology, 253–55
wikhegans, 65n. 14
Williams, Glyndwr, 90, 93
Williams, Roger, 162; *A Key into the Language of America,* 159
Wilson, Gilbert L., 80

Wind from an Enemy Sky (McNickle), 247
Windigo (Chief): Lake Nipigon map by, 52–53n. 51
Wingenund (Delaware warrior): map by, **12**
Winslow, Edward, 162
Winslow, Josiah, 174
Winston County, Mississippi: "mother mound" *(Nanih Waiya)* in, 249
Winthrop, John, Jr., 178
witnesses: and Indian deeds, 163
Women, Fire, and Dangerous Things: What Categories Reveal about the Mind (Lakoff), 277–78
Wood, Charles Erskine, 243
Wood, Denis, 147, 279

Wood, Raymond, 80, 81
wood and wood carvings: and pictographs, 47, 58
Woodward, David, 64; *History of Cartography*, vol. 1, 158; redefinition of concept of map by, 64
word maps: vs. graphical maps, 75
words: mapping by, 164–69, 172, 181
world: Eskimo concepts of, 73
worship: stored records on Ojibwa, 83
writing: on Quapaw/Arkansas painted buffalo hide, 195
written vs. cartographic information: Aztec lack of distinction between, 113

Yahey, Charlie (Dunne-za dreamer), 262, 281
Yoncomis: Weantinock deed by, 176
York, Erasmus. *See* Kalli-herua
York River: Southeastern Indian villages along, 213
Yuchi: four-directions motifs of, 250
Yurok (northwestern California), 250; cosmograms of, 254

Zacatepec: *lienzo* from, 124, **125**, 126
zenith: in American Indian cosmologies, 253
Zuñiga Map of 1608, 212
Zuni Pueblo (New Mexico): center shrine of, 250

DATE DUE

OCT 26 2004			
Nov 19 MOE			
NOV 23 REC'D			
OhioLINK			
MAY 0 3 REC'D			
OhioLINK			
JAN 2 6 REC'D			
OhioLINK			
APR 0 2 REC'D			

GAYLORD — PRINTED IN U.S.A.

E59.C25C37 1998
Cartographic encounters : perspectives on Native American mapmaking and map use